MOLECULAR BIOLOGY OF NUCLEOCYTOPLASMIC RELATIONSHIPS

MOLECULAR BIOLOGY OF NUCLEOCYTOPLASMIC RELATIONSHIPS

S. PUISEUX — DAO

Laboratoire de Biologie Cellulaire Végétale,
Université Paris VII,
75221 Paris Cédex 05, France

ELSEVIER SCIENTIFIC PUBLISHING COMPANY
AMSTERDAM / NEW YORK

ELSEVIER SCIENTIFIC PUBLISHING COMPANY
335 Jan van Galenstraat
P.O. Box 211, Amsterdam, The Netherlands

AMERICAN ELSEVIER PUBLISHING COMPANY, INC.
52 Vanderbilt Avenue
New York, New York 10017

Library of Congress Cataloging in Publication Data

Main entry under title:

Molecular biology of nucleocytoplasmic relation-
ships.

Symposium supported by the International Union
of Biological Sciences, et. al.
Includes index.
1. Cell nuclei--Congresses. 2. Cytoplasm--Con-
gresses. 3. Molecular biology--Congresses.
4. Cellular control mechanisms--Congresses.
I. Puiseux-Dao, S. II. International Union of
Biological Sciences.
QH595.M62 574.8'73 75-23166
ISBN 0-444-41343-X

Printed in The Netherlands

FOREWORD

One of the most important questions of Biology today is that of how cells differentiate. This process, which is obviously characteristic of both multicellular plants and animals, involves precise cellular mechanisms. Some unicellular organisms, however, also show readily distinguishable differentiation during their development; this is so for *Acetabularia* and certain strains of *Saccharomyces* and *Chlamydomonas* which exhibit two different functional and even morphological stages depending upon their conditions of culture.

Unicellular living material, when successfully cultured, is undoubtedly easier to work with. Moreover, the analysis of its cell physiology may be justifiably expected to be the simplest possible one. Indeed, the past few years have witnessed a burst of significant publications on nucleocytoplasmic relationships in unicellular organisms throwing light on fundamental mechanisms in eucaryotic cell physiology.

By far the most impressive results concern the interrelations between the nucleus and organelles containing extranuclear DNA, the mitochondria and plastids. A systematic use of all the modern tools of Cell Biology, involving genetics, biochemistry and electron microscopy, has gone to establish that two genomes with their own transcription and translation apparatus can cooperate to build a complex differentiated machinery. The first step is to know precisely how the genomes function, especially when compared to the procaryotic cell which has already yielded results that serve at present as the unique reference system. That, in fact, was one purpose of this Symposium, and some fine experimentation was reported both on the most detailed mechanisms known of viral and bacterial transcription and translation as well as on the functioning of mitochondrial and plastid RNA polymerases, ribosomes, tRNAs and translating factors in eucaryotes like *Saccharomyces* or *Chlorella*.

The mitochondrial genome is certainly the most studied in eucaryotic cells, especially in some strains of *Saccharomyces* and *Tetrahymena*. Restriction enzymes have been used to elucidate its linear and circular forms. Its mode of replication, mostly multidirectional and including exchanges of homologous strands, has been discussed in relation to the non-Mendelian transmission of some mitochondrial characters. Inhibitors and antibiotics have once again proved to be powerful investigational tools.

Another point discussed was the actual cooperation between the nucleus and extranuclear genomes in the building of mitochondrial or chloroplastic structures like ribosomes and membranes under different environmental conditions. The nuclear control of some ribosomal or membrane proteins is so specific that the philosophical question of the origin of the organelles containing extranuclear DNA was raised again.

Of course, nucleocytoplasmic relationships influence the fate of the cell as a whole, and a study of cell morphogenesis is one way to investigate the problem. The shape of a cell depends on the plasma membrane, whose growth and mobility are influenced by microtubules and microfilaments, principally in animal cells, and by the frame of the cell wall in the case of plants. Nuclear mutants affecting cell shape, movements of organelles and cell division in Ciliates seem to be very promising tools and may help in understanding the role of microfilaments and microtubules and of the membrane itself. Furthermore information was gathered concerning the nuclear control of cell-wall polysaccharide synthesis in *Acetabularia.* But it became clear during the meeting that mitochondrial and other DNA similar to that of plasmids can also play a role in specifying plasma membrane characteristics. The cell membrane thus has a very complicated architecture, as was indeed illustrated with the most refined techniques of electron microscopy.

It should also be borne in mind that the plasma membrane, in the living state, is an eternally mobile structure, a dissipative one which regulates the relations of the cell with both the external medium and other cells. Cyclic nucleotides and specific membrane receptors are involved in these processes in unicellular organisms. This was reported for *Dictyostelium*, but is evident for other species as well.

However, the mechanism by which exogenous information influences the physiology of the cell is still poorly known, and precise ideas as to the different levels at which control could be exerted are first required. The chain of biochemical events lying between nuclear DNA and cytoplasmic protein synthesis has been shown to be longer and more complicated in eucaryotes than in procaryotes. Unicellular organisms can help to unravel parts of that chain; *Acetabularia*, for instance, seems to provide indications on the transport of macromolecules from the nucleus to the cytoplasm.

Although investigations on the interrelation between the nucleus and the plasma membrane trail far behind those on the relationships between the nucleus and the mitochondria or plastids, a determined effort is now required to make up for this lag. We wish to express the desire that unicellular material be given its due place in this development for it will undoubtedly furnish data significant for all living organisms as postulated long ago by Professor Brachet in his comparison of *Acetabularia* with the Amphibian embryo, and for which he was most suited to introduce this Symposium.

We wish to thank him and all the participants for their pleasant collaboration, and hope that the present volume consisting of papers given at the Symposium will reflect its success.

S. Puiseux-Dao

LIST OF PARTICIPANTS

Alix, J.H.	Institut de Biologie Physicochimique, Paris, France.
Apel, K.	Harvard University, Cambridge, Mass., U.S.A.
Babloyantz, A.	Université Libre de Bruxelles, Bruxelles, Belgium.
Bannwarth, H.	Max-Planck-Institut, Wilhelmshaven, Germany.
Barat, M.	Université Paris XI, Orsay, France.
Beisson, J.	Centre Génétique Moléculaire, CNRS, Gif-sur-Yvette, France.
Benedetti, L.	Institut de Biologie Moléculaire du CNRS et de l'Université Paris VII, Paris France.
Berger, S.	Max-Planck-Institut, Wilhelmshaven, Germany.
Bernardi, G.	Institut de Biologie Moléculaire du CNRS et de l'Université Paris VII, Paris, France.
Bogorad, L.	Harvard University, Cambridge, Mass., U.S.A.
Boloukhere, M.	Université Libre de Bruxelles, Rhode-St-Genese, Belgium.
Bonot, C.	Université Paris XI, Orsay, France.
Bonotto, S.	Radiobiologie CEN SCK, Mol, Belgium.
Borghi, H.	Université Paris VII, Paris, France.
Borst, P.	Universiteit van Amsterdam, Amsterdam, The Netherlands.
Brachet, J.	Université Libre de Bruxelles, Rhode-St-Genese, Belgium.
Chanu, J.	Université Paris VII, Paris, France.
Chapeville, F.	Institut de Biologie Moléculaire du CNRS et de l'Université Paris VII, Paris, France.
Ciferri, O.	Istituto di Fisiologia Vegetale, Pavia, Italy.
Dazy, A.C.	Université Paris VII, Paris, France.
Dehm, P.	Max-Planck-Institut, Wilhelmshaven, Germany.
Delseny, M.	Centre Universitaire, Perpignan, France.
Dujon, B.	Centre de Génétique Moléculaire CNRS, Gif-sur-Yvette, France.
Dubacq, J.P.	Université Paris VI, Paris, France.
Dumitrescu, S.	INSERM, Hôpital St. Antoine, Paris, France.
Duntze, W.	Ruhr, Universität Bochum, Bochum, Germany.
Esnault, R.	Université Paris VII, Paris, France.
Felenbok, B.	Université Paris XI, Orsay, France.
Foury, F.	Université de Louvain, Louvain-la-Neuve, Belgium.
Frankel, J.	University of Iowa, Iowa, U.S.A.
Frontali, L.	Istituto di Fisiologia Generale, Roma, Italy.

Goffeau, A.	Université de Louvain, Louvain-la-Neuve, Belgium.
Gouhier, M.	Université Paris XI, Orsay, France.
Gros, F.	College de France, Paris, France.
Guignery, C.	Centre d'Etudes Atomiques, Saclay, France.
Herring, A.J.	Queen Mary College, London, England.
Hiernaux, J.	Université Libre de Bruxelles, Bruxelles, Belgium.
Hoober, J.K.	Temple University, Philadelphia, Penn., U.S.A.
Hoursiangou-Neubrun, D.	Université Paris VII, Paris, France.
Karakashian, M.	Max-Planck-Institut, Wilhelmshaven, Germany.
Kirchmann, R.	Radiobiologie CEN SCK, Mol, Belgium.
Kloppstesch, K.	Max-Planck-Institut, Wilhelmshaven, Germany.
Koop, H.U.	Freie Universität, Berlin, Germany.
Kretshmer, H.	Max-Planck-Institut, Wilhelmshaven, Germany.
Legros, F.	Laboratoire de Physiopathologie, Bruxelles, Belgium.
Leong, T.Y.	Max-Planck-Institut, Wilhelmshaven, Germany.
Levy, M.	Université Paris VII, Paris, France.
Loiseaux, S.	Université Grenoble I, Grenoble, France.
Lurquin, P.	Radiobiologie CEN SCK, Mol, Belgium.
Mache, F.	Université Grenoble I, Grenoble, France.
Margulies, M.M.	Smithsonian Institution, Rockville, U.S.A.
Mazza, A.	International Institute of Genetics and Biophysics, Napoli, Italy.
Miller, G.	Harvard University, Cambridge, Mass., U.S.A.
Morpurgo, G.	Istituto Superiore di Sanita, Roma, Italy.
Mounolou, J.C.	Université Paris XI, Orsay, France.
Muller, S.	Institut für Pflanzenphysiologie und Zellbiologie, Berlin, Germany.
Netrawali, M.S.	Bhabha Atomic Research Center, Bombay, India.
Niemann, R.	Max-Planck-Institut, Wilhelmshaven, Germany.
Novak, B.	Biozentrum der Universität, Basel, Switzerland.
Ohad, I.	The Hebrew University of Jerusalem, Jerusalem, Israel.
Pai, T.	Freie Universität, Berlin, Germany.
Pinon, H.	Université Paris XI, Orsay, France.
Pouchelet, M.	INSERM, Hôpital Saint-Antoine, Paris, France.
Prevost, F.	Université Paris XI, Orsay, France.
Prigogine, I.	Université Libre de Bruxelles, Bruxelles, Belgium.
Puiseux-Dao, S.	Université Paris VII, Paris, France.
Rahmsdorf, H.J.	Freie Universität, Berlin, Germany.
Raisonnier, A.	INSERM, Hôpital Saint-Antoine, Paris, France.
Robineaux, R.	INSERM, Hôpital Saint-Antoine, Paris, France.
Samarina, O.P.	Institute of Molecular Biology, Moscow, USSR.
Schael, U.	Freie Universität, Berlin, Germany.

Schatz, G.	Biozentrum der Universität, Basel, Switzerland.
Scherrer, K.	Institut de Biologie Moléculaire, CNRS, Paris VII, France.
Schweiger, H.G.	Max-Planck-Institut, Wilhelmshaven, Germany.
Schweyen, R.J.	Universität München, München, Germany.
Slonimski, P.	Centre de Génétique Moléculaire, CNRS, Gif-sur-Yvette, France.
Spring, H.	Cancer Research Center, Heidelberg, Germany.
Sussmann, M.	Hebrew University Migrash Harussim, Jerusalem, Israel.
Tait, A.	University of Edinburgh, Edinburgh, Scotland.
Tavlitsky, J.	Institut de Biologie Moléculaire du CNRS et de l'Université Paris VII, Paris, France.
Tiboni, O.	Istituto di Fisiologia Vegetale, Pavia, Italy.
Timofeeva, M.	Institute of Molecular Biology, Moscow, USSR.
Trendelenburg, M.	Cancer Research Center, Heidelberg, Germany.
Vanden Driessche, T.	Université Libre de Bruxelles, Rhode-St-Genese, Belgium.
Weil, J.H.	Institut de Biologie Moléculaire, Strasbourg, France.
Wilkie, D.	University College, London, England.
Williamson, D.	National Institute for Medical Research, London, England.
Woodcock, C.L.F.	University of Massachusetts, Amherst, Mass., U.S.A.
Yamakawa, M.	Max-Planck-Institut, Wilhelmshaven, Germany.
Zubarev, T.	Institute of Genetics and Selection of Industrial Microorganisms, Moscow, USSR.

Supported by:

The International Union of Biological Sciences
The European Cell Biology Organization
La Délégation générale à la Recherche Scientifique et Technique
Le Ministère des Affaires Etrangères et le Ministère de l'Education Nationale
Le Service d'Exploitation Industrielle des Tabacs et Allumettes

CONTENTS

PART ONE

Intergenomic cooperation in the cell

SECTION I

COMPARATIVE PROTEIN SYNTHESIS IN PROCARYOTES AND ORGANELLES CONTAINING EXTRANUCLEAR DNA

CHAPTER I.1

TRANSCRIPTION AND ITS REGULATION AT THE RIBOSOMAL LEVEL IN PROKARYOTIC SYSTEMS

F. GROS and M. CRÉPIN

Département de Biologie moléculaire, Institut Pasteur, Paris, France

One strikingly differentiative trait between eukaryotic organisms concerns the nature of the relationships existing in the cell between the two main steps of gene expression, namely transcription and translation.

In eukaryotic systems, as is well known, numerous processing events take place from the primary step of DNA transcription until the translation of the modified messenger chain by the ribosomal machinery. In other words, transcription and translation occur in different cell compartments and there are good reasons to believe that the mechanism of exporting messenger RNA from the nucleus towards the cytoplasm is performed by protein components of the type described by the Russian school (1) under the name of 'Informatin', rather than by the ribosomes themselves.

In contrast, in prokaryotes, and particularly in the bacterial cell, transcription and translation are *two closely coordinated processes*, to the extent that messenger RNA is never compartmentalized nor stored before it is read out by the ribosomal machinery. The most significant evidence of this is the now classical series of electron micrographic pictures taken by Hamkalo and Oscar Miller (2) of active chromosomal portions from a bacterial cell.

In fig. 1 one can see a low magnification picture of the extruded contents of osmotically shocked *Salmonella typhymurium* cells taken from a log phase culture. One can distinguish a network of DNase digestible fibers with attached ribonuclease sensitive polyribosomes.

Also quite convincing in this respect is the picture shown (fig. 2) of genetically active segments of *E. coli* chromosome with attached polyribosomes. The segments are 'clusters' of genes exhibiting coupled transcription and translation.

More recently, Antoinette Ryter at the Pasteur Institute has obtained evidence (3) that the *E. coli* nucleus from actively grown cells is, in fact, composed of two regions: one which appears as a dense vacuole and which is not transcribed under steady state conditions and another, not visible on electron micrography but which can be 'materialized' by autoradiographic methods, which is located at the periphery of the nucleus and is probably composed of extruded DNA loops that would 'stretch out' in the course of

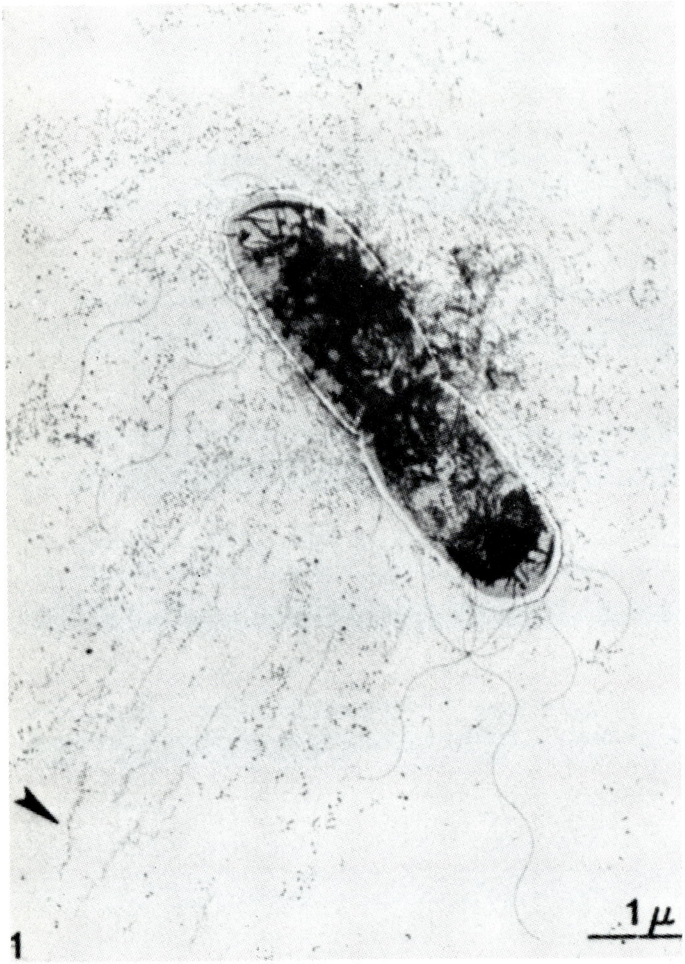

Fig. 1. Low magnification of the extruded contents of an osmotically shocked *Salmonella typhimurium* cell from a log phase, broth-grown culture. The arrow designates one of the 19 rRNA loci in the field.

transcription. The addition of puromycin — a typical translation inhibitor — causes a retraction of these loops, further suggesting that active translation is, in some sense, coupled with transcription as well as with chromosomal unfolding.

Although it is clear that transcription and translation are tightly coupled 'on time', the question arises as to whether this coupling is nothing but fortuitous. In other words, does the process of translation 'control', in

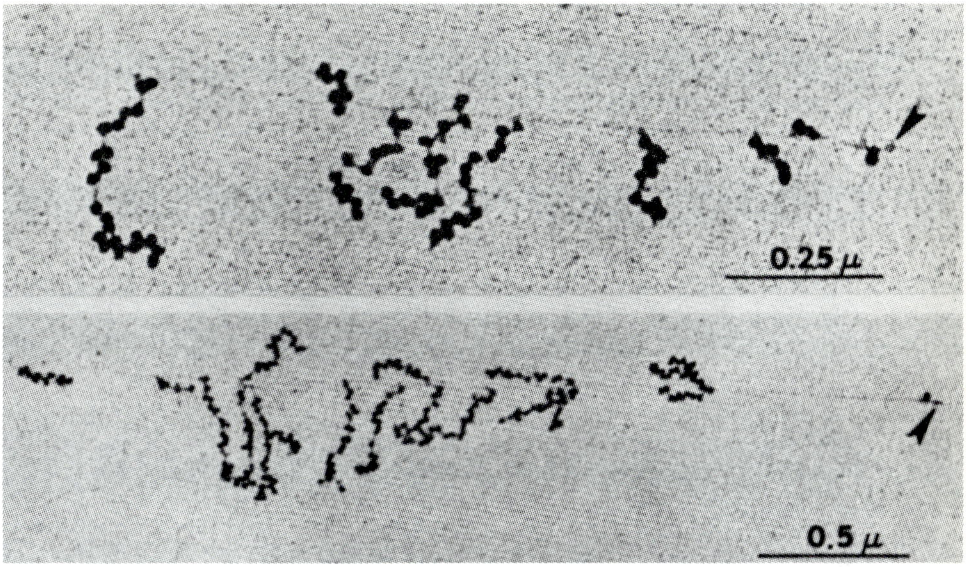

Fig. 2. Genetically active segments of *Escherichia coli* chromosome with attached polyribosomes. The arrows indicate RNA polymerase molecules on or very near the initiation sites for transcription.

some specific manner, the rate of transcription by the DNA-dependent RNA polymerase?

It was already suggested 10 years ago, by G. Stent (4), that ribosome movement along the messenger acts as a 'driving force' to 'release' the messenger chain from its template and that, in the absence of ribosome movement, chain propagation is somehow impaired. Many laboratories, including our own, have indeed observed that the rate of RNA transcription is proportional to the amount of ribosome per cell (5), but the exact nature of the coupling mechanism remained obscure until it was shown by Moldave (6) that ribosomes can stimulate DNA transcription in an in vitro reconstituted system. Later, Revel et al. (7,8) demonstrated that translation initiation factors — those protein fractions that are involved in the correct positioning of initiator tRNA on the ribosome—messenger complex — play a major role in this coupling since, in their absence, *E. coli* washed ribosomes only very weakly stimulated T_4 DNA transcription.

It was thus proposed that ribosome movement along the messenger is not the real "driving force" (as first suggested) but that it is the formation of a translation initiation complex which facilitates RNA chain release. However, given the crude state of purity of the factors — as then isolated — it was difficult to further specify which of them was actually responsible for the coupling effect.

In the present chapter we would like to report on experiments aimed at answering two questions:

1) How does the formation of a translation initiation complex affect the rate of transcription from a phage DNA template? Since there are numerous ligands and effectors involved in initiating protein synthesis, as well as many intermediary steps, this question remains fully relevant.

2) Can we define the type of ribosomal proteins that are specifically involved in the coupling between transcription and translation?

The approach to the first question was relatively simple: DNAs from bacteriophages of the λ or lambdoid family were copied in vitro with highly purified RNA polymerase plus the requisite nucleotide substrates in a radioactive state. The rate of RNA synthesis was measured by acid precipitation. To study the influence of the translation machinery, "NH_4Cl washed" ribosomal subunits were added to the system either alone or in the presence of part of or the whole effectors of translation initiation.

Firstly, as shown in table 1, "NH_4Cl washed" ribosomes, even when freed from initiation factors, display a small *but quite significant stimulating capacity* towards in vitro transcription. It is clear from this experiment that adding ribosomes prepared at high Mg^{2+} (i.e. enriched in 70S) causes a 2—3 fold stimulation. Purified 30S subunits are quite active as opposed to 50S, and when both subunits are present together, under conditions where a dimer can reform, the 30S activity is lost. This suggests, in passing, that the enhancing effect resulting from 70S addition might, in fact, be ascribed to contaminating 30S.

TABLE 1

Stimulation of ∅ 80 dlac DNA transcription by 70S ribosomes, 30S and 50S subunits. The reaction mixture contained, in a total volume of 0.10 ml: 0.04 M Tris—HCl, pH 7.9; 0.01 M $MgCl_2$; 0.1 mM EDTA; 0.1 M KCl; 2 mM dithiothreitol; 200 μM each of GTP, ATP, CTP and [^{3}H]UTP (800—5000 counts × min^{-1} × $nmole^{-1}$); 6 μg λ plac DNA or 4 μg ∅ 80 dlac DNA and 2 μg RNA polymerase purified as described by Burgess (19). The mixture was incubated for 20 min at 25°C in the presence or absence of 70S (100 μg) 50S (80 μg) or 30S ribosomes (60 μg) washed with a high salt buffer (Tris—HCl, pH 7.4, 0.02 M; $MgCl_2$, 0.01 M; NH_4Cl, 1 M; mercaptoethanol, 0.01 M). Samples were precipitated by 5% trichloroacetic acid at 0° C, filtered on Schleicher and Schüll Selectron membranes and counted in the Inter-technique scintillation counter.

Transcription reaction mixture	pmoles UMP incorporated into RNA
Control	420
+ 70S	1110
+ 30S	980
+ 50S	390
+ 30S + 50S	415

The ribosomes used in this experiment were "initiation factor free", and the small stimulation observed can be interpreted by saying that they can still attach to newly synthesized RNA, helping its release and thereby causing an increase in its rate of synthesis. We shall return to the nature of the ribosomal proteins that seem responsible for this effect later in this chapter.

What we should like to argue next is that this "primary" ribosomal effect is considerably enhanced when ribosomes have the capacity to form a stable initiation complex.

A first general indication that this is so is given by an experiment comparing ribosomes taken at each phase of the translation cycle. The 30S subunits used in the preceding experiment (table 1) were "derived", that is to say prepared artificially, by dissociating total ribosomes at low Mg^{2+}. After the cycle of translation has been completed, 70S ribosomes are released as so called "run off 70S"; they then become dissociated and the new 30S subunits combine with IF to form the so called "active 30S". It must be recalled that "derived" 30S and "run off" 70S are freed from initiation factors, these factors being released from polysomes at an early stage. It is clear from fig. 3 that "active 30S" are very efficient as compared with "derived" 30S or "run off" 70S. Therefore, 30S ribosomes carrying a complete set of factors are far more potent than are "factor-free" 30S, suggesting that *factor mediated* ribosomal complexes are the key elements in coupling.

In order to compare the *relative importance* of initiation factors in the ribosomal dependent coupling between transcription and translation, it was necessary to add them either *separately* or in *defined combination* to

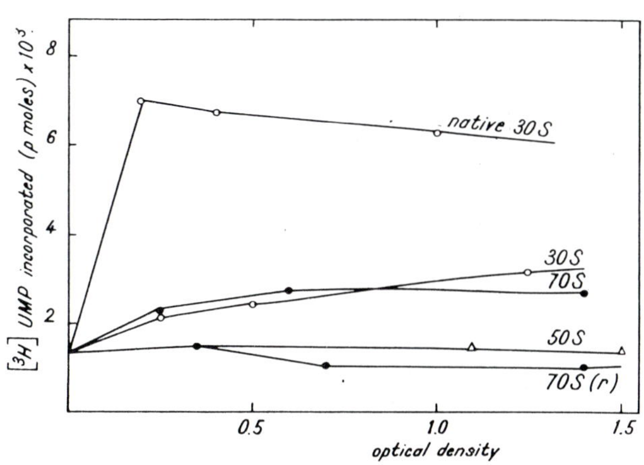

Fig. 3. Stimulation of λ plac DNA directed RNA synthesis by 30S and 50S ribosomal subunits or 70S ribosomes. To each reaction mixture various amounts of 30S, 50S or 70S ribosomes, washed with a high salt buffer (NH_4Cl, 1 M), were added as indicated. The same experiment was performed with "run off" 70S ribosomes designated 70S (r) and "native" 30S subunits isolated on sucrose gradient by zonal centrifugation.

TABLE 2

∅ 80 dlac DNA transcription in the presence of 70S ribosomes plus initiation factors. Saturating amounts of IF_2 (8 μg) and IF_3 (2 μg) factors were added to mixtures including 70S ribosomes (160 μg) and RNA synthesis was determined as described in the legend of table 1. Initiation factors were purified from the 1 M "NH_4Cl ribosomal wash fluid" obtained from 2 kg *E. coli* MRE 600. From this crude preparation initiation factors (IF_1, IF_2, IF_3) were precipitated by 80% saturation of ammonium sulphate. Further purification was achieved by chromatography on phosphocellulose (Hershey et al. (20)). IF_2 factor was further purified on DEAE Sephadex A-50 and IF_3 factor on Sephadex G-50.

Transcription reaction mixture	pmoles UMP incorporated into RNA
Control	530
+ 70S	1210
+ 70S + IF_2	2400
+ IF_2	560
+ 70S + IF_3	2840
+ IF_3	540

"NH_4Cl washed ribosomes" within a DNA—RNA polymerase transcription mixture. As seen in table 2, as well as in fig. 4, addition of IF_2 or IF_3 greatly enhances the ribosome dependent stimulation.

Yet, if one examines in more detail the kinetics of response with various factor concentrations, it is clear that IF_3 is far more active on a molar basis

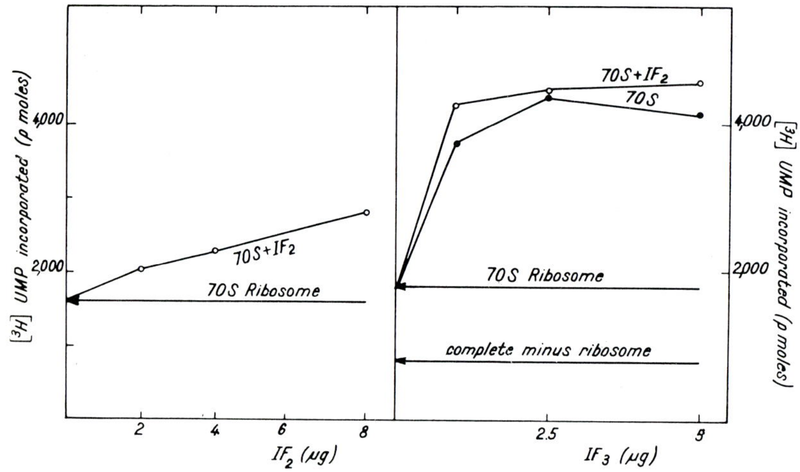

Fig. 4. Stimulation of λ plac DNA transcription by 70S ribosomes complemented with IF_2 factors. RNA synthesis was measured in incubation mixtures containing λ plac DNA (6 μg), RNA polymerase (2 μg), 70S ribosomes (120 μg) and either various amounts of IF_2 or various amounts of IF_3, alone or in the presence of IF_2 factor (8 μg).

than IF_2. Calculation shows that maximal stimulation is achieved for amounts of IF_3 corresponding to 0.3—0.5 molecule per ribosome.

This IF_3-dependent stimulation of the *coupling* is also observable when 30S are used instead of 70S. These results are consonant with the current views on translation initiation. It has been observed by Voorma and Bosch in Holland (9) that IF_3-dependent binding of 30S subunits to messenger precedes the interaction of the complex thus formed with IF_2 and fMet-tRNA.

$$70S \overset{+IF_3}{\rightleftarrows} 30S - IF_3 + 50S \quad (1)$$

$$30S - IF_3 + mRNA \rightarrow \underbrace{30S - IF_3 - mRNA}_{\text{complex I}} \quad (2)$$

$$\text{'Complex I'} + IF_2 + IF_1 + \text{fMet-tRNA} \rightarrow \underbrace{30S \begin{cases} IF_2 \text{—fMet-tRNA} \\ IF_1 \text{—mRNA} \end{cases}}_{\text{complex II}} + IF_3 \quad (3)$$

$$\text{'Complex II'} + 50S \xrightarrow{GTP} 70S \begin{cases} \text{fMet} \\ \text{mRNA} \end{cases} + IF_2 + IF_1 + GDP + P \quad (4)$$

The next question examined was whether formation of complex II in the presence of fMet-tRNA (step 3 above) causes a higher stimulation than formation of complex I.

That this is indeed the case is shown in table 3 and fig. 5. In table 3 it can be seen that addition of fMet-tRNA to the ribosomes and factors still further increases the coupling effect, maximal enhancement being achieved with 30S, IF_2, IF_3 and fMet-tRNA. It can be seen that under conditions where stable complex II is formed stimulation of transcription over the uncoupled system can reach up to 20-fold.

It could be argued that ribosomes, factors and tRNA do not stimulate transcription because they interact with newly synthesized RNA but rather in some direct fashion, perhaps at the level of the DNA-dependent RNA polymerase. Several arguments, which we cannot develop within the frame of this chapter, show that this is not so (10). One of them is based upon the use of an antibiotic, kasugamycin, which inhibits protein synthesis by causing release of initiator tRNA from the initiation complex. As expected, the addition of kasugamycin, which in itself has no direct effect on transcription, causes reduction in the degree of stimulation observed when initiation complex II is formed, the level of stimulation at kasugamycin excess corresponding to that that would be achieved in the presence of 30S + IF_3 alone (fig. 6).

TABLE 3

Transcription of λ plac DNA during formation of a translation initiation complex. Initiation factors were fractionated and purified as described in the legend of table 2. $tRNA_f^{Met}$ was aminoacylated and formylated as previously described (Revel et al. (21)). RNA synthesis was measured in the presence of $fMet\text{-}tRNA_f^{Met}$ (12 pmoles), saturating amounts of IF_1 (3 μg), IF_2 (8 μg) and IF_3 (2 μg) factors, plus 30S subunits (80 μg).

Transcription reaction mixture	pmoles UMP incorporated into RNA
Control	550
+ 30S	970
+ 30S + IF_3	1820
+ 30S + IF_2	1615
+ 30S + IF_3 + IF_2	2115
+ 30S + IF_3 + fMet-tRNA	3615
+ 30S + IF_3 + IF_2 + fMet-tRNA	3990
+ 30S + IF_2 + fMet-tRNA	2670
+ 30S + fMet-tRNA	1010

Taken together, these data clearly suggest that it is the step of formation of a translation initiation complex which is probably responsible for the large stimulation observed in the rate of RNA synthesis whenever DNA transcription is coupled with protein synthesis.

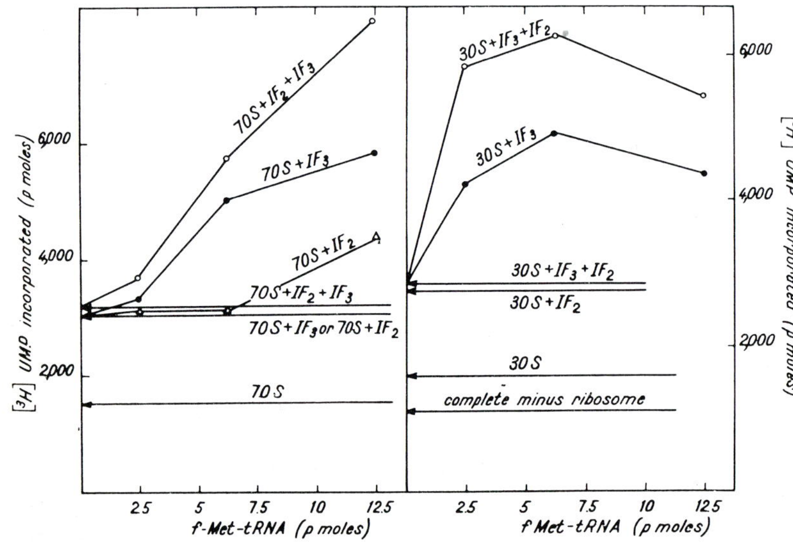

Fig. 5. fMet-tRNA dependent stimulation of λ plac RNA synthesis during formation of a translation initiation complex. Standard conditions were used for λ plac DNA transcription. Assays were carried out in the presence of different concentrations of fMet-$tRNA_f^{Met}$, 70S ribosomes and IF_2 or IF_3 factors.

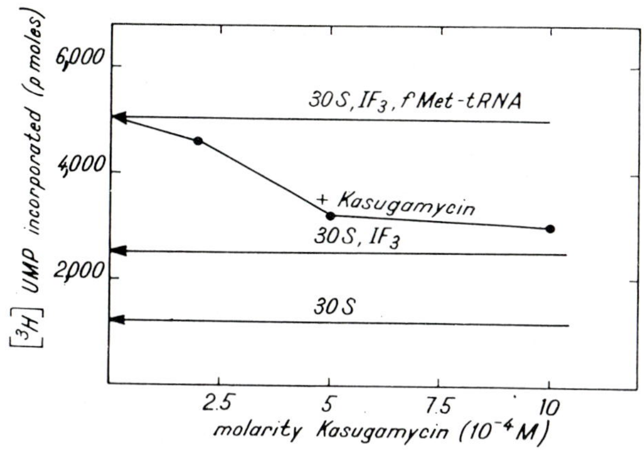

Fig. 6. Effect of kasugamycin on transcription in the presence of 30S, IF_3 and fMet-tRNA. After preincubation of λ plac DNA (6 μg) and RNA polymerase (2 μg) at 25° C, for 4 min, transcription was initiated by the combined addition of nucleoside triphosphates, in the presence of 30S subunits (80 μg), IF_3 factor (2 μg), fMet-$tRNA_f^{Met}$ (12 pmoles) and several different amounts of kasugamycin.

Before examining the second general question raised in our introduction it is important to discuss the stage of the overall transcription process which is sensitive to coupling. In brief, we have observed (10) that *both* the rate of propagation and the number of chains initiated per unit time are increased. This latter effect could be explained, particularly if one agrees with C. Thomas that newly transcribed RNA chains transiently remain associated with the DNA template, in a way similar to the one depicted in fig. 7 (taken from Thomas' work), this situation being responsible for the appearance of "DNA bubbles". By interacting with the "initiation region" of the messenger, ribosomes would facilitate release of the RNA product from the DNA and permit more polymerase molecules to find access to the promoter proximal region.

The role of ribosomal proteins in the coupling mechanism

Can one better define the role of the ribosome itself in the coupling mechanism?

More precisely, it is known that the small ribosomal particle contains 21 different structural proteins, each being present in a single copy. These proteins obviously have distinctive functions during RNA translation, particularly with respect to the messenger RNA binding, fixation of initiator or internal tRNAs, peptide synthesis, translocation etc.

It would be interesting to specify which among the 21 distinct protein species permit binding of 30S to a newly synthesized RNA transcript.

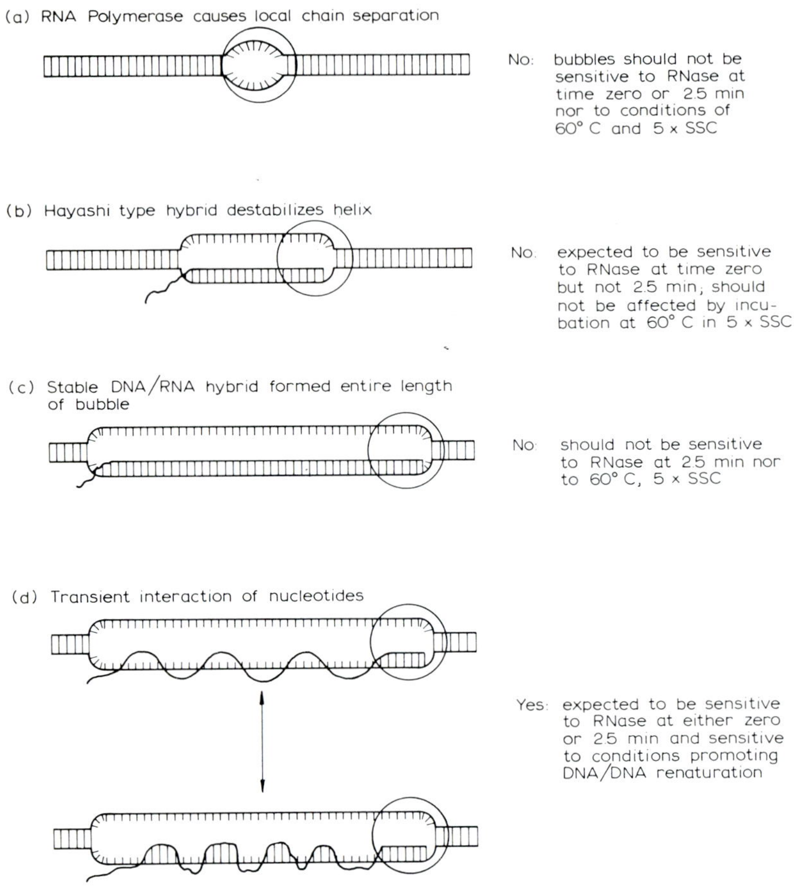

Fig. 7. Models to account for transcription bubbles. (According to C. Thomas.)

One way to tackle this problem is provided by the availability of specific monovalent antibodies directed against each individual 30S protein. Such "Fab" fractions have been obtained completely pure by workers in Stoeffler's group in Berlin (11, 12). Hence, we have undertaken to probe each one for its capacity to "uncouple" transcription and translation. To do this each purified Fab was added, at various concentrations, to a system in which the transcription of λ plac was stimulated by ribosomes plus initiation factors. Fig. 8 describes the effects of each antibody on the "differential", i.e. ribosomal dependent stimulation. It must be outlined that none of the tested antibodies showed any effect by itself on transcription (i.e. when tested in the absence of ribosomes). Moreover a non-immune antiserum prepared under the same conditions displayed very little activity.

In table 4 the results were computed by calculating the dose of Fab pro-

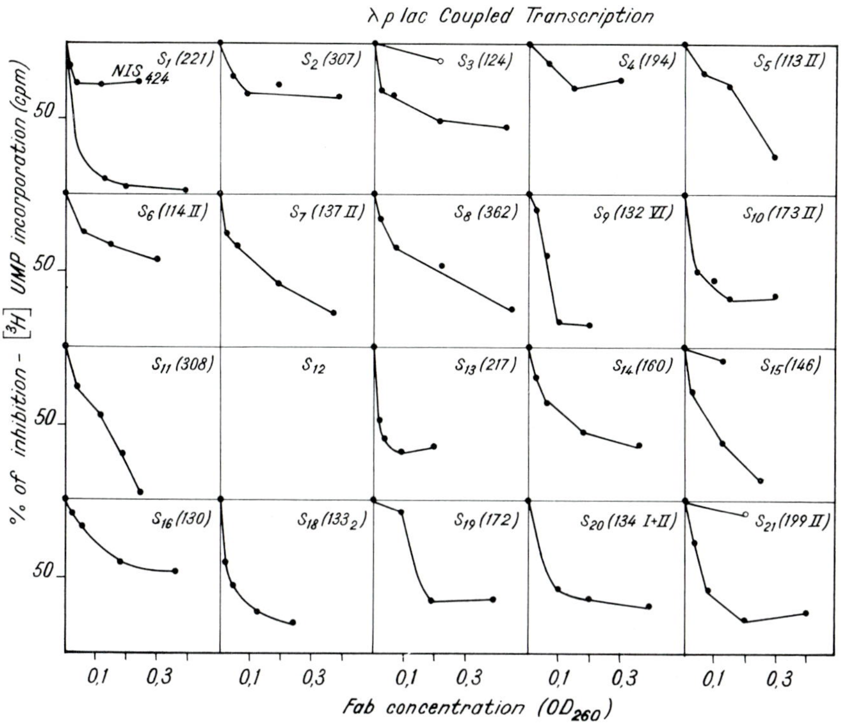

Fig. 8. Concentration dependence of Fab mediated inhibition of plac DNA "coupled" transcription in the presence of 30S ribosomal subunits.

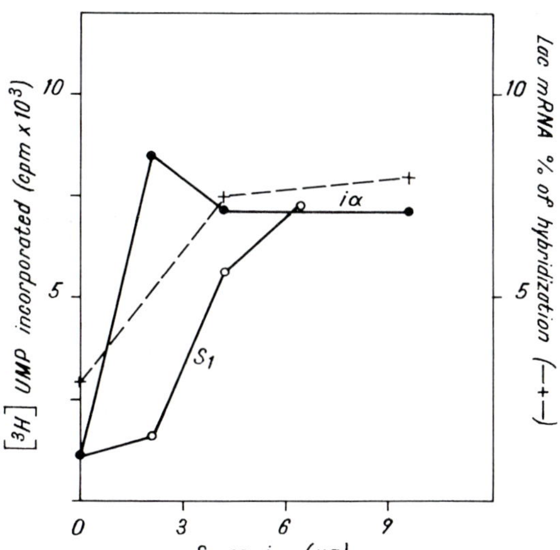

Fig. 9. Effect of S_1 and interference factor iα on the overall and lac specific in vitro RNA synthesis using a lac transducing DNA plus a conventional RNA polymerase system.

TABLE 4

Tentative classification of 30S proteins with regard to their involvement in ribosome dependent stimulation of in vitro phage DNA transcription.

Fabs	Relative amounts of Fab required for 50% inhibition (in OD_{280})	Extent of inhibition
NIS	>0.4	—
S_1	0.07	++
S_2	>0.4	—
S_3	0.2	—
S_4	>0.4	—
S_5	0.2	—
S_6	>0.4	—
S_7	0.15	+
S_8	0.2	—
S_9	0.06	++
S_{10}	0.05	++
S_{11}	0.13	+
S_{12}	ND	ND
S_{13}	0.03	++
S_{14}	0.15	+
S_{15}	0.19	—
S_{16}	>0.4	—
S_{17}	ND	ND
S_{18}	0.04	++
S_{19}	0.1	+
S_{20}	0.08	++
S_{21}	0.07	++

ducing 50% inhibition. From such a comparison one may derive the data from table 5 which allow the classification of the antibodies into 3 categories: i) those which cause *no inhibition* and which are directed against S_2, S_3, S_4, S_5, S_6, S_8, S_{15}, S_{16}; ii) those which have a *moderate action* they

TABLE 5

30S proteins involved in the coupling of transcription and translation according to immunochemical studies. Comparative effects of anti-30S protein Fabs on ribosome dependent stimulation of in vitro phage DNA transcription.

Major effect	S_1, S_9, S_{10}, S_{13}, S_{18}, S_{20}, S_{21}
Minor effect	S_7, S_{11}, S_{14}, S_{19}
No effect	S_2, S_3, S_4, S_5, S_6, S_8, S_{15}, S_{16}
Not tested	S_{12}, S_{17}

include antibodies against $S_7, S_{11}, S_{14}, S_{19}$; and iii) those which *strongly* inhibit coupling. The antigens corresponding to these Fabs are S_1, S_9, S_{10}, $S_{13}, S_{18}, S_{20}, S_{21}$.

Hence, it can be proposed that proteins from this latter group play a major role in the "recognition" of nascent mRNA.

Interestingly, protein S_1 falls within this group. This deserves special mention because it was reported some years ago by Van Duyn and Kurland (13) that S_1 has a key function in the binding of synthetic messengers to 30S subunits, and hence could be regarded as a universal device for "hooking" the triplets of the messenger chain onto ribosomes. This view concerning the key function of protein S_1 is reinforced by the observation according to which S_1 is one of the few proteins in the vicinity of which IF factors do bind (14).

It has also been shown that S_1 (which incidently is also one of the Qβ replicase subunits) is capable of interacting directly with mRNA chains even in the absence of ribosomes; this does not apply solely to Qβ or R_{17} phage RNAs (15) but to various types of messengers, including synthetic homopolymers (16).

These considerations prompted us to study whether S_1, when added in large amounts and in a free state, could stimulate in vitro transcription. As a control we tested protein S_9 which also belongs to the category of "coupling proteins".

The results in fig. 9 are quite clear in showing that addition of S_1 alone

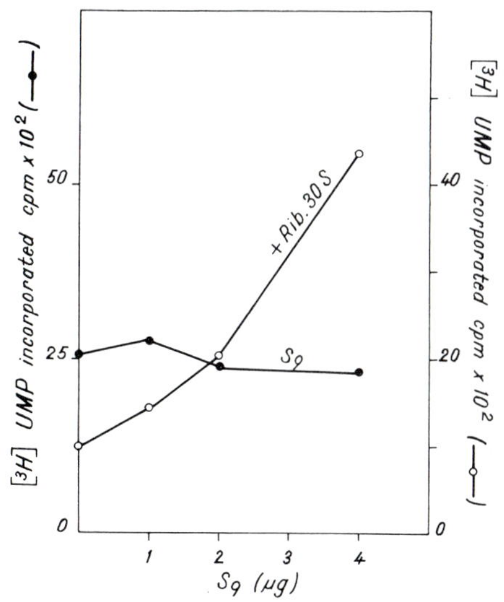

Fig. 10. Effect of protein S_9 on the overall rate of transcription of a lac transducing DNA, in the presence or absence of washed 30S subunits.

(or of iα, which is supposedly the same entity) stimulates in vitro transcription of λ h80 dlac DNA whereas free S_9 does not (fig. 10).

It cannot be excluded that S_1, which is known to exist in both "bound" and "free" states in the cell sap, could act directly at the polymerase level (17), but we prefer to regard this S_1-dependent stimulation as a reflection of the fact that it is capable of functioning as a "linker" between nascent mRNA and ribosomal subunits, thereby possessing enough affinity for the "nascent RNA" to help its release from the DNA template. The S_1 effect on transcription constitutes an excellent argument in support of the view that coupling involves conformational changes of the nascent RNA rather than actual protein synthesis. Similar data with S_1 have been reported by Leavit et al. using a different system.

Conclusion

In conclusion, the sequence of events underlying the "coupling" between translation and DNA transcription is complex, but, schematically, we think it is based upon interaction of the newly synthesized RNA chain with the small ribosomal subunit, a process primarily involving protein S_1 plus some "satellite" proteins such as S_9, S_{10}, S_{13}, S_{18}, S_{20}, S_{21}, etc.

i) This "factor free" ribosome messenger interaction leads to formation of unstable complexes and, consequently, causes only weak effects on the rate of transcription initiation.

ii) Reinforcement of these complexes would be mediated by initiation factor IF_3, the role of which is to permit the small subunit to select with great accuracy the portion of the mRNA chain where protein synthesis begins (18) (a portion that can be designated as the "initiation region").

iii) Further reinforcement of the 30S—messenger complex would then be achieved by the IF_2-dependent positioning of the initiator tRNA.

We suggest that it is the formation of this stable "initiation complex" which, in a manner that remains to be elucidated, would stimulate "release" of the messenger 5′ proximal portion from the DNA template.

This situation illustrates the degree of sophistication involved in what can be regarded as the nuclear cytoplasmic interactions occurring among prokaryotic systems.

References

1. O.P. Samarina, E.M. Lukanidin and G.P. Georgiev, 6th Karolinska Symp. Research Methods in Reproductive Endocrinology, May 21 (1973) 130.
2. O.L.E. Jr. Miller, B.R. Beatty, B.A. Hamkalo and C.A. Jr. Thomas, Cold Spring Harbor Symp. Quant. Biol. 35 (1970) 505.
3. A. Ryter, personal communication.
4. G.S. Stent, Science 144 (1964) 816.

5. S. Naono, J. Rouvière and F. Gros, Biochim. Biophys. Acta 129 (1966) 271.
6. D. Shin and K. Moldave, J. Mol. Biol. 21 (1966) 231.
7. M. Revel and F. Gros, Biochem. Biophys. Res. Comm. 27 (1967) 92.
8. M. Revel, M. Herzberg, A. Becarevic and F. Gros, J. Mol. Biol. 33 (1968a) 231.
9. H.O. Voorma and L. Bosch, personal communication.
10. M. Crepin, J.C. Lelong and F. Gros, 6th Karolinska Symp., Research Methods in Reproductive Endocrinology, May 21—23 (1973) 33.
11. G. Stöffler, R. Masenbank, M. Lütgenaus, R. Maschler, C.A. Morrison, H. Zeichhardt and R.A. Garrett, Mol. Gen. Genetics 127 (1973) 89.
12. J.C. Lelong, D. Gros, F. Gros, A. Bollen, R. Maschler and G. Stöffler, Proc. Nat. Acad. Sci. U.S. 71 (1974) 248.
13. J. Van Duyn and P.H. Van Knippenberg, J. Mol. Biol. 84 (1974) 185.
14. A. Bollen, L. Kahan, R.L. Heimark, A. Cozzone, R.R. Traut and J.W.B. Hershey (J. Biol. Chem., sous presse).
15. R. Kempfer, personal communication.
16. M. Tal, M. Avivam, A. Kanarck and A. Weiss, Biochim. Biophys. Acta 281 (1972) 381.
17. J.C. Leavitt, K. Moldave and D. Nakada, J. Mol. Biol. 70 (1972) 15.
18. M. Revel, H. Aviv, Y. Groner and Y. Pollack, FEBS Letters 9 (1970) 213.
19. R.R. Burgess, J. Biol. Chem. 244 (1969) 6160.
20. J.W.B. Hershey, E. Romold-O'Donnell, D. Kolakofsky, K.F. Dewey and R.E. Thach, in Moldave K. and L. Grossmar (Eds.) Methods in Enzymology, Vol. 20, Academic Press, N.Y. (1971) p. 235.
21. M. Revel, G. Brawerman, J.C. Lelong and F. Gros, Nature 219 (1968b) 1016.

Dr. OHAD:

Is it possible that some of the antibodies against the 30S specific proteins do not recall the proteins located inside the core?

Dr. GROS:

I do not think so. This point has been examined with great caution by the group in Berlin (Stöffler, Maschler and Wittmann). Using several independent criteria (such as gel diffusion techniques especially adapted to ribosomes, studies of the hydrodynamic properties of antibody-ribosome complexes and, more recently — in situ detection by electron microscopic techniques), they could show that all the anti-30S protein antibodies readily interact with their cognate antigen on the ribosome. This conclusion is further supported by the fact that virtually all the tested Fab species inhibit poly-U-directed polyphenylalanine synthesis, even though the extent of inhibition varies with the kind of antibody in use.

Dr. BOGORAD:

What about the transcription -- translation coupling for nontranslated RNAs? (i.e., rRNAs, tRNAs).

How much free S_1 is there in the *E. coli*?

Could this affect the control of transcription without actual coupling to translation?

Dr. GROS:

Concerning transcription coupling for nontranslated RNAs there is for the time no compelling evidence that such a situation does in fact exist, although many obvious models could be proposed (such as the possible interaction of ribosomal proteins, or aminoacyl RNA synthetases with their cognate polynucleotides, in statu nascendi . . . etc.).

The question as how much S1 exists free in the cell sap is still controversial. If there is a substantial amount the possibility of S1 directly interacting with the DNA dependent RNA polymerase cannot be ruled out (see 17).

But one of the major difficulties in interpreting *any* modification which is introduced into a ribosome by some kind of chemical, physical or immunological reagent can be biased by the nature of the cooperative forces involved in the maintenance of the organelle. It is interesting to notice however that, according to Zimmerman et al. (1972), S_7, S_9, S_{10} are located on or near the 3′ end portion of the 16S RNA, together with other protein components and yet anti-S_9 and S_{10} show a strong inhibition effect whereas anti-S_7 does not, suggesting that topological orientation with respect to S_1 does not suffice to explain all the observed effects.

CHAPTER I.2

tRNA STRUCTURE IN VIRAL GENOMES AND ITS POSSIBLE ROLE IN GENE REPLICATION AND TRANSLATION

A. PROCHIANTZ, C. BÉNICOURT, P. YOT and A.-L. HAENNI

Laboratoire de Biochimie du Développement, Institut de Biologie Moléculaire du C.N.R.S. et de l'Université Paris VII, 2 Place Jussieu, 75005 Paris

Introduction

Ever since it was shown that tRNAs are present in all cells, and that they play a key role in protein biosynthesis, they have been extensively studied, and to date they are the best known of the macromolecules of the translation machinery. Their small size — about 25,000 daltons — their great stability and their importance in deciphering the genetic code explain the wide interest which these molecules have aroused. Moreover, because they are readily available in pure and often in crystalline form, tRNA molecules represent unique models for the study of nucleic acid-protein interactions. Indeed, tRNAs are recognized by a series of very specific enzymes, many of which can now be obtained in purified form.

tRNA specific enzymes

Presented below, is a brief survey of some of these tRNA specific enzymes and of the reactions they catalyze.

1) The numerous modified bases found in all tRNAs result from the action of specific enzymes (methylases, for example) that recognize not only certain bases, but also certain positions of these bases within the tRNA molecule.

2) The 3′ terminus of all tRNAs is composed of the nucleotide sequence -CCA, which is metabolically less stable then the rest of the molecule. The tRNA nucleotidyltransferase is an enzyme capable of specifically adding back the two CMPs and the AMP to tRNA molecules lacking the intact 3′ terminal sequence.

3) Recently, the demonstration of the existence of 5S tRNA precursors, that is tRNA molecules containing some 40 extra bases at their 5′ end, has led to the isolation of a specific endonuclease, RNase P, which cleaves off these extra bases to yield the 4S tRNA molecule.

4) Of all the enzymes that recognize tRNAs, the aminoacyl-tRNA synthetases are undoubtedly the most specific: every cell contains 20 aminoacyl-

tRNA synthetases, each one being specific for a given amino acid and responsible for the esterification of that amino acid onto its cognate tRNA.

5) In turn, the aminoacyl-tRNAs are recognized by a soluble elongation factor of protein synthesis, and it is in the form of a complex with this factor and GTP that aminoacyl-tRNAs are correctly positioned on the ribosome for peptide synthesis.

6) Finally, peptidyl-tRNAs, or aminoacyl-tRNAs with a blocked amino group on the amino acid are recognized by another specific enzyme of tRNAs, the peptidyl-tRNA hydrolase which cleaves off the peptide or the N-acyl amino acid moiety from the tRNA.

This rapid survey shows us the multiplicity of the possible interactions that tRNA molecules can undergo with these various specific enzymes.

Regulation at the level of tRNAs

A regulation of translation could in theory operate at the level of any of these interactions.

It has for instance been postulated that the peptidyl-tRNA hydrolase could regulate protein synthesis by regenerating tRNAs from peptidyl-tRNAs that had accidentally left the ribosomal machinery. Likewise, the modification of a pre-existing aminoacyl-tRNA synthetase by bacteriophage infection of *E. coli* also suggests that these enzymes could regulate translation. Another example is the modification of a pre-existing tRNA — such as by methylases — which could lead to a more active, or to a less active tRNA species for protein synthesis.

Other types of regulation by tRNAs exist. The DNA bacteriophage T_4, for instance, possesses in its genome the information for the synthesis of several tRNAs not present in the uninfected cell, and whose anticodons correspond to codons frequently represented in the messenger RNA (mRNA) of the phage; therefore, the presence of these tRNAs favors the synthesis of phage-specific proteins. A similar situation could be one wherein a given cistron contained many codons corresponding to a rare tRNA species in the cell; in these conditions, the rate of protein synthesis would be regulated by the intracellular concentration of that tRNA.

One should not neglect another possible primordial role afforded by tRNAs. Cytokinins are modified bases found in tRNAs. During tRNA degradation, these cytokinins are released and appear to act as plant hormones.

Functions of tRNAs

As aminoacyl-tRNAs, the main function of tRNAs is to donate their amino acid for protein synthesis in a system requiring the complete translation machinery including ribosomes, and which characteristically is dependent on the presence of an mRNA as template. For a long time this was considered the sole function of tRNAs. More recently it has become apparent, however, that tRNAs are pleiotropic, and operate at various levels in cell development.

Aminoacyl-tRNAs have been shown to donate their amino acid for peptide synthesis in the absence of ribosomes and mRNA. This is the case of certain aminoacyl-tRNAs that donate their amino acid for bacterial cell wall synthesis. It is also the case of the N-terminal addition of amino acids to performed proteins, which also operates via aminoacyl-tRNAs. In both of these situations, the macromolecular species with which the aminoacyl-tRNAs interact are very different from those they encounter in the classical protein synthesizing system.

Aminoacyl-tRNAs can also function as regulatory molecules without donating their amino acid. Thus, histidyl-tRNA acts as a co-repressor of the histidine operon.

Recently, tRNAs have appeared to possess still another very different function from those of donating amino acids for protein synthesis. As uncharged molecules, they reportedly act as primers of DNA synthesis. Indeed, studies carried out with certain oncogenic RNA viruses such as the Rous Sarcoma Virus (RSV) and the Avian Myeloblastosis Virus have revealed that these viruses are capable of encapsidating host tRNAs. What is more, of these tRNAs, some can easily be removed from the viral genome, whereas others are retained very strongly and can only be released by such drastic treatments as heating the RNA in the presence of EDTA. The work pursued with the RSV RNA strongly suggests that a tRNA of host origin and specific for tryptophan is the molecular species tightly retained by the viral RNA, and that it is the primer molecule for the synthesis of DNA complementary to the viral RNA catalyzed by the RNA-dependent DNA polymerase.

In the examples of tRNA functions presented so far, the tRNAs are always operative as well-defined small molecular species. However, in the last few years, evidence has accumulated that tRNA structures when they are integrated into much larger RNA molecules might also be operative. Thus, several viral RNAs appear to possess a tRNA structure at their 3′ end as an integral part of their viral genome, judging from the fact that these genomes are recognized by several enzymes mentioned above and normally thought to be specific of tRNAs (1—14).

This intriguing feature will be illustrated here using the RNA of Turnip Yellow Mosaic Virus (TYMV) as an example; indeed studies with this RNA have brought to date the most conclusive evidence that it possesses a tRNA structure. The fact that this RNA of 2×10^6 daltons interacts with all the tRNA specific enzymes examined to date suggests that its corresponding tRNA structure, which is not masked by 98% of the remaining RNA, must be of primordial importance for virus development.

Results

Esterification of TYMV RNA by valine

Using an *E. coli* cell-free extract devoid of tRNAs, we have shown that valine is esterified to the 3′ end of TYMV RNA in the conditions that allow the esterification of valine onto tRNA (1,2). No other amino acid could be bound to the RNA; the ratio of valine esterified per RNA molecule of 2×10^6 daltons varied with the batch of RNA between 0.3 and 0.8, but never exceeded 1.

The following four methods enabled us to define the nature of the linkage between valine and the RNA of TYMV.

1) The half-life time of TYMV valyl-RNA was determined by incubation at 37°C and pH 8.6: it was of 60 min, which corresponds to the half-life time of valyl-tRNA in the same conditions, suggesting an ester linkage between valine and the viral RNA which is hydrolyzed at that pH.

2) When TYMV valyl-RNA was acetylated chemically, acetylvalyl-RNA was formed, indicating that valine in valyl-RNA possesses a free amino group.

3) In the presence of the peptidyl-tRNA hydrolase, acetylvaline was liberated from TYMV acetylvalyl-RNA, as it is from acetylvalyl-tRNA.

4) Direct analysis of the base in the viral RNA to which valine was bound was performed by treatment of TYMV valyl-RNA by pancreatic RNase and snake venom phosphodiesterase, and yielded adenosine.

We concluded from these experiments that in TYMV RNA, valine is attached to the 3′ position of a terminal adenosine by an energy rich ester bond.

RNA species to which valine is bound

In order to define if the valine acceptor might be a contaminating tRNA molecule of host origin that co-purified with TYMV RNA, attempts were made to liberate such a putative tRNA using treatments with sodium dodecylsulfate, urea, EDTA, or dimethylsulfoxide and formaldehyde. However, after all these treatments, valine still co-chromatographed with high molecular weight RNA (1,2).

We also excluded the possibility that a $tRNA^{Val}$ of host origin had become covalently linked to the 3′ end of the viral genome by an RNA ligase (2). To this effect, tRNA was extracted from the uninfected host (Chinese cabbage leaves), charged with ^{14}C-valine, submitted to T_1 RNase digestion together with ^3H-valyl-RNA of TYMV, and the valyl-oligonucleotides formed were analyzed using a DEAE-cellulose column in the presence of urea. At least two valyl-oligonucleotides originated from the host valyl-tRNA and must correspond to at least two tRNA acceptors of valine in the host cells. On the other hand, valyl-RNA of TYMV yielded only one valyl-oligonucleotide, and its position of elution differed from the position of those pertaining from host valyl-tRNA.

These experiments thus demonstrated that valine forms an ester bond with the adenosine at the 3′ end of the viral genome itself and not of a contaminating tRNA.

Recognition by valyl-tRNA synthetase and the tRNA nucleotidyltransferase

In order to define which enzyme in the *E. coli* cell-free extract was responsible for the esterification reaction, the extract was replaced by highly purified valyl-tRNA synthetase. Surprisingly, in these conditions, no esterification of valine was observed, and only if tRNA nucleotidyltransferase was included could the charging reaction be restored (2,3). It was further demonstrated that the viral RNA binds AMP covalently in the presence of tRNA nucleotidyltransferase (15).

Therefore, the RNA as it is extracted from the virion is devoid of a 3′ terminal adenosine whose presence is mandatory for charging to occur, and further, valyl-tRNA synthetase is the enzyme responsible for the esterification of valine onto TYMV RNA.

Having defined that three enzymes normally thought to be specific of tRNAs, namely the tRNA nucleotidyltransferase, valyl-tRNA synthetase and the peptidyl-tRNA hydrolase, recognize the 3′ region of TYMV RNA, it appeared of interest to examine the effect of other tRNA specific enzymes in such a system.

Recognition by an endonuclease

Incubation of TYMV RNA (charged or not with valine) with an enzyme fraction behaving similarly to RNase P revealed that there resulted the formation of a 4.5S RNA fragment corresponding to the 3′ region of TYMV RNA, indicating a great similarity between the structures in the 5S tRNA precursor and in the 3′ region of TYMV RNA (9). Since the isolated 4.5S fragment could also be charged by valine, it was concluded that the integrity of the viral genome is not required for esterification to proceed, and that other parts of the viral genome do not participate in the tRNA-like structure.

Recognition by the ribosomal system

Using an *E. coli* protein synthesizing system, TYMV RNA as mRNA, a mixture of tRNAs, all the amino acids except valine, and either TYMV valyl-RNA or the 4.5S RNA fragment charged with valine, it was found that some of the valine was recovered in the hot TCA precipitate (8,14). By including in the incubation mixture a competitive inhibitor of valyl-tRNA synthetase, namely valinol-AMP, we were able to verify that the incorporation observed corresponded to the direct transfer of valine from TYMV RNA, and not to incorporation after a transacylation process. Experiments are under way to define whether a further maturation or modification of the valyl-RNA is a prerequisite prior to the incorporation of valine into peptide chains.

Recognition by the elongation factor EF-T

The fact that valyl-RNA donates its amino acid moiety for protein synthesis, led us to investigate whether it could form a complex with GTP and the *E. coli* elongation factor EF-T. Ternary complex formation had already been reported using the corresponding wheat germ elongation factor (7).

Experiments carried out using highly purified EF-T (EF-Tu and EF-Ts) demonstrated that both valyl-RNA and valyl-4.5S RNA of TYMV could form such a ternary complex in conditions identical to those leading to the formation of a valyl-tRNA · GTP · EF-T complex (14). As with tRNAs, the uncharged RNA did not lead to complex formation.

Discussion

The results of the experiments described above suggested the model presented below (fig. 1) for TYMV valyl-RNA (2,14). The tRNA-like structure is encompassed within the 3′ region of the viral RNA. As it is extracted from the virion, the RNA terminates with the sequence -CC; a terminal adenosine must be bound to the RNA prior to charging with valine.

Since these experiments were undertaken, other viral RNAs have been reported to accept an amino acid. The RNAs of other Tymoviruses, such as the Okra Mosaic Virus and the Eggplant Mosaic Virus, also accept valine (4,10); the RNAs of the Bromoviruses as well as Cucumber Mosaic Virus accept tyrosine (5,13); Tobacco Mosaic Virus binds histidine (6,7). With respect to animal systems, Mengovirus RNA has been shown to accept an

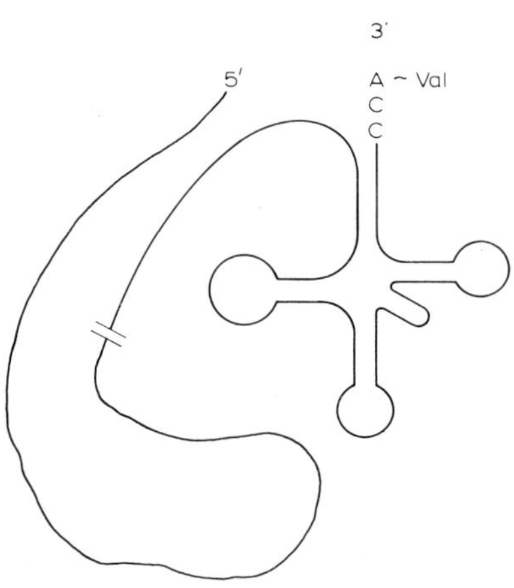

Fig. 1. Possible model for TYMV valyl-RNA.

amino acid, namely histidine (11), and the RNA extracted from Encephalomyelocarotidis Virus appears to be charged with serine (N. Carey et al., personal communication).

These experiments might indicate that other viral RNAs contain a tRNA-like structure in their 3′ region. It should, however, be pointed out that the RNA of several animal viruses are terminated at their 3′ end by a poly(A) sequence. This nevertheless does not exclude the possibility that the poly(A) sequence follows a tRNA-like structure in the genome; it is known that tRNAs are efficiently used as primers by the poly(A) synthetase.

As for RNA bacteriophages, they are known to possess in their genome the 3′ terminal sequence -CCA common to all tRNAs. Moreover, in the case of Qβ at least, it has been shown that two of the four subunits of the replicating system are the elongation factors EF-Tu and EF-Ts (16). These indications prompted us to look for tRNA-like structures in these RNAs. All attempts at charging the genome of RNA bacteriophages with an amino acid have been unsuccessful to date. However, we have observed that after treatment of Qβ, R17 and MS2 RNA with commercial snake venom phosphodiesterase such that it removes the 3′ terminal -CCA sequence from tRNAs, the tRNA nucleotidyltransferase adds back two CMPs for one AMP in conditions in which it is highly specific for tRNAs. It therefore appears that in the genome of RNA phages, parts of tRNA structures might be present.

The physiological role of tRNA-like structures in viral genomes is as yet unknown. At least two possible levels at which such structures could be involved in the regulation of virus development can be envisaged.

1) The tRNA-like structure could be involved in the recognition of the genome by the RNA replicase. Were the replicating system of these viruses to contain the elongation factors as subunits, then the tRNA structure could represent the binding site for the replicase, and could thereby regulate the replication process.

2) The tRNA-like structure could play a regulatory role at the level of translation, either by preferentially favoring the synthesis of virus-specific proteins, or conversely, by interfering with host-specific protein synthesis.

The fact that several viral RNAs contain a tRNA-like structure raises the question of the origin of these structures. It will be interesting to compare the primary sequence of the tRNA-like structure with the primary sequence of the corresponding cytoplasmic, mitochondrial and chloroplastic isoacceptor tRNAs in the host cell. Should significant similarities be found between the tRNA-like structure and one of the isoacceptors, it will be tempting to draw conclusions as to the possible origin of these viruses.

Acknowledgements

This work was carried out in Dr. F. Chapeville's laboratory; we wish to thank him for his constant encouragement and for many stimulating discussions.

These studies were supported by research grants from NATO (No. 769) and from the ATP Différenciation Cellulaire (No. 4.910), Centre National de la Recherche Scientifique.

References

1. M. Pinck, P. Yot, F. Chapeville and H.M. Duranton, Nature 226 (1970) 954.
2. P. Yot, M. Pinck, A.L. Haenni, H.M. Duranton and F. Chapeville, Proc. Nat. Acad. Sci. U.S. 67 (1970) 1345.
3. S. Litvak, D.S. Carré and F. Chapeville, FEBS Letters 11 (1970) 316.
4. M. Pinck, S. Chan, M. Genevaux, L. Hirth and H.M. Duranton, Biochimie 54 (1972) 1093.
5. T.C. Hall, D.S. Shih and P. Kaesberg, Biochem. J. 129 (1972) 969.
6. B. Öberg and L. Philipson, Biochem. Biophys. Res. Commun. 48 (1972) 927.
7. S. Litvak, L. Tarrago-Litvak and J.E. Allende, Nature New Biol. 241 (1973) 88.
8. A.L. Haenni, A. Prochiantz, O. Bernard and F. Chapeville, Nature New Biol. 241 (1973) 166.
9. A. Prochiantz and A.L. Haenni, Nature New Biol. 241 (1973) 168.
10. M. Pinck, M. Genevaux and H.M. Duranton, Biochimie 56 (1974) 423.
11. R. Salomon and U.Z. Littauer, Nature 249 (1974) 32.
12. C. Benicourt and A.L. Haenni, FEBS Letters 45 (1974) 228.
13. R.J. Kohl and T.C. Hall, in press.
14. A.L. Haenni, A. Prochiantz and P. Yot, in D. Richter (Ed.) Lipmann Symposium: Energy, Biosynthesis and Regulation in Molecular Biology, De Gryter & Co., Berlin, New York (1974).
15. S. Litvak, L. Tarrago-Litvak and F. Chapeville, J. Virol. 11 (1973) 238.
16. T. Blumenthal, T.A. Landers and K. Weber, Proc. Nat. Acad. Sci. U.S. 69 (1972) 1313.

Dr. WEIL:

Is the viral RNA, deprived of the 3′ terminal tRNA like structure, still infectious?

Drs. PROCHIANTZ and HAENNI:

Such an experiment is difficult to perform in so far as the endonuclease preparations used to liberate the 4.5S RNA fragment are still contaminated with other RNases which fragment the viral RNA elsewhere, and we therefore do not have an intact RNA deprived of its 4.5S tRNA-like structure that would allow us to answer this question. However, it is very likely that the viral RNA deprived of its tRNA-like structure is not infectious. This presumption is based on experiments carried out with TMV RNA and with R17 and Qβ RNA: elimination of the 3′ terminal adenylate from TMV RNA causes a marked loss in infectivity*; on the other hand the infectivity of R17 and Qβ RNA does not require the presence of the terminal adenylate, but is lost on removal of the penultimate cytidylate**,***. It should be stressed that in the case of RNA phages, the terminal adenylate is probably added after replication.

* A. Steinschneider and H. Fraenkel-Conrat, Biochemistry 5 (1966) 2735.
** K. Kamen, Nature 221 (1969) 321.
*** U. Rensing and J.T. August, Nature 224 (1969) 853.

CHAPTER I.3

CONSIDERATIONS ON THE PROBLEM OF THE JOINT EVOLUTION OF TWO DIFFERENT TRANSLATION APPARATUSES WITHIN THE SAME CELL

J. NINIO

Laboratoire de Biochimie du Développement, Institut de Biologie Moleculaire du C.N.R.S. et de l'Université Paris VII, 2 Place Jussieu, 75005 Paris

Introduction

The following problem was suggested by L.E. Orgel. Consider within the framework of the endosymbiotic theory for the origin of the eukaryotic cell, an initial situation in which there are two translation apparatuses working in parallel. One of them is cytoplasmic. The other one functions in the ancestors of the mitochondria or the chloroplasts. Initially, there is a certain amount of incompatibility between the two systems. If their components (tRNAs and activating enzymes for instance) can diffuse through the membrane of the organelle, mischarging and miscoding errors will be made, of the type which is observed in heterologous systems. One can foresee that the evolution of the translation machineries will obey two conflicting trends: a) reducing the number of genes corresponding to "duplicate" functions; b) maintaining an overall accurate translation in both systems.

Is it possible to deduce from these premices certain general rules (concerning for instance the patterns of presence and absence of organelle-specific tRNAs and aminoacyl-tRNA ligases) that could be confronted to the available data?

Misacylation and mistranslation in heterologous systems

When the components of two translation apparatuses from two different species are mixed, functional errors occur. Extensive studies have been devoted to the phenomenon of misacylation in heterologous systems, that is, the incorrect charging of tRNAs from one organism by aminoacyl-tRNA ligases from another (1). Giégé et al. (2) studied in a systematic fashion the acylation of purified tRNAs from yeast or *Escherichia coli* by *Bacillus subtilis* valyl-tRNA ligase. 17 purified tRNAs from *E. coli* and 10 purified tRNAs from yeast were tested in the classical incubation conditions that are in use for the homologous reaction. *E. coli* and yeast $tRNA^{Val}$ are acylated to a high level (100 and 86% respectively, compared to *B. subtilis* $tRNA^{Val}$).

However, other tRNAs may be acylated to an appreciable extent, chiefly *E. coli* $tRNA^{Ile}$ and $tRNA_f^{Met}$ (78 and 20%) and yeast $tRNA^{Phe}$ and $tRNA_f^{Met}$ (83 and 11%).

A similar phenomenon holds at the ribosomal level. This was shown in two independent papers (3,4). In the first of these studies, *E. coli* ribosomes were used, and R17 mRNA was translated using *E. coli* tRNA and competing with it, preacylated yeast glycyl-$tRNA^{Gly}$. The coat protein synthesized in vitro was isolated and analyzed by thin-layer mapping of its tryptic digests. Bergquist, Burns and Plinston showed that for the most part, the glycine codons were translated with fidelity by the yeast glycyl-tRNAs. However, one of the peptides had more glycine incorporated into it from yeast $tRNA^{Gly}$ than would be expected from the composition of the peptide, implying that a codon for another amino acid in the R17 RNA was being translated as glycine by the yeast glycyl-$tRNA^{Gly}$ acting on *E. coli* ribosomes (3). Hunter and Jackson (4) studied the synthesis of rabbit globin in the reticulocyte transfer system using either homologous or *E. coli* tRNA. Again, a detailed analysis of the peptides revealed miscoding errors to which little attention had been paid in previous studies (5,6).

The occurrence of errors in heterologous systems cannot be explained satisfactorily by the traditional qualitative geometrical conception of recognition (7—10). The phenomenon becomes quite natural when one starts to understand enzyme or ribosomal discrimination in quantitative kinetic terms (11—14).

Heterologous sub-cellular system from a similar cell

Suppose we mix the translation apparatus from the cytoplasm and from the chloroplast or the mitochondrion of a same species. Should we expect errors to be made?

Mistranslation

Cell-free systems for protein synthesis have been isolated from mitochondria (15—17) and chloroplasts (17—18). The accuracy of translation in these systems has not been studied yet. There is at least one cytoplasmic tRNA species which might be expected to produce mistranslation errors on chloroplastic ribosomes. Upon fractionation of chloroplastic tRNA from *Phaseolus vulgaris* on a RPC-5 column, Guillemaut et al. (19) found five or six peaks of leucine acceptance. They demonstrated (personal communication) that the chloroplasts contain two leucyl-tRNA ligases, one of which (A) is identical to the cytoplasmic ligase, the other one (B) being specific of the chloroplasts. Two of the major peaks of chloroplastic $tRNA^{Leu}$ are acylatable by ligase B exclusively, and are not detected among the cytoplasmic $tRNA^{Leu}$. The two other major peaks of chloroplastic $tRNA^{Leu}$ are acylatable by both ligases A and B, and are also found in the cytoplasm.

Consider that the purely chloroplastic tRNAs read the leucine codons a, b, c, d while the other tRNAs would read the leucine codons c, d, e, f. There should exist a cytoplasmic $tRNA^{Leu}$ responding to codon a or b. Such a cytoplasmic tRNA would be absent from the chloroplast in spite of the fact that, i) it is acylatable by a ligase present in the chloroplast, ii) it reads codons that are used on chloroplastic mRNAs. Therefore, it would be of interest to determine why it is not being used by the chloroplastic translation machinery. A plausible explanation is that it would sometimes respond to non-leucine codons if used on chloroplastic ribosomes.

Misacylation

There is no detailed study on the acylation of a tRNA specific for one amino acid and present in a given subcellular fraction (cytoplasm or organelle) by a ligase specific for another amino acid and originating from a different subcellular fraction. Very often, the available data describe the cross-acylation of total tRNA from one fraction by the total enzyme from the other fraction. This is exemplified in table 1.

The first line of this table, involving serine, is typical of the intuitive expectations in the case where the chloroplastic and the cytoplasmic ligases are different. As a general rule, such are the "clean" situations from which more detailed studies have stemmed (table 3). The second line in table 1 shows a reversed pattern. Chloroplastic tRNA accepts five times more asparagine when it is donated by the heterologous enzyme than when it is transferred by the homologous ligase. It may be that the chloroplastic ligase is loosing its activity during the extraction or that, being strongly attached to the membrane, very little of it is recovered. Whatever the case, it appears that there is a cytoplasmic ligase which is potentially able to donate important amounts of asparagine to the chloroplastic tRNA, and yet, it would not be used to that effect. A likely possibility is that this ligase, if introduced in the chloroplast, would acylate a chloroplastic tRNA specific for an amino

TABLE 1

The attachment of two aminoacids on tRNAs from the chloroplast or the cytoplasm of *Phaseolus vulgaris*, catalyzed by aminoacyl-tRNA ligases from either source is expressed in nmoles of aminoacid attached per mg of tRNA (from (32)).

Amino acid	Cytoplasmic tRNAs		Chloroplast tRNAs	
	Cytoplasmic enzyme	Chloroplast enzyme	Chloroplast enzyme	Cytoplasmic enzyme
Serine	0.49	0.14	1.24	0.69
Asparagine	0.61	0.18	0.19	1.01

acid other than asparagine. It is even plausible that some of the observed excessive incorporation might be due to such an effect.

Ways to reduce the errors

Let us now describe the initial situation, within the framework of the endosymbiotic theory. The ancestor of the organelle is trapped within a larger cell. If initially the two species were very closely related, it would be difficult to distinguish between the consequences of the endosymbiotic theory, and those of the cluster-clone alteration theory. Here, we assume that the two species which have fused are distantly related so that mixing their translation apparatus would cause too many errors. We will consider that the membrane of the organelle has a low permeability towards macromolecules. Too high a permeability would result in over-ambiguous translation. No permeability at all would be inconsistent with a further evolution during which there is a takeover of a large fraction of the organelle genome by the central apparatus. From now on, there are many conceivable ways (which may or may not be used) through which errors could be reduced, for instance:

a) The elongation factors of one apparatus evolve in such a way as to become inefficient towards the tRNAs from the other one. Concommitantly, the potentially ambiguous tRNAs evolve towards being discriminated against by the heterologous elongation factors.

b) The "dangerous" aminoacyl-tRNA ligases would gain specific determinants for their intracellular localization: attachment to a membrane, participation to multienzymatic complexes.

c) The evolution of nucleases, proteases or other components that would detect and destroy tRNAs or ligases found in a wrong place; the evolution of enzymes that may label some of the tRNA species making them easily recognizable and degradable should they invade the other translation system.

d) A general evolution of the tRNAs and the ligases resulting in a diminished level of errors in the heterologous reactions. It may prove easier for a ligase to discriminate against *all* heterologous tRNAs than to work efficiently and accurately in both compartments.

Suppression of superfluous genes

So far, we have focused on potential errors and the ways to reduce them. There will also be cases where a component of one apparatus may be replaced without dammage to accuracy by the corresponding component from the other apparatus. Some of these situations would arise at a second stage of evolution, after a preliminary adjustment of the two apparatus has occurred. Since there is no clue in our possession concerning the various interactions that were possible initially, it is difficult to make predictions and say

TABLE 2

Two cases of differences in codon recognition patterns between a prokaryote and eukaryote (21-24).

		E. coli	Yeast (cytoplasm)
Isoleucine codons	AUU	○ (A)	○ (C)
	AUC	○ (A)	○ (C)
	AUA	○ (B)	○ (C)
Valine codons	GUU	○ (E) ○ (D)	○ (F) ○ (G)
	GUC	○ (D)	○ (F)
	GUA	○ (E)	○ (F) ○ (G)
	GUG	○ (E)	○ (G)

what particular gene in the organelle genome was the most likely to be deleted without endangering the translation process. There are, however, two general ordering notions which it may be useful to consider: the overlap in the codon recognition patterns, and the practical multiplicity of the tRNAs.

Overlap in the codon recognition patterns

Since one tRNA may read more than one codon, there are several sets of anticodons which may correspond to the degenerate codons of a given amino acid. Indeed, different species have different codon recognition patterns (20). Table 2 shows some recent and reliable data (21—24) concerning valine and isoleucine. A word of caution is necessary: the interpretation of the experiments of the binding of tRNAs to ribosomes in the presence of triplets has been changing in recent years, mainly with the increase in knowledge on the chemical modifications of the bases located in position I of the anticodons (25). We have no example yet of differences in codon recognition patterns either within prokaryotes, or within eukaryotes. Apart from some indications on the behaviour of *Tetrahymena pyriformis* $tRNA^{Leu}$ (26), we know almost nothing about the codon response of organelle tRNAs.

If initially the two translation apparatus were making use of different codon recognition patterns, some of the tRNAs would have been difficult to eliminate. Suppression of a tRNA of type C in table 2 would have required its simultaneous replacement by the two tRNAs A and B. Replacement of A or B by C would be plausible, but the replacement of A *and* B by C could raise accuracy problems. As a consequence we would expect the codon recognition patterns of organelle tRNAs to display an unusual amount of overlap, i.e., the proportion of the codons that are read by two or more isoaccepting tRNAs differing in their codon specificities, should be relatively large.

Practical multiplicity of the tRNAs

We will assume that initially there was one aminoacyl-tRNA ligase per amino acid and per translation apparatus. The "practical multiplicity" shall be defined as the number of different anticodons corresponding to one amino acid in a given translation apparatus. This number indicates the minimum number of tRNA molecules of different sequences that the ligase must recognize and distinguish from all other tRNAs. In general when two wild-type isoaccepting tRNAs have different anticodons, their primary structures differ in at least five positions. Two tRNAs chargeable with different amino acids such as *E. coli* $tRNA_3^{Gly}$ and $tRNA_1^{Val}$ may be more similar in sequence than two isoaccepting tRNAs (for instance, *E. coli* $tRNA_1^{Val}$ and $tRNA_{2a,b}^{Val}$) (27,28). In all known cases, the practical multiplicity for phenylalanine is one (29). There are at least four different anticodons for leucine in *E. coli* (20). The smaller the practical multiplicity, the easier it would be for

TABLE 3

Organelle-specific acyl tRNA ligases.

Amino acid	Organism	Organelle	Presence of specific ligase	Criteria	References
Arginine	Phaseolus vulgaris	C	?	c	(32)
Aspartic acid	Neurospora crassa	M	+	b	(35)
Isoleucine	Euglena gracilis	C	+	a, b	(33)
	Euglena gracilis	C	+	a	(34)
	Euglena gracilis	M	—	a	(34)
Leucine	Phaseolus vulgaris	C	+	a, b	(19, 32)
	Tetrahymena pyriformis	M	+	a	(26)
	Neurospora crassa	M	+	b	(35)
	Rat (liver)	M	+	a	(36)
	Nicotiana tabacum	M	+	a	(37)
Lysine	Phaseolus vulgaris	C	+	c	(32)
Methionine	Neurospora crassa	M	+	a	(38)
Phenylalanine	Euglena gracilis	C	+	a, b	(33)
	Neurospora crassa	M	+	b	(35)
Tyrosine	Rat (liver)	M	+	a	(36)
Valine	Phaseolus vulgaris	C	+	a, c	(32)

C, chloroplasts; M, mitochondria; a, ability to acylate resolved peaks of tRNA; b, chromatographic behaviour; c, response to chemical agents.

the aminoacyl-tRNA ligase to evolve. Therefore, one would expect that the organelles are using their own ligases for charging the tRNAs of high practical multiplicity such as those for serine, leucine and arginine, while the organelles could possibly make use of a cytoplasmic ligase for the acylation of low multiplicity tRNA species such as those for phenylalanine, tyrosine and tryptophan among others. Of course, since during evolution, organelle tRNAs are suppressed and replaced by cytoplasmic species, the situation for the ligase replacements is changing with time.

The data compiled in table 3 suggest that, whenever the cytoplasmic and the organelle ligases have been compared thoroughly, differences are detected. There is only one negative result. Thus, it may be that the organelles possess a complete set of specific aminoacyl-tRNA ligases in which case, the criterium of practical multiplicity would be of no value. On the other hand, what appears as an organelle-specific ligase may be the result of a post-translational modification of the cytoplasmic ligase. Indeed, it was shown by Weeks and Gross (30) that two mutations in the nuclear gene leu5 could affect both cytoplasmic and mitochondrial leucyl-tRNA ligases of *Neurospora.* On the other hand, this may reflect a situation where the organelle contains a cytoplasmic-derived ligase in addition to the organelle-specific one, as discussed above.

Discussion

When discussing the reasons for the universality of the genetic code, one has to take into consideration the possibility for primitive organisms with different codes to exchange genetic and translational material. Brenner (quoted in Crick (31)) suggested that the exchange of genes may have been important for the establishment of a universal code. The boundaries of primitive organisms were probably less defined than nowadays, and one has to take seriously into account the possibility that some macromolecules implied in translation were not univoquely linked to one organism. Therefore, the problem of the joint evolution of two translation apparatus within a same cell that we have discussed in this lecture has bearing on another very important evolutionary problem.

Owing to the scarcity of the data, we did not attempt to discuss fully all possibilities with all their ramifications. At several places we have been consciously schematic, our main purpose being to point out the type of experimental data which appears to us to be particularly relevant to the problem, and therefore should deserve closer attention and further investigations.

Acknowledgements

I am indebted to Prof. L.E. Orgel for formulating the problem discussed here, and to Professor F. Chapeville for allowing me to pursue some investigations on that subject in his laboratory.

References

1. K.B. Jacobson, Progr. Nucleic Acid Res. Mol. Biol. 11 (1971) 461.
2. R. Giegé, D. Kern, J.P. Ebel, H. Grosjean, S. De Henau and H. Chantrenne, Eur. J. Biochem. 45 (1974) 351.
3. P.L. Bergquist, D.J.W. Burns and C.A. Plinston, Biochemistry 7 (1968) 1751.
4. A.R. Hunter and R.J. Jackson, Eur. J. Biochem. 15 (1970) 381.
5. I.B. Weinstein, S.M. Friedman and S. Ochoa, Cold Spring Harbor Symp. Quant. Biol. 31 (1966) 671.
6. B. Weisblum, F. Gonano, G. van Ehrenstein and S. Benzer, Proc. Nat. Acad. Sci. U.S. 53 (1965) 328.
7. F.H.C. Crick, J. Mol. Biol. 19 (1966) 548.
8. W. Fuller and A. Hodgson, Nature 215 (1967) 817.
9. C.R. Woese, Nature 226 (1970) 817.
10. B. Roe, M. Sirover and B. Dudock, Biochemistry 12 (1973) 4146.
11. J. Ninio, J. Mol. Biol. 56 (1971) 63.
12. J. Ninio, Progr. Nucleic Acid Res. Mol. Biol. 13 (1973) 301.
13. J. Ninio, J. Mol. Biol. 84 (1974) 297.
14. J.P. Ebel, R. Giegé, J. Bonnet, D. Kern, N. Beffort, C. Bollack, F. Fasiolo, J. Gangloff and G. Dirheimer, Biochimie 55 (1973) 547.
15. A.H. Scragg, H. Morimoto, V. Villa, J. Nekhorocheff and H.O. Halvorson, Science 171 (1971) 908.
16. H.C. Blossey and H. Küntzel, FEBS Letters 24 (1972) 335.
17. N.G. Avadhani and D.E. Buetow, Biochem. J. 128 (1972) 353.
18. N. Chua, G. Blobel and P. Siekevitz, J. Cell. Physiol. 57 (1973) 798.
19. P. Guillemaut, A. Steinmetz, G. Burkard and J.H. Weil, C.R. Soc. Biol. France 167 (1973) 961.
20. C.T. Caskey, A. Beaudet and M. Nirenberg, J. Mol. Biol. 37 (1968) 99.
21. F. Harada and S. Nishimura, Biochemistry 13 (1974) 300.
22. M. Yarus and B.G. Barrell, Biochem. Biophys. Res. Commun. 43 (1971) 729.
23. S. Takemura, M. Murakami and M. Miyazaki, J. Biochem. 65 (1969) 553.
24. K. Takeishi, T. Takemoto and S. Nishimura, Biochem. Biophys. Res. Commun. 47 (1972) 746.
25. S. Nishimura, Progr. Nucleic Acid Res. Mol. Biol. 12 (1972) 49.
26. N. Chiu, A.O.S. Chiu and Y. Suyama, J. Mol. Biol. 82 (1974) 441.
27. C. Squires and J. Carbon, Nature New Biol. 233 (1971) 274.
28. M. Yaniv and B.G. Barrell, Nature New Biol. 233 (1971) 113.
29. G. Keith, F. Picaud, J. Weissenbach, J.P. Ebel, G. Petrissant and G. Dirheimer, FEBS Letters 31 (1973) 345.
30. C.O. Weeks and S.R. Gross, Biochem. Genet. 5 (1971) 505.
31. F.H.C. Crick, J. Mol. Biol. 38 (1968) 367.
32. G. Burkard, P. Guillemaut and J.H. Weil, Biochim. Biophys. Acta 224 (1970) 184.
33. B.J. Reger, S.A. Fairfield, J.L. Epler and W.E. Barnett, Proc. Nat. Acad. Sci. U.S. 67 (1970) 1207.
34. N. Kislev, M.I. Selsky, C. Norton and J.M. Eisenstadt, Biochim. Biophys. Acta 287 (1972) 256.
35. W.E. Barnett, D.H. Brown and J.L. Epler, Proc. Nat. Acad. Sci. U.S. 57 (1967) 1775.
36. C.A. Buck and M.M.K. Nass, Proc. Nat. Acad. Sci. U.S. 60 (1968) 1045.
37. R.H. Guderian, R.L. Pulliam and M.P. Gordon, Biochim. Biophys. Acta 262 (1972) 50.
38. J.L. Epler, L.R. Shugart and W.E. Barnett, Biochemistry 9 (1970) 3575.

Dr. LURQUIN:

Is there evidence showing organ specificity for tRNAs and/or aminoacyl-tRNA ligases either at the cytoplasmic or at the organelle level? If this is the case, how would that fit into an evolutionary scheme?

Dr. NINIO:

Differences in the chromatographic behaviour of tRNAs from different tissues or of tRNAs extracted at different stages of development have been reported. Alterations in elution profiles may be related to changes in the relative amounts of various isoacceptors, or changes in methylation or other post-transcriptional modifications of tRNA. No clear-cut case of differing sequences has yet been reported (see references in U.Z. Littauer and H. Inouye, Ann. Rev. Biochem. 42, 439). I am not aware of studies bearing on tissue specificity of aminoacyl-tRNA ligases.

CHAPTER I.4

tRNA AND AMINOACYL-tRNA SYNTHETASES FROM WILD-TYPE AND PETITE YEAST MITOCHONDRIA

G. BALDACCI, C. FALCONE, L. FRONTALI, G. MACINO and C. PALLESCHI

Centro di Studio per gli Acidi Nucleici, C.N.R., Istituto di Fisiologia Generale, Università di Roma, Rome, Italy.

tRNA and aminoacyl-tRNA synthetases from yeast mitochondria were studied and the localization of the genetic information for these constituents investigated.

The rate and the extent of acylation of mitochondrial tRNA was studied in homologous (mitochondrial-mitochondrial) and heterologous (mitochondrial-cytoplasmic) synthetase-tRNA mixtures.

Results show that mitochondrial phenylalanyl-tRNA synthetase is different from the cytoplasmic counterpart and that the same enzyme is present in a petite mutant completely lacking mitochondrial DNA.

Moreover, hybridization experiments were performed between several tRNAs and fragments of mitochondrial DNA of different buoyant density obtained by sonication and fractionation on a CsCl gradient. The hybridization patterns indicate that these genes are not clustered and suggest that more than one gene exists for $tRNA_{Ser}$ and $tRNA_{His}$.

Introduction

Results reported in this paper refer to yeast mitochondrial tRNA and aminoacyl-tRNA synthetases as a part of the mitochondrial machinery for macromolecular synthesis, whose degree of autonomy is still not fully established.

As far as aa-tRNA synthetases are concerned, the existence of specifically mitochondrial enzymes different from the cytoplasmic counterparts has been observed in rat liver (1) and in *Neurospora crassa* (2) and immunologically demonstrated in *Tetrahymena pyriformis* (3). However, the localization of the genetic information for these mitochondrial constituents had not been investigated. On the contrary, several mit-tRNAs are known to be coded for by mitochondrial genome in HeLa cells (4,5) in *Xenopus* (6) and in yeast (7—11) but the problem of the number and function of tRNA coded for by mit-DNA is still open.

Materials and methods

A wild-type strain (DM) was used, whose characteristics have been previously described (12) and a petite mutant called G5, obtained from the former by treatment with ethidium bromide (40 μg/ml for 40 generations). The absence of mit-DNA was repeatedly checked by CsCl analytical gradients. Culture media contained 1% yeast extract, 2% glucose and 2% pepton.

The methods for the preparation of mitochondrial and supernatant tRNA and aminoacyl-tRNA synthetases have been previously described (11).

Acylation studies were performed in the following conditions: 5 μmoles ATP, 15 μmoles $MgCl_2$, 40 μmoles Tris—HCl buffer pH 7.4, 100—200 μg cytoplasmic or mitochondrial tRNA, 0.1—0.5 mg cytoplasmic or mitochondrial enzyme proteins, 0.12 μmoles ^{3}H aminoacid (spec. act. 83 μCi/μmole) in a final volume of 0.5 ml. Reaction was stopped at various intervals by addition of 3 ml 67% ethanol plus 0.5 M NaCl; the samples were kept for at least 1 h at —20°C. The precipitate was washed 3 times with 67% ethanol plus 0.5 M NaCl and then dissolved in 0.6 ml of 0.2 M triethylamine. Aliquots of each sample were counted for radioactivity after addition of 10 ml Bray scintillation solution in a Beckman liquid scintillator. For the preparation of aa-tRNA for the hybridization experiments, the acylation mixture contained in 1 ml: 50 μmoles Tris—HCl buffer (pH 7.4), 0.2 mg tRNA, 350 μCi ^{3}H amino acid (Amersham products, spec. act. 10—50 Ci/mmole), 10 μmoles ATP, 30 μmoles $MgCl_2$, 0.5 mg mitochondrial enzyme protein.

DNA was prepared, sonicated and fractionated as previously reported (13). The hybridization procedure was essentially that reported by Casey et al. (8) and contained 33% formamide and 70 μg/ml aminoacyl-tRNA.

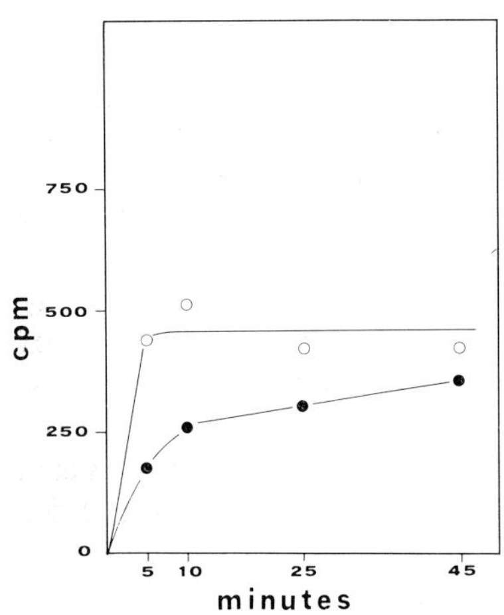

Fig. 1. Acylation rates of tRNA in homologous and heterologous mixtures. Standard assay mixture containing ^{3}H phenylalanine, 100 μg mitochondrial enzyme protein, and 200 μg mitochondrial (○) or 250 μg supernatant (●) tRNA.

Results and discussion

To test the specificity of aa-tRNA synthetases from yeast mitochondria, we investigated the acylation rate as well as the acylation extent in homologous (mitochondrial-mitochondrial) or heterologous (mitochondrial-cytoplasmic) mixtures. The acylation rate was studied in the presence of low enzyme concentrations, by comparing the rates of acylation by the mitochondrial synthetase of the homologous and the heterologous tRNA; the total amounts of these tRNAs were chosen as to have in both mixtures about the same amount of tRNA accepting the tested aminoacid. Results for the case of phenylalanine, reported in fig. 1, were obtained with an enzyme preparation from G5 mitochondria, but the same results are obtained with wild-type mitochondria. Supernatant synthetases were also tested and in this case too the acylation rate is higher in homologous mixtures.

The acylation extent was studied under different conditions, in which enzyme was in excess and the amount of tRNA bound radioactivity was only dependent on the amount of tRNA in the mixture. In this case the same tRNA preparation was assayed in the presence of two different (mitochondrial and cytoplasmic) enzyme preparations.

Though $tRNA_{Phe}$ is certainly coded for by mit-DNA (see later), results reported in fig. 2 show that this tRNA is hyperacylated in heterologous mixtures, while hyperacylation does not occur in the mixture containing

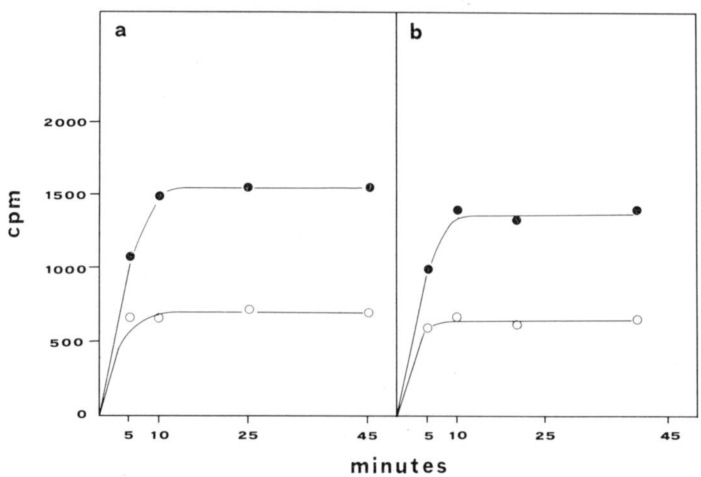

Fig. 2. Extent of acylation of mitochondrial (a) or supernatant (b) tRNA in the presence of homologous and heterologous synthetases. (a) Standard assay mixture containing ^{3}H phenylalanine, 100 μg mitochondrial tRNA and 400 μg mitochondrial (○) or cytoplasmic (●) enzyme protein. (b) Standard assay mixture containing ^{3}H phenylalanine, 200 μg supernatant tRNA and 400 μg cytoplasmic (●) or mitochondrial (○) enzyme protein.

cytoplasmic tRNA (fig. 2b). The explanation of this result is not clear (we ruled out trivial explanations like the presence in mitochondrial preparations of enzymes having a lytic activity on tRNA) but similar cases have been observed in several heterologous mixtures (14) and also in chloroplast-cytoplasm mixtures (15). In one case heterologous mischarging has been demonstrated (16).

Also in this case the same results are obtained with enzymes from wild-type and from G5 mitochondria, therefore the results concerning the acylation rate as well as the acylation extent show the existence of mitochondrial synthetases both in the wild-type and in the DNA_0 mutant. The localization of the genetic information for these mitochondrial constituents is therefore certainly nuclear. In the case that mitochondrial and cytoplasmic synthetases might contain common subunits, all of them should be coded for by nuclear genome.

As far as the information for tRNA is concerned, hybridization studies have demonstrated the existence of 12 genes for tRNA in HeLa cells (5), of 15 in *Xenopus* (6) and of 20 in *Saccharomyces cerevisiae* (7); these low figures raise a very interesting problem: does mit-DNA contain the information for the complete set of tRNA (one for each aminoacid incorporated into protein in mitochondria) or are some tRNAs imported from the cytoplasm? The way to solve this problem seems to be the identification of the individual tRNAs coded by the mitochondrial genome, the study of their response to codons and of the possible degeneracy of the mitochondrial genetic code. The technique by Weiss (17) has been used to identify tRNAs coded by mit-DNA in yeast, and 14 different aa-tRNAs have been found to hybridize with mit-DNA (10).

We report here the hybridization of five aa-tRNAs performed by the same technique on mit-DNA of *S. cerevisiae* fragmented by sonication and fractionated on CsCl density gradient. Mit-DNA from yeast has actually been shown (18,13) to contain sections of widely different base composition, and hybridization on the different fractions of a preparative CsCl density gradient can indicate the localization of the genes for tRNA on heavy or light fragments of mit-DNA. Results of these experiment are reported in fig. 3.

The hybridization profiles are different for the different tRNAs tested and indicate the following:

1) On the same mit-DNA preparation seryl- and phenylalanyl-tRNA hybridize on fragments of widely different buoyant density: seryl-tRNA mainly on heavy fragments and phenylalanyl-tRNA almost exclusively on fragments lighter than bulk DNA. This shows that the genes for tRNA are not clustered on mit-DNA from *S. cerevisiae.*

2) Three aa-tRNA (valyl-, phenylalanyl-, isoleucyl-tRNA) give an hybridization peak displaced on the lighter region of the gradient, on the contrary histidyl- and seryl-tRNA hybridize both on the heavier and the lighter region

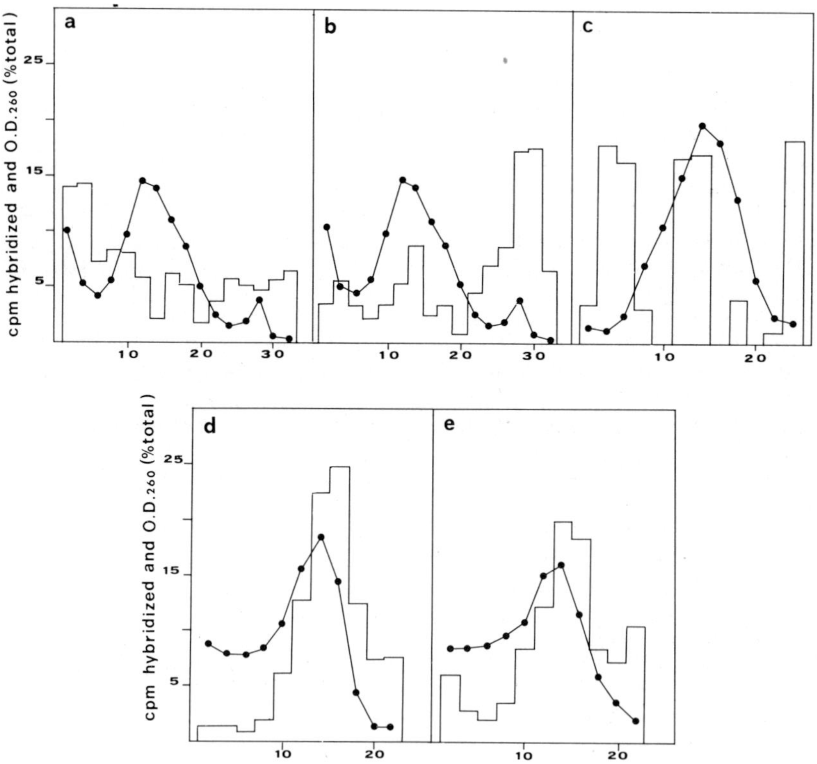

Fig. 3. Fractionation on CsCl density gradient of sonicated mit-DNA from w.t. strain and annealing of the fractions with aa-tRNAs. Fractionation and annealing conditions as previously reported (11). (a) w.t. mit-DNA (0.55 O.D. units) and ^{3}H seryl-tRNA (2000 cpm/μg); total DNA bound radioactivity 985 cpm. (b) w.t. mit-DNA (0.55 O.D. units) and phenylalanyl-tRNA (3000 cpm/μg); total DNA bound radioactivity 1130 cpm. (c) w.t. mit-DNA (0.51 O.D. units) and histidyl-tRNA (2714 cpm/μg); total DNA bound radioactivity 1653 cpm. (d) w.t. mit-DNA (1.3 O.D. units) and isoleucyl-tRNA (3300 cpm/μg); total DNA bound radioactivity 560 cpm. (e) w.t. mit-DNA (1.4 O.D. units) and valyl-tRNA (1700 cpm/μg); total DNA bound radioactivity 497 cpm. The average value of five blanks with nuclear DNA was subtracted from the data (130 cpm in a, 70 in b, 350 in c, 70 in d, 110 in e) O.D. (●——●) and radioactivity (height of the bars) are expressed as fraction of the total.

of the gradient (hybridization on light and heavy regions, where the optical density is very low, implies hybridization levels at least 100 times higher than hybridization in the region of the peak).

The reasons for this peculiar hybridization profiles remain to be investigated and the possibility of the existence of more than one gene for $tRNA_{Ser}$ and $tRNA_{His}$ is to be checked.

References

1. C.A. Buck and N.M.K. Nass, Proc. Nat. Acad. Sci. U.S. 60 (1968) 1045.
2. W.E. Barnett and D.M. Brown, Proc. Nat. Acad. Sci. U.S. 57 (1967) 142.
3. O.S.A. Chiu and Y. Suyama, Biochim. Biophys. Acta 229 (1973) 557.
4. Y. Aloni and G. Attardi, J. Mol. Biol. 55 (1971) 271.
5. M. Wu, N. Davidson, G. Attardi and Y. Aloni, J. Mol. Biol. 71 (1972) 81.
6. I.B. Dawid, J. Mol. Biol. 63 (1972) 201.
7. L. Reijnders and P. Borst, Biochem. Biophys. Res. Commun. 47 (1972) 126.
8. J. Casey, M. Cohen, M. Rabinowitz, H. Fukuara and G.S. Getz, J. Mol. Biol. 63 (1972) 431.
9. M. Cohen and M. Rabinowitz, Biochim. Biophys. Acta 281 (1972) 192.
10. M. Rabinowitz, J. Casey, P. Gordon, J. Locker and H.J. Hsu, in A.M. Kroon and C. Saccone (Eds.) Symposium on Mitochondrial Biogenesis, Academic Press, New York (1974).
11. F. Carnevali, C. Falcone, L. Frontali, L. Leoni, G. Macino and C. Palleschi, Biochem. Biophys. Res. Commun. 51 (1973) 651.
12. G. Bernardi, F. Carnevali, A. Nicolajeff, G. Piperno and G. Tecce, J. Mol. Biol. 37 (1968) 493.
13. F. Carnevali and L. Leoni, Biochim. Biophys. Res. Commun. 47 (1972) 1322.
14. K.B. Jacobson, Progr. Nucleic Acid Res. Mol. Biol. 11 (1971) 461.
15. G. Burkard, P. Guillemaut, J.H. Weil, Biochim. Biophys. Acta 224 (1970) 184.
16. K.B. Jacobson, S. Nishimura, W.E. Barnett, R.J. Mano, P. Cammarano and G.O. Novelli, Biochim. Biophys. Acta 91 (1964) 305.
17. S.B. Weiss, W.T. Hsu, T.W. Foft and N.H. Sherberg, Proc. Nat. Acad. Sci. U.S. 61 (1968) 114.
18. C. Christiansen, G. Christiansen and A.L. Back, J. Mol. Biol. 84 (1974) 65.

CHAPTER I.5

EFFECT OF DIFFERENT ANTIBIOTICS ON THE PRESENCE OF A CHLOROPLAST-SPECIFIC ELONGATION FACTOR IN *CHLORELLA VULGARIS*

O. CIFERRI, O. TIBONI and A.R. AMILENI

Institute of Plant Physiology, University of Pavia, Pavia, Italy

In *Chlorella vulgaris* the chloroplast-specific elongation factor G (EF-G_{chl}) appears to be present only if the alga is grown under conditions that allow development of photosynthesis. Substantial amounts of EF-G_{chl} may be isolated from dark-grown cells illuminated for at least 24 hr. Treatment of dark-grown cells with either chloramphenicol (CAP, 3 mg/ml) of rifampicin (RIF, 0.2 mg/ml) greatly reduces the amount of EF-G_{chl} that may be isolated from cells exposed subsequently to light for 24 hr. No significant decrease was observed when cycloheximide (CHI, 0.1 mg/ml) was employed.

Introduction

Light-grown cells of *Chlorella vulgaris* contain three different elongation factors endowed with translocase activity. One of such factors, elongation factor 2 (EF-2), is active on the cytoplasmic (80S) ribosomes while the other two, elongation factors G (EF-G), catalyze protein synthesis in vitro on ribosomes of the 70S type such as those from *Escherichia coli* (1). Evidence was presented indicating that one of the latter factors is associated with the chloroplasts of the alga (EF-G_{chl}) while the other is that presumably present in the mitochondria (EF-G_{mit}). The chloroplast-specific EF-G appears to be present only if the alga is grown under conditions that allow development of photosynthesis and of the chloroplast. Indeed EF-G_{chl} is absent or undetectable in cells of a non-photosynthetic mutant or in those of the wild type strain grown in the dark in the presence of glucose (1,2). On the other hand, the mitochondrion-specific factor (EF-G_{mit}) is present in cells from both strains grown under any of the tested experimental conditions. Furthermore, such a factor appears to be produced in the cytoplasm of the cell rather than in the organelle itself since it is present in cells of the non-photosynthetic mutant grown in the presence of antibiotics, such as chloramphenicol and erythromycin, that inhibit mitochondrial protein synthesis (2).

The present paper reports preliminary experiments aimed at establishing the cellular site of synthesis of the chloroplast-specific elongation factor G.

Material and methods

Chlorella vulgaris was grown mixo- or heterotrophically in the media and under the experimental conditions already reported (1,2). Preparation of cell-free extracts, purification of the elongation factors and assay conditions were previously described (1,2) except than in certain experiments the "analytical" DEAE-Sephadex columns were twice as long (60 cm) and the volume of the elution buffer increased to a total of 200 ml (100 ml of 0.24 M KCl and 100 ml of 0.35 M KCl).

Results

Effect of light on the presence of EF-G_{chl}

If dark-grown cells of *C. vulgaris* are exposed to light for different periods of time, EF-G_{chl} is detected in substantial amounts only after a light treatment for approximately 24 hr (table 1). The presence of small amounts of EF-G_{chl} in cells exposed to light for shorter periods of time (e.g. 3 hr) seems to indicate that this factor may be present even in dark-grown cells. However its concentration is probably so reduced to make its detection problematic. In addition, as compared to EF-G_{mit}, EF-G_{chl} appears to be less stable so that only after prolonged light treatment the chloroplast-specific factor may be consistently detected in substantial amounts in the cell extracts.

Effect of different antibiotics on the presence of EF-G_{chl}

Treatment of cells of *C. vulgaris* with CAP or CHI produces strikingly different effects on the activity of the chloroplast-specific factor. As depicted in fig. 1 in cells treated with CAP one hr after exposure to light and then illuminated for a further 23 hr, EF-G_{chl} was undetectable in the ex-

TABLE 1

Effect of light treatment on the presence of chloroplast- and mitochondrion-specific elongation factors G. Cells were grown in the dark to a concentration of 5×10^6 cells/ml. The cultures were then illuminated (1000 lux) for different time intervals. The organelle-specific elongation factors G were isolated and purified from approximately 3 g (wet weight) of cells. One unit of EF-G activity was defined as the amount of protein that catalyzes the incorporation of 1 pmole of phenylalanine/15 min at 30° C.

Light treatment (hr)	EF-G_{chl}, total units	EF-G_{mit}, total units
0	0	2346
3	25	2544
6	0	4490
24	510	2647
48	735	2181

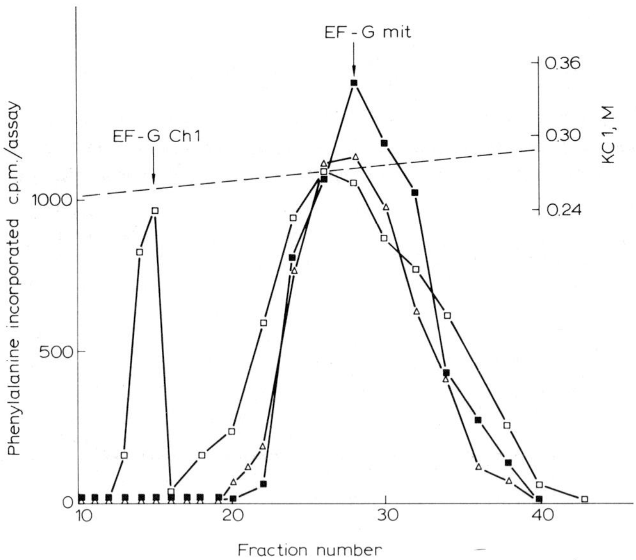

Fig. 1. Effect of chloramphenicol and cycloheximide on the presence of the chloroplast- and mitochondrion-specific elongation factors G. Cells were grown in the dark to a concentration of 1×10^7 cells/ml (approximately 48 hr). After illumination for one hr, chloramphenicol (CAP, 3 mg/ml) or cycloheximide (CHI, 0.1 mg/ml) were added and the cells , except in the control experiment, illuminated for a further 23 hr. The organelle-specific elongation factors G were isolated, purified and assayed as already reported (1,2). ■——■ control (1 hr of light); △——△ CAP (24 hr of light); □——□ CHI (24 hr of light).

tracts as in the control cells exposed to light for one hr only. On the other hand, EF-G_{chl} could be isolated from cells exposed to CHI, an inhibitor of cytoplasmic protein synthesis. Such findings were confirmed in the experiment II (table 2) showing that addition CHI 3 hr before illumination still allowed the isolation of EF-G_{chl} from the cell-free extracts. Again, in cells treated with CAP hardly any EF-G_{chl} was detected. It may be added that in separate experiments it was found that the concentration of CHI employed was sufficient to block within 5 min almost completely (more than 95% inhibition) the incorporation of aminoacids by intact cells.

The role of the chloroplast protein synthetic activity on the production of EF-G_{chl} was confirmed in an experiment employing rifampicin (table 2, exp. III), an antibiotic reported to inhibit selectively in *C. pyrenoidosa* the synthesis of chloroplast ribosomal RNA (3). As compared to control cells, those exposed to the antibiotic 6 hr before light treatment showed a 10-fold reduction in the amount of EF-G_{chl} while no significant reduction in the amount of the EF-G_{mit} was observed.

TABLE 2

Effect of chloramphenicol (CAP), cycloheximide (CHI) and rifampicin (RIF) on the presence of the chloroplast- and mitochondrion-specific elongation factors G. Experimental conditions: I) see legend to fig. 1; II) as in fig. 1 except that CAP and CHI were added to the cultures 3 hr before illumination; III) as in fig. 1 except that RIF was added 6 hr before illumination. In each experiment, the elongation factors were isolated and purified from 3 to 6 g (wet weight) of cells.

Experiment	Addition	Cell concentration cells $\times 10^7$/ml		Organelle-specific elongation factors G, percent	
		before light period	at the end of light period	$EF\text{-}G_{chl}$	$EF\text{-}G_{mit}$
I	none (control, 1 hr of light)	1.2	1.2	0	100
	CAP (3 mg/ml)	1.2	1.9	0	100
	CHI (0.1 mg/ml)	1.2	1.2	7	93
II	none (control, 24 hr of light)	0.5	2.5	14	86
	CAP (3 mg/ml)	1.0	2.8	4	96
	CHI (0.1 mg/ml)	0.9	1.4	27	73
III	none (control, 24 hr of light)	0.3	2.1	11	89
	RIF (0.2 mg/ml)	0.4	4.6	1	99

Discussion

It is too early to draw any conclusion on the cellular site of synthesis of the chloroplast-specific elongation factor G. It appears quite likely, however, that its synthesis requires the activity of the chloroplast protein synthetic machinery. It remains to be established whether the role of the chloroplast is direct and $EF\text{-}G_{chl}$ is synthesized within the organelle. Alternatively, a product of the chloroplast protein synthetic activity may regulate the production of $EF\text{-}G_{chl}$ in a different cell compartment or it may convert an inactive precursor, synthesized elsewhere, to the active form present in the organelle. Whichever the mechanism, the data so far obtained indicate that the presence of $EF\text{-}G_{chl}$ is correlated with the presence and the activity of a functional organelle. Such results are at variance with those reported for the mitochondrion-specific factor in *C. vulgaris* (2), yeast (4,5,6) and *Neurospora crassa* (7). In the latter two organisms $EF\text{-}G_{mit}$ appears to be produced in the cytoplasm and to be present in cells in which most of the mitochondrial functions, including protein synthesis, are impaired or absent.

Acknowledgement

The work was supported by a grant from the Laboratorio di Genetica Biochimica ed Evalozionistica of the Consiglio Nazionale delle Ricerche.

References

1. O. Ciferri and O. Tiboni, Nature 245 (1973) 209.
2. O. Ciferri, O. Tiboni, G. Lazar and J. Van Etten in A.M. Kroon and C. Saccone (Eds.) Biogenesis of Mitochondria, Academic Press (1974) 107.
3. G. Galling, C. Salzmann and E. Spiess, Planta 114 (1973) 269.
4. B. Parisi and R. Cella, FEBS Letters 14 (1971) 209.
5. A.H. Scragg, FEBS Letters 17 (1971) 111.
6. D. Richter, Biochemistry 10 (1971) 4422.
7. Z. Barath and H. Kuntzel, Proc. Nat. Acad. Sci. U.S. 69 (1972) 1371.

Dr. WILLIAMSON:
Is your mutant coded by the chloroplast DNA?

Dr. CIFERRI:
We do not know.

CHAPTER I.6

MEMBRANE-BOUND RIBOSOMES IN CHLOROPLASTS. POSSIBLE ROLE IN MEMBRANE BIOSYNTHESIS

A. MICHAELS and M.M. MARGULIES

Radiation Biology Laboratory, Smithsonian Institution, 12441 Parklawn Drive, Rockville, Maryland 20852, U.S.A.

When actively growing cells of *Chlamydomonas reinhardi* are treated with chloramphenicol, a large portion of chloroplast ribosomes are isolated bound to chloroplast membranes. The membrane-bound chloroplast ribosomes occur as polysomes. A portion of these ribosomes are bound to the membranes by nascent protein chains. The membrane-polyribosome complex carries out amino acid incorporation into protein. The reaction has the properties expected for an in vitro protein synthesis system carried out on chloroplast ribosomes. Incorporation involves elongation and completion of protein chains, but not initiation. Both nascent and completed proteins appear to be present at the end of the incorporation reaction.

Introduction

Reports from a number of laboratories have provided evidence that some chloroplast ribosomes occur bound to chloroplast membranes in both algae and higher plants (1—6). Maximum yields of membrane-bound ribosomes are obtained when cells of *Chlamydomonas reinhardi* are treated with chloramphenicol before they are homogenized (2,4). Over half of the chloroplast ribosomal RNAs are found in the 15,000 *g* pellet from homogenates of chloramphenicol-treated cells (4). The chloroplast ribosomal RNAs of the 15,000 *g* supernatant are largely chloroplast monomeric ribosomes (4). Only small amounts are found as polyribosomes (Michaels, unpublished). Our investigations were carried out to determine if chloroplast ribosomal RNAs in the 15,000 *g* pellet of homogenates occurred as ribosomes bound to chloroplast membranes, and to characterize the ribosome-membrane association.

Materials and methods

Cells of *C. reinhardi* (Arg-1, mating type plus) were grown, and chloroplast membranes were prepared as already described (4). Electron microscopy, density gradient centrifugations, release of ribosomal particles from membranes with 1 mM puromycin and 500 mM KCl, and analyses were carried out essentially as previously described with exceptions noted in the figure legends. Amino acid incorporation into protein was carried out essentially as described by Spencer and Wildman (7). Samples were prepared for counting and were counted as described by Mans and Novelli (8). A cell extract fraction (CE) was necessary for amino acid incorporation. CE was prepared by passing a 2 hr 150,000 *g* supernatant of homogenate through a column of Sephadex G-25, and then combining the peak of ultraviolet-absorbing material which eluted with the void volume.

The following compounds were obtained from Sigma Chemical Co., St. Louis, Mo., U.S.A.: actinomycin D; ATP, disodium salt; CTP, sodium salt; deoxyribonuclease, pancreatic type I (E.C. 3.1.4.5); phospho(enol) pyruvate, trisodium salt; pyruvate kinase type II, (ATP: pyruvate phosphotransferase, E.C. 3.1.4.22). Cycloheximide and D-threo chloramphenicol were obtained from Calbiochem., La Jolla, Calif., U.S.A. Rifamycin was obtained from Dow Chemical Co., 2020 Dow Center, Midland, Mich., and streptogramin A,B from Merck Chemical Div., Merck and Co., Inc., Rahway, N.J., U.S.A. Tritiated L-amino acid mixture (catalogue No. NET 250) was obtained from New England Nuclear Co., Boston, Mass., U.S.A. It had an average specific radioactivity of 12 Ci/mmole.

Results and discussion

Membranes from chloramphenicol-treated cells of *C. reinhardi* contain mostly 16S and 23S chloroplast ribosomal RNAs (fig. 1). Particles of the size and staining properties of ribosomes are seen, when membranes are examined by electron microscopy (fig. 2). The arrangement of some of the ribosomes suggests that they are polyribosomes.

The form in which ribosomes occur bound to membranes and the nature of the linkage of the ribosomes to the membranes was studied. The detergent

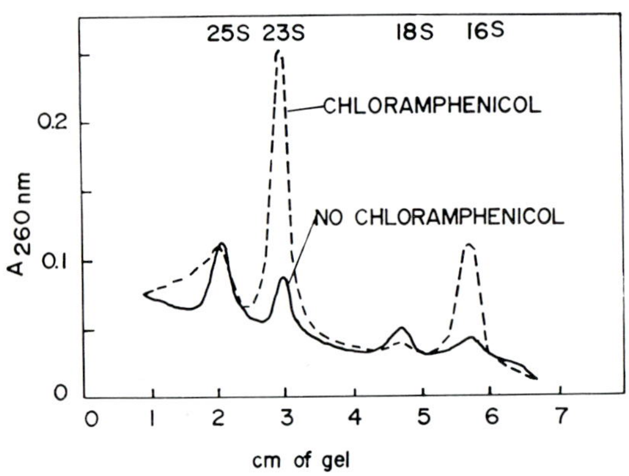

Fig. 1. Electrophoretogram showing RNAs from equal amounts of chloroplast membranes from untreated (———) and chloramphenicol-treated (- - - - - -) cells.

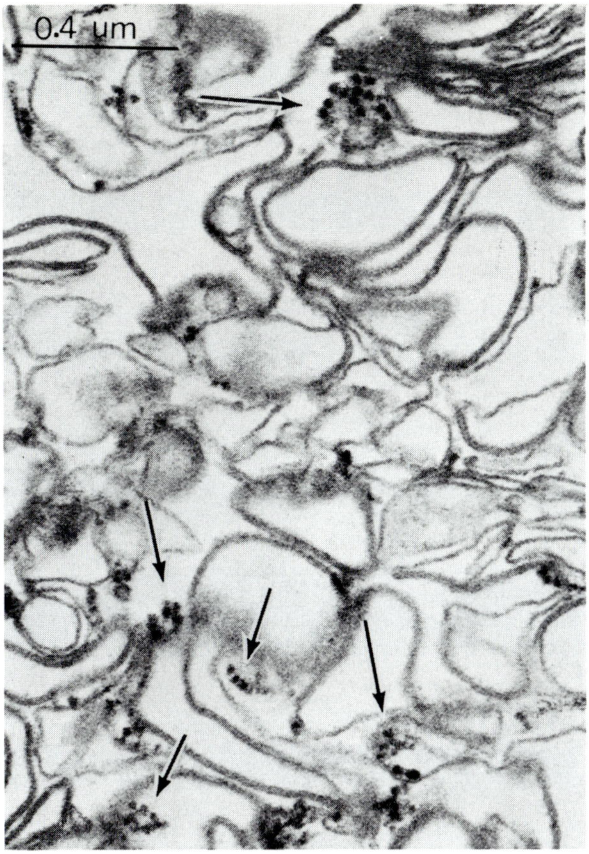

Fig. 2. Electron micrograph of isolated membranes from chloramphenicol-treated cells. Arrows indicate the location of polysome-like arrangements.

Nonidet P-40 releases polysomes from the membranes. Profiles typical of polyribosomes are obtained when detergent-treated membranes from chloramphenicol-treated cells are analyzed by density gradient centrifugation (fig. 3). Equal amounts of membranes from cells that are not treated with chloramphenicol do not yield polyribosome profiles when they are treated with detergent. Conversion of polyribosomes to chloroplast monomeric ribosomes occurs on treatment with RNase, but only if trypsin and chymotrypsin are also included during RNase digestion (Margulies, unpublished). A large portion of ribosomes are released from membranes of chloramphenicol-treated cells, as ribosomal subunits, when membranes are treated with 1 mM puromycin and 500 mM KCl (fig. 4). Both puromycin and KCl are required for maximum release of subunits. The requirement for puromycin indicates that the nascent protein chain is important in the binding of polyribosomes

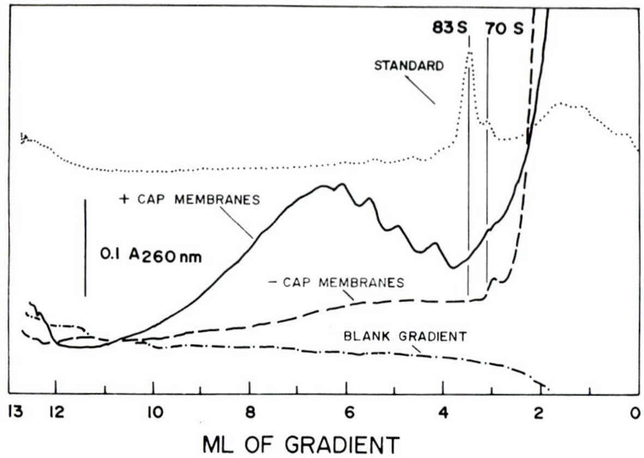

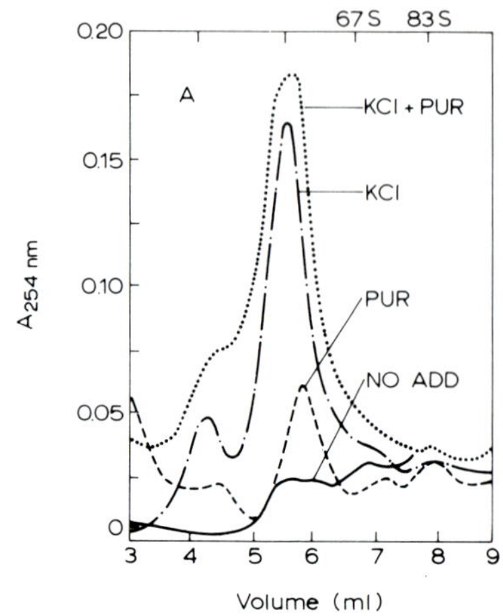

Fig. 3. Polysomes from membranes of chloramphenicol-treated cells. Membranes from chloramphenicol-treated cells (———) containing 325 μg of chlorophyll in 0.3 ml of buffer (4) and 0.25 M sucrose were treated with 10 μl of Nonidet P-40 (from Particle Data Laboratories, Elmhurst, Ill., U.S.A.). The 0.3 ml sample was layered over 12 ml of an exponential sucrose gradient (4) and overlaid with 0.7 ml of buffer. A sample of membranes from untreated cells containing 290 μg of chlorophyll was treated in the same way (- - - - - -). The gradients were centrifuged for 1 hr at 40,000 rpm in a type SW 40 Ti Spinco rotor, and scanned at 254 nm.

Fig. 4. Release of ribosomal subunits from membranes of chloramphenicol-treated cells by puromycin and KCl. Membranes containing 350 μg of chlorophyll were incubated for 15 min at 0°C followed by 15 min at 37°C in 1 ml of medium containing 125 mM sucrose, 25 mM $MgCl_2$, 5 mM dithiothreitol, 25 mM Tris pH 7.5. In addition, reaction mixture contained: 25 mM KCl (———); 1 mM puromycin and 25 mM KCl (- - - - - -); 500 mM KCl (- · - · -); 1 mM puromycin and 500 mM KCl (· · · · · ·). Reaction mixtures were layered on exponential sucrose density gradients (4) containing 25 mM KCl, 25 mM $MgCl_2$, 25 mM Tris pH 7.5, and 5 mM dithiothreitol. The gradients were centrifuged for 4 hr at 40,000 rpm in a type SW 40 Ti Spinco rotor, and scanned at 254 nm.

to the membranes. The presence of polyribosomes with nascent protein on the membranes indicates that these ribosomes probably function in protein synthesis.

Amino acid incorporation into protein by membranes from chloramphenicol-treated cells was studied to determine if the membrane-bound ribosomes actually function in protein synthesis. Membranes with attached ribosomes carry out incorporation of a mixture of tritiated amino acids into hot trichloroacetic acid insoluble material. Incorporation is linear for approximately 30 min, and continues for at least 60 min (fig. 5). Rates of incorporation of up to 10 pmole amino acid per mg chlorophyll per hour are obtained. Incorporation is inhibited by chloramphenicol, but not by cycloheximide (table 1). Thus, incorporation is carried out by 70S ribosomes (9,10). Other properties of the system rule out the possibility that incorporation is carried out by contaminating bacteria (table 1). Incorporation is completely dependent on CE, ATP and membranes. It is inhibited by RNase, and is partially dependent on GTP. Incorporation is dependent on the amount of ribosomes attached to membranes, since less incorporation is obtained by an equal amount of membrane from cells that have not been treated with chloramphenicol. Incorporation does not require concomitant RNA synthesis, since it is not inhibited by DNase, actinomycin D, rifamycin, nor by omission of UTP and CTP. Thus, incorporation is carried out on preexisting mRNA.

The antibiotic inhibitor of initiation of protein synthesis by 70S ribosomes, streptogramin (11), does not inhibit amino acid incorporation by chloroplast membranes. Streptogramin appears to be an inhibitor of initiation in vivo in *C. reinhardi* (4). Thus, it is concluded that amino acid incorporation by membranes results from continuation of translation on a preformed membrane-ribosome-mRNA complex.

Over 80% of the radioactive protein in amino acid incorporation reaction mixtures is recovered in membranes that are washed once by centrifugation.

TABLE 1

Characteristics of protein synthesis by chloroplast membranes. The data represent the summary of a number of different experiments using different membrane preparations. Incorporation was carried out with 0.3 to 0.6 mg chlorophyll/ml and 4 μCi of a mixture of tritiated amino acids. Incorporation varied from 250 to 1400 cpm/0.05 ml of complete reaction mixture at the 30 min time points selected here. Results in one experiment were referred to incorporation obtained with complete reaction mixture in that experiment. Each value is the average (± average deviation) of two experiments. In each duplicate samples were taken.

Source of membranes	Conditions of incorporation	Incorporation (% of complete)
CAP-treated cells	Complete	100
	Minus ATP and ATP generating system	13 ± 3
	Minus UTP, CTP and GTP	44 ± 1
	Minus GTP	50 ± 4
	Minus UTP, CTP	94 ± 15
	Minus CE	3 ± 1
	Minus membranes	3 ± 2
	Plus chloramphenicol, 100 μg/ml	33 ± 3
	Plus cycloheximide, 100 μg/ml	108 ± 1
	Plus deoxyribonuclease, 80 μg/ml	91 ± 13
	Plus ribonuclease, 40 μg/ml	7 ± 3
	Plus rifamycin, 5 μg/ml	110 ± 26
	Plus actinomycin D, 20 μg/ml	130 ± 24
	Plus streptogramin, 100 μg/ml	102 ± 28
Untreated cells	Complete	32 ± 1

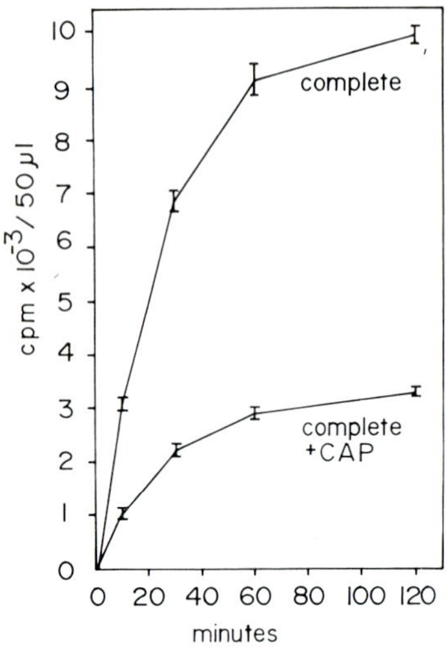

Fig. 5. Effect of chloramphenicol on amino acid incorporation. The reactions were carried out as described in Methods. Each ml of reaction mixture contained 1 mg of chlorophyll and 40 µCi of tritiated amino acids. Dispersion is average deviation for duplicate samples.

When washed membranes are sedimented in a density gradient that is layered over a cushion of 2 M sucrose, all radioactive protein and chlorophyll is found at the density gradient-cushion interface (fig. 6). If membranes are first treated with Nonidet P-40 (fig. 6A), some radioactivity, but no chlorophyll is found at the cushion-density gradient interface. This radioactivity might represent membrane that is not completely disrupted, but from which chlorophyll is removed. Some radioactivity is found in the polysome region of the gradient. This result indicates that some radioactive protein in the membranes is in nascent protein, and is confirmed by the observation that ribosomal subunits released with puromycin and KCl are not radioactive (Margulies, unpublished). Some radioactivity is also found at the top of the gradient. This last observation suggests that some radioactive protein is attached to the membrane, but is no longer attached to ribosomes.

Conclusions

The data show that functional polyribosomes occur attached to the chloroplast membranes of *C. reinhardi.* The data also suggest that these ribosomes are synthesizing membrane protein, and that this protein is inserted into the membrane as it is synthesized. The thylakoid polyribosome complex offers a promising tool to study a portion of membrane biosynthesis in vitro.

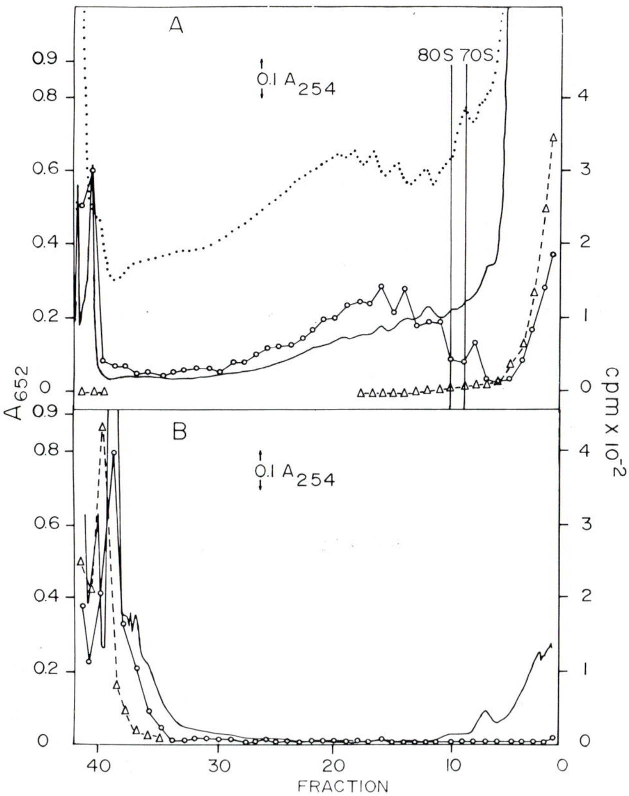

Fig. 6. Analysis of products of amino acid incorporation by density gradient centrifugation. Membranes from chloramphenicol-treated cells were labeled (fig. 5) and they were recovered from the reaction mixture and washed once by centrifugation. A sample of labeled membranes (17,000 cpm and 0.17 mg chlorophyll) in 0.5 ml of buffer that lacked sucrose (4) was treated with 50 μl of 10% Nonidet P-40 (A) or was not treated with detergent (B). The samples were layered on top of exponential sucrose gradients of 11.5 ml under which had been placed 0.5 ml of 2.0 M sucrose in buffer (4). Samples were centrifuged 1 hr at 40,000 rpm in a type SW 40 Ti Spinco rotor. Gradients were scanned at 254 nm (———) and 0.3 ml fractions collected. Radioactive protein (○—○—○) was determined on a single 0.1-ml sample from each fraction. Relative amounts of chlorophyll were determined as A_{652} (△—△—△) of 80% acetone extracts of 0.1 ml samples. A profile obtained with detergent-treated unlabeled membranes is shown for comparison (· · · ·).

Acknowledgements

This work was aided in part by a Grant number 430059 from the Smithsonian Research Foundation and the capable technical assistance of H. Lee Tiffany and is published with the approval of the Secretary of the Smithsonian Institution. Figs. 1, 2 and 4 are published with permission from the Rockefeller University Press.

References

1. I. Chen and S. Wildman, Biochim. Biophys. Acta 209 (1970) 207.
2. N. Chua, G. Blobel, P. Siekevitz and G.E. Palade, Proc. Nat. Acad. Sci. U.S. 70 (1973) 1554.
3. H. Falk, J. Cell Biol. 42 (1969) 582.
4. M.M. Margulies and A. Michaels, J. Cell Biol. 60 (1974) 65.
5. K.L. Tao and A. Jagendorf, Biochim. Biophys. Acta 324 (1973) 518.
6. I.I. Phillipovich, A.M. Tongur, B.A. Alina and A.I. Oparin, Exp. Cell Res. 62 (1970) 399.
7. D. Spencer and S.G. Wildman, Biochemistry 3 (1964) 954.
8. R.J. Mans and G.D. Novelli, Arch. Biochem. Biophys. 94 (1961) 48.
9. D. Boulter, Ann. Rev. Plant Physiol. 21 (1970) 91.
10. N.G. Avadhani and D.E. Buetow, Biochem. J. 128 (1972) 353.
11. S. Pestka, Ann. Rev. Microbiol. 25 (1971) 487.

Dr. MACHE:

Are the contaminant 80S ribosomes possibly attached to the outer membrane of the chloroplast's envelope?

Dr. MARGULIES:

The 80S ribosomes might be attached to pieces of the outside of the chloroplast envelope. This possibility should be considered, since cytoplasmic ribosomes are attached to the outer envelope of mitochondria*. However, the 80S ribosomes might be attached to pieces of endoplasmic reticulum. At present, we have no evidence to decide between these possibilities.

* Kellems, R.E. and Butow, J. Biol. Chem. 249 (1974) 3304.

CHAPTER I.7

POLYSOMES ASSOCIATED WITH YEAST MITOCHONDRIA: EVIDENCE FOR A CELL-SAP ORIGIN*

A.F.M. MOORMAN and L.A. GRIVELL

Section for Medical Enzymology, Laboratory of Biochemistry, University of Amsterdam, Eerste Constantijn Huygensstraat 20, Amsterdam, The Netherlands

Mitochondria from all organisms so far studied contain ribosomes distinct from their cell-sap counterparts (1). In a number of cases, careful washing and fractionation procedures have permitted isolation of these ribosomes virtually free of cell-sap ribosome contamination. An essential element in such procedures has undoubtedly been the use of EDTA-containing media for mitochondrial isolation (2).

Yeast mitochondria prepared under these conditions are highly active in the synthesis of specific mitochondrial proteins (3), yet paradoxically, on subsequent analysis, are singularly poor in polysomes (fig. 1a). This profile was not significantly improved by the use of various detergents (Triton X-100, Brij 58, deoxycholate) for mitochondrial lysis or gradient analysis, or by the inclusion of ribonuclease inhibitors (heparin, polyvinylsulphate, rat-liver post-microsomal supernatant) in media. Direct analysis of mitochondrial lysates on sucrose gradients was also negative. Treatment of yeast protoplasts to a period of recovery in osmotically stabilized growth medium followed by rapid cooling, which improves the protein synthetic ability of mitochondrial preparations (3) was without effect, as was also the termination of the recovery period with chloramphenicol or erythromycin as inhibitors of polysome run-off.

The minor peak in fig. 1a probably represents an aggregate of 74S particles, as reported previously (4).

In contrast, as fig. 1b shows, mitochondria prepared in Mg-containing media do yield polysomes in agreement with previously published results (5—7). At the same time, the RNA content is increased by a factor of 4—5 (cf. (8)). In view of this and the fact that magnesium has been reported to promote aggregation of microsomal membranes (9,10), leading to increased association of what is possibly a special class of cell-sap ribosomes (11,12), we considered it necessary to verify whether such polysomes were indeed mitochondrial.

---

* Presented by P. Borst.

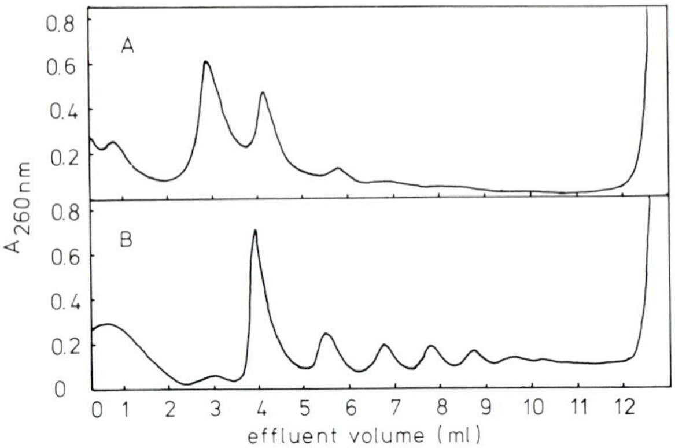

Fig. 1. Sucrose gradient analysis of polysome preparations isolated from EDTA- and Mg-washed mitochondria. *Saccharomyces carlsbergensis* NCYC-74 was grown in semi-synthetic lactate medium (21). Labelling was performed by growing the cells for several generations in the same medium, containing 0.3 mCi [2,4-^{3}H]adenine per liter or in phosphate-free semi-synthetic lactate medium, containing 1.7 mCi sodium [^{32}P]orthophosphate per liter. Mitochondria were prepared essentially as described previously (4), from protoplasts which had been incubated for 1 hr at 28°C in growth medium, osmotically stabilized by the addition of 1 M sorbitol (25 g cells per liter medium). EDTA-washed mitochondria were washed in 0.6 M mannitol, 0.1 mM EDTA (pH 6.7), while Mg-washed mitochondria were prepared by washing in 0.6 M mannitol, 10 mM $MgCl_2$ (pH 6.7). Lysis of the mitochondria and ribosome purification were accomplished as described previously (4). Polysomes were concentrated by pelleting through 1.5 M sucrose and 10 mM Mg-acetate, 10 mM Tris—HCl, 100 mM NH_4Cl, 6 mM 2-mercaptoethanol (pH 7.5). Centrifugation was then for 16 hr in the Spinco-30 rotor at 30,000 rev/min and at 2°C. Pellets were rinsed with 10 mM Mg-acetate, 10 mM Tris—HCl 50 mM NH_4Cl, 6 mM 2-mercaptoethanol (pH 7.5), resuspended in the same buffer, in which the NH_4Cl concentration was increased to 100 mM, and clarified by centrifugation at 10,000 *g* for 10 min. Preparations were analysed directly for polysomal content on 15—25% isokinetic gradients (4), containing 10 mM Mg-acetate, 10 mM Tris—HCl, 50 mM NH_4Cl, 6 mM 2-mercaptoethanol (pH 7.5), in the Spinco SW-41 rotor. Centrifugation was for 105 min at 40,000 rev/min and at 5°C. Absorbance profiles were monitored continuously by upward displacement through a Zeiss PMQII spectrophotometer, modified to accept an MR ID flow-through cuvette. A) Polysome preparation from EDTA-washed mitochondria. B) Polysome preparation from Mg-washed mitochondria.

RNA from the two preparations was, therefore, analysed by gel electrophoresis under conditions which separate mitochondrial and cell-sap RNA with high resolution (13) and with labelled cell-sap RNAs as internal marker. Figs. 2a and 2b show that whereas EDTA-washed mitochondria are virtually free of cell-sap contamination and yield typical mitochondrial RNAs (14), Mg-washed mitochondria contain an 80:20% mixture of cell-sap and mitochondrial RNAs. Evidently the polysomes present in Mg-washed mitochondria are largely — if not entirely — of cell-sap origin.

Cell-sap ribosomes associated with Mg-washed mitochondria are apparent-

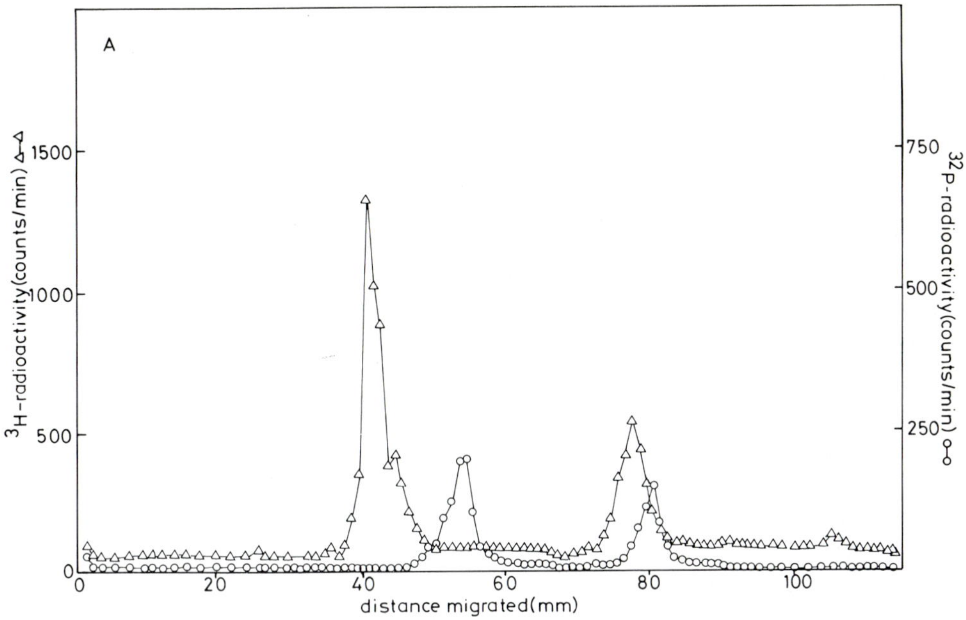

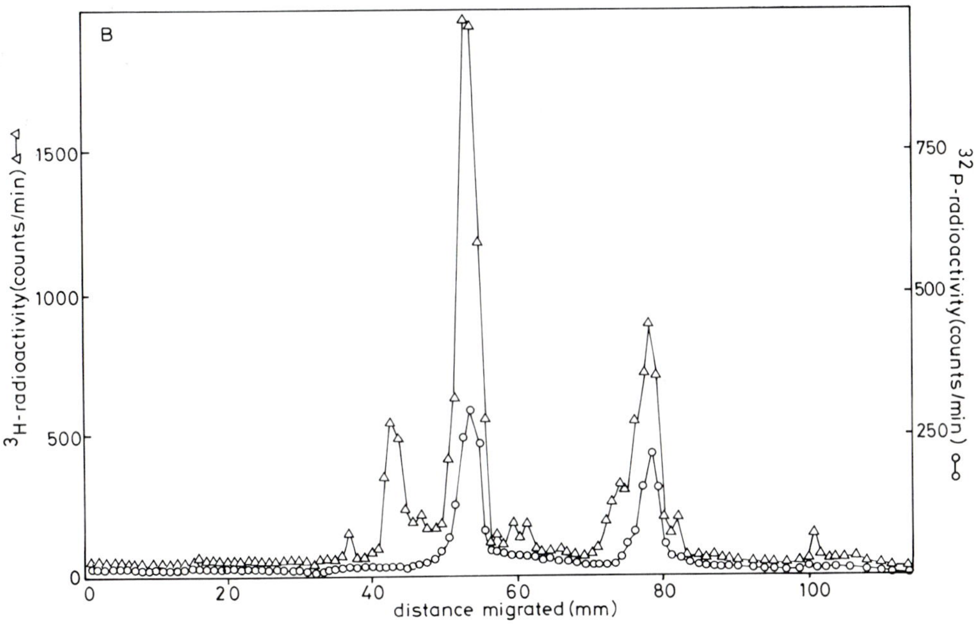

Fig. 2. Polyacrylamide gel electrophoresis of ^{3}H-labelled RNA associated with EDTA- and Mg-washed mitochondria. RNA was extracted directly from a mitochondrial suspension in 15 mM NaCl, 1.5 mM Na-citrate (pH 7.0), according to a modified Kirby procedure (22) as described in (13), in which is also described the procedure for isolation of marker cell-sap RNA. Gel electrophoresis was carried out in 2.4% pre-run gels at 8°C using the buffer system of Peacock and Dingman (23). Gel electrophoresis was for 2.5 hr at 2.5 mA/gel. Transfer RNA has run off the gel. For radioactivity determinations frozen gels were sliced, solubilized with 0.5 ml protosol (New England Nuclear) and counted in PPO-POPOP-toluene in a liquid scintillation system. A) RNA from EDTA-washed mitochondria. B) RNA from Mg-washed mitochondria.

TABLE 1

Antibiotic sensitivity of peptidyl transferase by mitochondrial ribosomes. Peptidyl transferase was measured according to De Vries et al. (25), using acetyl-^{3}H-leucyl-tRNA as substrate (input 7000 cpm/assay; 1 A_{260} unit/assay). Values are the means of two experiments and have been corrected for blanks without ribosomes.

Additions or omissions	Percentage activity in ribosomes washed in	
	EDTA	Mg
None	100*	100**
Puromycin omitted	2	2
Chloramphenicol (100 μg/ml)	7	12
Anisomycin (60 μg/ml)	98	91

* 4598 cpm/A_{260} unit.
** 1994 cpm/A_{260} unit.

ly poorly active (cf. (11)). Table 1 shows that antibiotic sensitivity of peptidyltransferase activity by ribosomes isolated from the two preparations is identical and typically mitochondrial. Antibiotic sensitivity of protein synthesis by the isolated mitochondria was also found to be identical (table 2). This is to be expected, however, since assay conditions used are probably sub-optimal for measurement of cell-sap ribosome activity. Clearly, functional tests alone are insufficient criteria for assay of mitochondrial purity.

Do typical polysomes exist in yeast mitochondria? Electron microscopy has shown a ribosome alignment which may be interpreted as polysomal structures (15). Polysomal structures from EDTA-washed mitochondria have been reported (16,17). From the unusual resistance to degradation by ribo-

TABLE 2

Antibiotic sensitivity of protein synthesis in isolated mitochondria. Assay of amino acid incorporation by intact mitochondria in vitro was carried out as described by Grivell et al. (24). Incubation was for 30 min at 30° C in a total volume of 0.5 ml. L-[U-^{14}C]Leucine (56 mCi/mmole; Radiochemical Centre, Amersham, Great Britain) was 143 μM. Values given have been corrected for zero time control (less than 3%) and are the means of two experiments. Protein determination was according to Lowry et al. (26).

Additions	Percentage activity in mitochondria washed in	
	EDTA	Mg
None	100*	100**
Chloramphenicol (100 μg/ml)	12	15
Cycloheximide (200 μg/ml)	98	99

* 67 pmole L-[U-^{14}C]leucine incorporated/mg protein per 30 min.
** 30 pmole L-[U-^{14}C]leucine incorporated/mg protein per 30 min.

nuclease it has been concluded that these polymeric forms probably were non-specific aggregates of either monosomes or polysomes. All our attempts at isolation of mitochondrial polysomes from yeast have so far failed. One possibility is that the highly hydrophobic nature of mitochondrial translation products results in aggregation, which hampers or prevents polysome isolation. The possibility also remains that, as in chloroplasts (18) and proposed for mitochondria (19), mitochondrial ribosomes are sufficiently firmly attached to the mitochondrial membrane to prevent the isolation of mRNA-ribosome complexes as intact entities. Prospects that mitochondrial polysomes may serve as a source of specific mitochondrial mRNA (all the more important because mitochondrial RNA in yeast does not contain poly(A) (20)) are, therefore, slender.

Acknowledgements

This work was supported in part by a grant from The Netherlands Foundation for Chemical Research (S.O.N.) with financial aid from The Netherlands Organization for the Advancement of Pure Research (Z.W.O.).

References

1. P. Borst and L.A. Grivell, FEBS Letters 13 (1971) 73.
2. W. Kleinow, W. Sebald, W. Neupert and Th. Bücher, in N.K. Boardman, A.W. Linnane and R.M. Smillie (Eds.), Autonomy and Biogenesis of Mitochondria and Chloroplasts, North-Holland, Amsterdam (1970) p. 140.
3. G.S.P. Groot and R.O. Poyton, Abstr. 9th FEBS Meeting, Budapest (1974) p. 271.
4. L.A. Grivell, L. Reijnders and P. Borst, Biochim. Biophys. Acta 247 (1971) 91.
5. N.G. Avadhani and D.E. Buetow, Biochem. J. 128 (1972) 353.
6. C.S. Cooper and C.J. Avers, in A.M. Kroon and C. Saccone (Eds.), The Biogenesis of Mitochondria, Academic Press, New York (1974) p. 289.
7. E. Agsteribbe, R. Datema and A.M. Kroon, in A.M. Kroon and C. Saccone (Eds.), The Biogenesis of Mitochondria, Academic Press, New York (1974) p. 305.
8. R. Datema, E. Agsteribbe and A.M. Kroon, Biochim. Biophys. Acta 335 (1974) 386.
9. B. Attardi, B. Cravioto and G. Attardi, J. Mol. Biol. 44 (1969) 47.
10. G.H. Hogeboom, W.C. Schneider and G.E. Pallade, J. Biol. Chem. 172 (1948) 619.
11. R.E. Kellems and R.A. Butow, J. Biol. Chem. 247 (1972) 8043.
12. G. Hallermayer and W. Neupert, FEBS Letters 41 (1974) 264.
13. L.A. Grivell, L. Reijnders and P. Borst, Eur. J. Biochem. 19 (1971) 64.
14. L. Reijnders, C.M. Kleisen, L.A. Grivell and P. Borst, Biochim. Biophys. Acta 272 (1972) 396.
15. P.V. Vignais, J. Huet and J. André, FEBS Letters 3 (1969) 177.
16. D. Ojala and G. Attardi, J. Mol. Biol. 65 (1972) 273.
17. R. Michel and W. Neupert, Eur. J. Biochem. 36 (1973) 53.
18. N.-H. Chua, G. Blobel, P. Siekevitz and G.E. Palade, Proc. Nat. Acad. Sci. U.S. 70 (1973) 1554.

19. C.L. Bunn, C.H. Mitchell, H.B. Lukins and A.W. Linnane, Proc. Nat. Acad. Sci. U.S. 67 (1970) 1233.
20. G.S.P. Groot, R.A. Flavell, G.J.B. Van Ommen and L.A. Grivell, Nature (1974) 167.
21. T. Ohnishi, K. Kawaguchi and B. Hagihara, J. Biol. Chem. 241 (1966) 1797.
22. J.H. Parish and K.S. Kirby, Biochim. Biophys. Acta 129 (1966) 554.
23. A.C. Peacock and C.W. Dingman, Biochemistry 7 (1968) 668.
24. L.A. Grivell, L. Reijnders and H. De Vries, FEBS Letters 16 (1971) 159.
25. H. De Vries, E. Agsteribbe and A.M. Kroon, Biochim. Biophys. Acta 246 (1971) 111.
26. O.H. Lowry, N.J. Rosebrough, A.L. Farr and R.J. Randall, J. Biol. Chem. 193 (1951) 265.

Dr. MACHE:

It could be that poly A was not found in mitochondria because poly A tails are too short.

Dr. BORST:

This was checked in two ways: a) Mitochondrial RNA was degraded with pancreatic RNase and the digest was electrophoresed on acrylamide gels. Whereas the 50-nucleotide poly(A) segment from cell-sap RNA was resolved, no such component was found in mtRNA. A poly(A) tract of less than 30 nucleotides would not have been detected, because of the high background of oligonucleotides. b) mtRNA was run through poly(U) Sepharose and none was bound under conditions where a poly(A) tract of 25 is still retained (Jacobson et al., J. Mol. Biol. 82 (1974) 213).

Together these results show that the poly(A) in yeast mtRNA, if present, must be shorter than 25 nucleotides.

Dr. MACHE:

There is another way to look for the possibility of poly A in mitochondria: has anybody looked for a poly A (or poly X) polymerase?

Dr. BORST:

Jacob et al. (Biochim. Biophys. Acta 361 (1974) 312) have reported that rat-liver mitochondrial preparations contain a poly(A) polymerase, but I do not know of a similar finding in primitive eukaryotes. However, even if yeast mitochondrial preparations were shown to contain poly(A) polymerase activity, it would still be difficult to prove that this enzyme is truly mitochondrial and that it adds A residues onto mtRNA species. I find our approach more direct and, therefore, more conclusive.

Dr. MARGULIES:

Has anyone demonstrated real mitochondrial polymerase?

Dr. BORST:

Even yeast mitochondria isolated in EDTA-containing solutions, yield on lysis some material that sediments faster than monomer ribosomes. This could be polysomal material. In all cases, however, where typical polysomal profiles were obtained from mitochondrial lysates, contamination with cell-sap polysomes was not rigorously excluded in my opinion.

Dr. SCHATZ:

Have you looked for mitochondrial polysomes in yeast by pulse-labeling mitochondria or cells with labeled amino acids, extracting the mitochondria with detergents, and checking for rapidly sedimenting, RNase sensitive species in the extract?

Dr. BORST:

We have pulse-labelled isolated mitochondria with leucine in the presence of cycloheximide and analysed the distribution of nascent protein in mitochondrial lysates on sucrose gradients. About 90% of the leucine incorporation was inhibited by chloramphenicol (CAP). The fraction of CAP-sensitive radioactivity sedimenting beyond the monosome peak was 7% in EDTA-washed mitochondria and 34% in Mg^{2+}-washed mitochondria. The RNase-sensitivity of this more rapidly sedimenting material was not checked.

Dr. WILLIAMSON:

S.G. Oliver in my lab has evidence that petite mutagenesis by 5 fluorouracil (5FU) is caused by incorporation of 5FU into mitochondrial RNA. This mutagenic effect is blocked by cordycepin. This argues that cordycepin has at least some inhibitory effect on yeast mitochondrial RNA synthesis.

CHAPTER I.8

DIFFERENTIATION AND THE EVOLUTIONARY PERSISTENCE OF SEMI-AUTONOMOUS ORGANELLES

G. MORPURGO

Orto Bonatico dell'Università di Roma, Largo Cristina di Svezia, Roma, Italia

In this paper an hypothesis on the meaning of the persistence in the eukaryotic cells of autonomous organelles with a bacterial type machinery for protein synthesis is discussed. According to the hypothesis, organelles were maintained during evolution to conserve for some functions a system which permits rapid and efficient adaptation to a changing environment. Rapid adaptation at the transcription level needs 70 S ribosomes and "naked" DNA to permit transcription-coupled translation. The change in the system of adaptation in eukaryotes was determined by the evolution of differentiative systems of control. The need of a different system of adaptation determined the evolution toward 80 S ribosomes which probably are better suited to perform adaptation at the translational level.

One of the main problems in the biology of the organelles of the eukaryotic cell is to understand what is the advantage for the cell to have two or three independent systems of protein synthesis. Actually both mitochondria and chloroplasts have their own system of protein synthesis with 70 S type ribosomes and their own DNA. According to the most recent results (based largely on studies on yeasts) on the mit-DNA there are the genes for the ribosomal RNA, for the transfer RNAs (of the mit-protein synthetic system) and probably for three polypeptides of the cytochromoxidase, for one of the cytochrome *b*, for 4 polypeptides of the mit-ATPase and for the coenzyme QH_2-cytC reductase (1,2). Surely there is still some function to be discovered but the number of genes that may be codified in the mit-DNA is very small. Still less clear is the situation in chloroplasts. Besides the chl RNAs the genes for cytochrome *b* and *f* and perhaps for RUDP carboxylase and for some proteins of the thylakoids are probably encoded in the DNA of the organelle (3). In both organelles most or all of the proteins of the ribosomes are codified in the nucleus and synthetized in the cytoplasm.

The question to set up is: what is the meaning for the economy of the cell to have two different systems of protein synthesis, one with 80 S and one with 70 S ribosomes? The latter synthetizes very few proteins but it is essential for the life of the cell: the interruption of its functioning with antibiotics that block protein synthesis at the level of the 70 S system is always lethal except for yeasts that can live anaerobically.

In the evolution from the prokaryotic to the eukaryotic cell one of the most important physiological changes is that in the control system of the adaptive protein synthesis. In prokaryotic organisms a large part of the genes are grouped in operons and are repressed and derepressed in a coordinate way directly from the changing environment. The level of the protein may fluctuate thousand times in the shift from the repressed to the derepressed condition. On the contrary in the eukaryotic cells, also in the most primitive eukaryotes, most of the proteins are constitutive and the adaptive proteins very rarely, as a consequence of a direct external influence, change by a factor more than ten.

The difference in the adaptive system among prokaryotes and eukaryotes may be attributed to the increasing importance of the differentiative control in the eukaryotes. Actually differentiation means activation and inactivation of genes in different moments of the development following a precise order determined by an internal clock. The construction of stable differentiated structures is incompatible with a system of regulation very sensitive to the external stimuli like that of prokaryotic organisms and therefore with a continuous change in the level of most of the proteins.

One of the best instances of the necessity of a temporal control in the differentiated systems is given by the hemoglobins (Hb). In all the vertebrates (except lamprey) the Hb molecule is a tetramer. The adult Hb (A) has two α- and two β-chains ($\alpha_2\beta_2$) while the foetal Hb is $\alpha_2\gamma_2$. The minor adult Hb (A_2) is $\alpha_2\delta_2$. The α-chain is always present while the γ-chain disappears concomitantly with the synthesis of β- and δ-chains. If we examine the linkage relationship of the structural genes involved in the production of the Hb chains we see that γ, β, δ genes are linked to each other while α gene is unlinked. Because all the genes arose by duplication of a single ancestral gene it must be concluded that α gene which is always active was translocated in a different portion of the genome probably to permit an independent regulation. γ, β, δ structural genes may be contiguous because a single switch is sufficient to shift from one gene to the others.

The rule that protein synthesis is scarcely adaptive in eukaryotic cells holds true for most of the enzymatic systems, notably with the exception of some mitochondrial and chloroplastic synthesis. In *Saccharomyces cerevisiae* the shift from anaerobiosis to aerobiosis induces the synthesis ex novo of cytochrome *b* and of cytochromoxidase. Unfortunately most of the data on the mit adaptive synthesis are limited to yeasts because these only may live both in aerobiosis and in anaerobiosis. It is known however that cytochromoxidase is adaptive also in mammalian cells in culture (4). More difficult to interpret is the situation in chloroplasts because it is different in the various organisms. Cytochromes *b* and *f* are fully adaptive in bean (5) and a similar situation exists in *Euglena* (3). In *Chlamydomonas* the situation is rather different because the chloroplastic synthesis is completely constitutive except in the mutant Y-1 (3). It is impossible here to discuss this complex situation but again it must be stressed that the only complete adaptive synthesis of proteins in the eukaryotic cells are linked to organelles with 70 S type ribosomes and that probably most of the structural genes of the organelles are fully adaptive. An interesting point is that also some proteins which are totally coded in the nucleus and synthetized in the cytoplasmic ribosomes, but are localized in the mitochondria, are adaptive also if the relative increase, following adaptation, is generally lower than that of the mit coded proteins. It seems in this case that the organelle may regulate also the synthesis of proteins codified in the nucleus and synthetized in the cytoplasm. It is unknown if this regulation takes place at the transcription or at the translation. The idea can be forwarded that the adaptation of these proteins takes place at the translation and is determined by the presence of the proteins encoded by the organelle chromosome regulated at the transcriptional level. These considerations suggest the following possibilities: 1) that strongly adaptive synthesis of proteins is linked to a bacterial type system, 2) that during the course of evolution the cell has preserved organelles with a partial genetic autonomy and their own protein synthesizing machinery because it was useful to maintain for some functions an efficient system of adaptation like that of bacteria.

The next question is the following: why a bacterial system is necessary for an efficient adaptive synthesis? Apparently 70S and 80S ribosomes should work equally well in the adaptive synthesis because adaptation takes place at the transcription while ribosomes operate during the translation of the genetic message. However some recent papers (6) has shown the validity of the theory of Stent that in bacteria transcription and translation are tightly coupled. The interruption of the translation interrupts at the same time the process of transcription. The linkage between translation and transcription is possible in prokaryotes but not in eukaryotes because protein synthesis occurs in the cytoplasm while transcription occurs in the nucleus and the two processes are physically separated. It is likely that the linkage between translation and transcription needs 70 S ribosomes and a "naked" bacterial type chromosome. According to this hypothesis the bacterial type structure of the organelles could be useful to the cell to conserve the bacterial type system of adaptation of protein synthesis. It is not known if in organelles transcription and translation are physically linked but, according to Fukuhara (7) during respiratory adaptation in yeasts there is a strong increase in the RNA synthesis in the mitochondria that, in this respect, behave like a bacterium in derepressed condition. It would be interesting to know if chloramphenicol, as in bacteria (8) interrupts the adaptive synthesis of messenger RNA during respiratory adaptation in yeasts.

A further question is the following: if 70 S ribosomes are particularly suited to perform transcriptional control (linked to the translational process) why the eukaryotic cells that lack this type of control in the evolution have developed larger and more complicated ribosomes? A possible answer is that eukaryotes have evolved a different type of control, i.e. a control at the translation. Actually the translational control may play a dominant role in protein synthesis in eukaryotes. Conconi (9) very recently has shown that one type of thalassemia (a disease of man in which βHb chains are not produced) is due to a defect at the translation because RNA messenger of β-chains is regularly produced but is not translated. Monroy (10) has shown that in the development of sea urchin egg "frozen" messenger RNA are present waiting to be activated by fertilization. Also the work of Tomkins (11) on the induction of tyrosintransaminase in hepatoma cells shows that the level of the enzyme after induction with corticosteroids is largely determined by a translational control.

I believe in summary that the transition from 70 S to 80 S ribosomes in the course of the evolution was selectively determined by the necessity for the cell to change the system of control of adaptive protein synthesis in relation to the formation of the differentiated systems. The organelles (on their origin we may equally well accept the symbiotic or the asymbiotic theory) initially preserved 70 S ribosomes and naked DNA to maintain a rapid and efficient adaptation of some functions (the energy production) for which was still useful to have a direct adaptation to the variable conditions. With the further progress of the evolution the chromosomes of the organelles have lost most of their genes: the structural genes that are still there are however regulated like bacterial genes and have probably acquired a central role in controlling the concentration of other proteins of the organelle which are codified in the nucleus and synthetized in the cytoplasm.

References

1. E. Ross, E. Ebner, R.O. Poyton, T.L. Mason, B. Ono and G. Schatz in A.M. Kroon and G. Saccone (Eds), The Biogenesis of Mitochondria, Academic Press N.Y. and London (1974) p. 477.
2. D.S. Beattie, L.F. Lin and R.N. Stuchell, in A.M. Kroon and G. Saccone (Eds.) The Biogenesis of Mitochondria, Academic Press N.Y. and London (1974) p. 465.
3. R. Sager, Cytoplasmic Genes and Organelles, Academic Press, N.Y. and London (1972).
4. N. Hakami and D.A. Pious, Nature 216 (1967) 1087.
5. J.W. Bradbeer, in X.X. Milborrow (Ed), Biosynthesis and its Control in Plants, Academic Press, N.Y. and London (1972) p. 279.
6. F. Gros and M. Crepin, Transcription and Regulation at the Ribosomal Level, this Symposium (1975) ch. I.1.
7. H. Fukuhara, Proc. Nat. Acad. Sci. U.S. 58 (1967) 1065.
8. F. Imamoto and S. Tani, Nature New Biol. 240 (1972) 172.
9. F. Conconi, P.T. Rowley, L. del Senno, S. Pontremoli, S. Volpato, Nature New Biol. 238 (1972) 83.
10. A. Monroy, R. Maggio and A.M. Rinaldi, Proc. Nat. Acad. Sci. U.S. 54 (1965) 107.
11. G.M. Tomkins, E.B. Thompson, S. Hayashi, T. Gelehrter, D. Cramer and B. Peterkofsky, Cold Spring Harbor Symp. Quant. Biol. XXI (1966) 349.

SECTION II

EXTRANUCLEAR DNA IN ORGANELLES OF EUCARYOTES

CHAPTER II.1

LES FONCTIONS GENETIQUES MITOCHONDRIALES CHEZ *SACCHAROMYCES CEREVISIAE*

B. DUJON

Centre de Génétique Moléculaire du C.N.R.S., 91190 Gif-sur-Yvette, France

L'organisation de l'information génétique à l'intérieur des cellules eucaryotes constitue l'un des aspects importants pour l'étude des *relations nucléocytoplasmiques*, thème de ce colloque. En effet il est maintenant clairement établi que si une quantité importante de l'information génétique dont dispose la cellule est portée par les chromosomes du noyau, il existe aussi une autre partie de cette information génétique qui est portée par l'ADN des organelles: mitochondries et chloroplastes. Ainsi dans chaque cellule eucaryote il existe au moins deux systèmes génétiques différents: le noyau et les mitochondries, ou même trois dans le cas des cellules possédant des chloroplastes. Ce bref exposé sera consacré à une discussion de quelques unes des questions qui concernent l'information génétique mitochondriale: en particulier on montrera que les mitochondries possèdent une information génétique qui leur est spécifique tant par la nature des fonctions codées que par les règles de la transmission héréditaire de cette information.

Deux types de mutations peuvent apparaître dans le génome mitochondrial de la levure *Saccharomyces cerevisiae*. Le premier type est représenté par les mutants à déficience respiratoire (petite ou ρ^-) dont l'ADN mitochondrial présente d'importantes réorganisations moléculaires: délétions de longs segments de séquences de bases et répétitions régulières d'autres segments restaurant ainsi la longueur initiale des molécules et la masse totale d'ADN mitochondrial (1—5). Le deuxième type est représenté par des mutations conferrant la résistance envers des inhibiteurs spécifiques des fonctions mitochondriales (chloramphénicol, érythromycine, spiramycine, paromomycine, oligomycine) (6—22). Ces mutations sont du type ponctuel.

Par une étude basée sur les fréquences de recombinaison entre les mutations un début de carte factorielle a pu être proposé. Trois loci, génétiquement liés, portent les mutations conferrant la résistance au chloramphénicol (locus R_I), à l'érythromycine, à la spiramycine ou aux deux à la fois (loci R_{II} et R_{III}) (22). Par l'étude de l'activité in vitro des ribosomes mitochondriaux purifiés, Grivell et al. (23) ont montré que ces trois loci devaient correspondre à l'information génétique nécessaire à la synthèse d'au moins un des éléments de la grosse sous-unité des ribosomes mitochondriaux et constituent ainsi un segment ribosomal. Un autre locus (P_I), génétiquement non lié aux précédents, porte les mutations conferrant la résistance à la paromomycine, un autre inhibiteur spécifique de l'activité des ribosomes mitochondriaux (20). Deux loci (O_I et O_{II}) génétiquement non liés entre eux ainsi que non liés au segment ribosomal et au locus P_I, portent les mutations conferrant la résistance à l'oligomycine (14). Ces mutations sont responsables de changements de l'ATPase liée à la membrane interne mitochondriale (24).

Ainsi le génome mitochondrial porte l'information génétique correspondant au moins à deux types de fonctions différentes. D'une part les mutations oligomycine-résistantes montrent qu'une partie du génome mitochondrial est impliquée dans la synthèse de certains polypeptides des complexes enzymatiques liés à la membrane interne mitochondriale tels que l'ATPase. Il est possible que ceci soit le cas aussi pour d'autres complexes liés à la membrane interne et dont certaines sous-unités sont traduites par les ribosomes mitochondriaux comme par exemple la cytochrome oxydase (25). D'autre part les mutations chloramphénicol, érythromycine, spiramycine, et paromomycine résistantes montrent que certains des éléments de la machinerie de synthèse protéique spécifique des mitochondries sont codés par l'ADN mitochondrial. Ceci est confirmé par les expériences d'hybridations moléculaires ARN/ADN qui montrent que les deux ARN ribosomaux mitochondriaux ainsi que plusieurs ARN de transfert mitochondriaux sont codés par l'ADN mitochondrial (26, 27, 28).

Grâce à l'étude de très nombreux croisements multifactoriels utilisant les mutations conferrant la résistance aux antibiotiques (8, 9, 10, 14, 20, 22) les traits essentiels de la recombinaison et de la transmission héréditaire des gènes mitochondriaux ont pu être dégagés. La descendance d'un croisement impliquant des mutations mitochondriales est caractérisée par une *assymétrie* de la fréquence des allèles des deux parents et par une *polarité* de recombinaison (14, 29). Un locus spécifique de l'ADN mitochondrial (locus ω) lié étroitement au locus R_I du segment ribosomal, gouverne l'existence de la polarité de recombinaison dans ce segment. Chaque cellule porte soit l'allèle ω^+ soit l'allèle ω^- dans son génome mitochondrial (9). Le segment ribosomal lié à ω est un segment polaire tandis que d'autres régions du génome mitochondrial sont non polaires (20, 29).

Les règles de la transmission héréditaire et de la recombinaison des génomes mitochondriaux ont été récemment décrites dans un modèle général pour la ségrégation et la recombinaison des molécules d'ADN mitochondrial (29). Ce modèle insiste sur le fait que la génétique mitochondriale est la génétique d'une population de molécules d'ADN mitochondrial. En cela la génétique des mitochondries à l'intérieur des cellules eucaryotes présente des similitudes avec la génétique des bactériophages dans les cellules procaryotes (30). Les idées principales de ce modèle seront brièvement résumées ici. Ce modèle suppose que, dans chaque cellule, existent de multiples copies du génome mitochondrial et que ces copies sont génétiquement équivalentes.

Cette idée est en accord avec le fait que le nombre de molécules d'ADN mitochondrial par cellule est élevé (environ 50 pour les cellules haploides et 100 pour les cellules diploïdes (31)). Ces copies forment un "pool" commun de génomes mitochondriaux dans lequel les événements d'appariemment et de recombinaison se font au hasard. La population de molécules d'ADN mitochondrial est donc panmictique. Cette population est caractérisée par une fraction appelée "input" c'est à dire par la contribution relative des molécules d'ADN mitochondrial de chacun des parents qui rentrent dans ce "pool" génétique. Ce rapport peut varier dans de larges proportions selon les souches impliquées dans le croisement et également selon les conditions expérimentales dans lesquelles est fait ce croisement. Les événements d'appariemment et de recombinaison des génomes mitochondriaux ont lieu continuellement. Ils sont non synchrones. D'autre part ils se produisent aussi bien dans les cellules dont le noyau est haploïde que dans celles dont le noyau est diploïde, et aussi bien dans les cellules ρ^+ que dans les cellules ρ^- (32). Ces événements ne sont génétiquement décelables que si la cellule dans laquelle ils ont lieu est mixte (hétéroplasmon) c'est à dire contient plusieurs types de molécules d'ADN mitochondrial portant des marqueurs génétiques différents. L'hétérogénéité peut être introduite par croisement ou par mutagenèse $\rho^+ \rightarrow \rho^-$. Dans le premier cas deux cellules, différant entre elles par leur génotype mitochondrial mais contenant chacune un seul type de molécules d'ADN mitochondrial (homoplasmon), fusionnent pour former un zygote qui contiendra ainsi des molécules des deux types parentaux. Dans le second cas l'hétérogénéité est introduite au sein d'une cellule homoplasmon après action d'un mutagène (induisant des mutations ρ^-) qui provoque de larges altérations de la séquence des bases dans certaines des molécules. Les phénomènes de ségrégation des molécules dans les cellules-filles (bourgeons) se produisent à chaque génération cellulaire et tendent rapidement à produire des cellules ne contenant qu'un seul type d'ADN mitochondrial et dans lesquelles les événements moléculaires d'appariemment et de recombinaison ne seront plus décelables. Il a été possible de calculer qu'une molécule subit en moyenne plusieurs (environ 4) événements d'appariemment et de recombinaison avant que les phénomènes de segrégation ne l'entraînent dans une cellule pure. Les événements élémentaires de recombinaison sont du type non réciproque (du type conversion), chaque événement produisant autant de molécules de type parental que de molécules de type recombiné et produisant seulement l'un des deux types recombinés possibles entre deux marqueurs. Les deux types de recombinés sont produits dans la population finale avec statistiquement la même fréquence ou au contraire avec des fréquences très différentes selon les régions dans lesquelles se trouvent les marqueurs considérés (polaire ou non polaire) et selon la nature des allèles des deux parents au locus ω. La fréquence des divers types cellulaires observés dans la descendance d'un croisement est donc la résultante d'un grand nombre d'événements moléculaires de recombinaison et de ségrégation.

Ce modèle rend parfaitement compte de tous les traits génétiques rencontrés dans les croisements mitochondriaux. D'autre part les prédictions quantitatives qu'il permet, notamment sur les variations de la fréquence des différents allèles et celle des recombinés en fonction des variations de l'input et sur la polarité de recombinaison, sont en accord avec l'ensemble des résultats expérimentaux des croisements multifactoriels (9, 10, 14, 15, 16, 17, 19, 20, 22, 33).

En conclusion il apparait que l'information génétique portée par l'ADN mitochondrial est spécifique, c'est à dire n'est pas une copie d'une partie de celle du noyau, et obéit à des règles de transmission héréditaire qui lui sont propres, et différentes de celles du noyau. La quantité d'information (longueur des séquences uniques) que représente le génome mitochondrial vis à vis du génome nucléaire est petite. Par contre il existe une forte redondance intermoléculaire de cette information mitochondriale puisqu'il existe de nombreuses copies du génome mitochondrial à l'intérieur d'une même cellule pour une seule copie du génome nucléaire dans les cellules haploïdes et deux dans les cellules diploïdes. Puisque chaque molécule d'ADN mitochondrial porte un cistron pour chacun des deux ARN ribosomaux mitochondriaux il y a donc une forte redondance des cistrons d'ARN ribosomaux mitochondriaux dans une même cellule. Or dans le noyau les séquences codant pour les ARN ribosomaux cytoplasmiques montrent aussi un haut degré de répétition. Peut être faut-il chercher dans la nécessité d'avoir un nombre élevé de copies des gènes ARN ribosomaux par cellule l'une des raisons de l'existence d'un génome mitochondrial autonome.

Bibliographie

1. J. Deutsch, B. Dujon, P. Netter, E. Petrochilo, P.P. Slonimski and M. Bolotin-Fukuhara, Genetics 76 (1974) 195.
2. G. Faye, H. Fukuhara, C. Grandchamp, J. Lazowska, F. Michel, J. Casey, G. Getz, J. Locker, M. Rabinowitz, M. Bolotin-Fukuhara, D. Coen, J. Deutsch, B. Dujon, P. Netter and P.P. Slonimski, Biochimie 55 (1973) 779.
3. J. Lazowska, F. Michel, G. Faye, H. Fukuhara and P.P. Slonimski, J. Mol. Biol. 85 (1974) 393.
4. F. Michel, J. Lazowaka, G. Faye, H. Fukuhara and P.P. Slonimski, J. Mol. Biol. 85 (1974) 411.
5. M. Rabinowitz, J. Casey, P. Gordon, J. Locker, H.S. Hsu and G.S. Getz, in Kroon and Saccone (Eds.) Biogenesis of Mitochondria Academic Press (1974) pp. 89—105
6. A.W. Linnane, G.W. Saunders, E.B. Gingold and H.B. Lukins, Proc. Nat. Acad. Sci. U.S. 59 (1968) 903.
7. D.Y. Thomas and D. Wilkie, Biochem. Biophys. Res. Commun. 30 (1968) 368.
8. D. Coen, J. Deutsch, P. Netter, E. Petrochilo and P.P. Slonimski, Mitochondrial Genetics. I. Methodology and Phenomenology. Symp. Soc. Exp. Biol. XXIV, Cambridge University Press (1970) 449.

9. M. Bolotin, D. Coen, J. Deutsch, B. Dujon, P. Netter, E. Petrochilo and P.P. Slonimski, Bull. Inst. Pasteur 69 (1971) 215.
10. R.A. Kleese, R.C. Grotbeck and J.R. Snyder, J. Bacteriol. 112 (1972) 1023.
11. C.H. Mitchell, C.L. Bunn, H.B. Lukins, A.W. Linnane, Bioenergetics 4 (1972) 363.
12. G.H. Rank and N.T. Bech-Hansen, Genetics 72 (1972) 1.
13. K. Suda and A. Uchida, Japan, J. Genetics 47 (1972) 441.
14. P.R. Avner, D. Coen, B. Dujon and P.P. Slonimski, Mol. Gen. Genetics 125 (1973) 9.
15. P.R. Avner and D.E. Griffiths, Eur. J. Biochem. 32 (1973) 312.
16. N. Howell, M.K. Trembath, A.W. Linnane and H.B. Lukins, Mol. Gen. Genetics 122 (1973) 37.
17. G.H. Rank, Heredity 30 (1973) 265.
18. M.K. Trembath, C.L. Bunn, H.B. Lukins and A.W. Linnane, Mol. Gen. Genetics 121 (1973) 35.
19. K. Wakabayashi and S. Kamei, FEBS Letters 33 (1973) 263.
20. K. Wolf, B. Dujon and P.P. Slonimski, Mol. Gen. Genetics 125 (1973) 53.
21. K. Suda and A. Uchida, Mol. Gen. Genetics 128 (1974) 331.
22. P. Netter, E. Petrochilo, P.P. Slonimski, M. Bolotin-Fukuhara, D. Coen, J. Deutsch and B. Dujon, Genetics 78 (1974) 1063.
23. L.A. Grivell, P. Netter, P. Borst and P.P. Slonimski, Biochim. Biophys. Acta 312 (1973) 358.
24. M. Somlo, P.R. Avner, J. Cosson, B. Dujon and M. Krupa, Eur. J. Biochem. 42 (1974) 439.
25. E. Ebner, T.L. Mason and G. Schatz, J. Biol. Chem. 248 (1973) 5369.
26. J. Casey, M. Cohen, M. Rabinowitz, H. Fukuhara and G.S. Getz, J. Mol. Biol. 63 (1972) 431.
27. G. Faye, C. Kujawa and H. Fukuhara, J. Mol. Biol. 88 (1974) 185.
28. H. Fukuhara, G. Faye, F. Michel, J. Lazowska, J. Deutsch, M. Bolotin-Fukuhara and P.P. Slonimski, Mol. Gen. Genetics 130 (1974) 215.
29. B. Dujon, P.P. Slonimski and L. Weill, Genetics 78 (1974) 415.
30. N. Visconti and M. Delbrück, Genetics 38 (1953) 5.
31. D.H. Williamson, Soc. Exp. Biol. Symp. 24 (1970) 247.
32. G. Michaelis, E. Petrochilo and P.P. Slonimski, Mol. Gen. Genetics 123 (1973) 51.
33. B. Dujon, A. Kruszewska, P.P. Slonimski, M. Bolotin-Fukuhara, D. Coen, J. Deutsch, P. Netter, E. Petrochilo and L. Weill, Mol. Gen. Genetics (in press).

CHAPTER II.2

INTRODUCTION FOR ROUND TABLE DISCUSSION ON MITOCHONDRIAL AND PLASTID DNAs

P. BORST

Section for Medical Enzymology, Laboratory of Biochemistry, University of Amsterdam, Eerste Constantijn Huygensstraat 20, Amsterdam, The Netherlands

In this Round Table Discussion we shall deal with some of the open questions on organelle DNAs. As introduction I have prepared two tables that summarize some of the available facts.

Table 1 shows that chloroplast DNAs are circular and all rather large. This makes these DNAs more difficult to handle and renders the task of fully elucidating their genetic function a formidable one. This may explain why most speakers in this Round Table stick to mtDNAs. The few chloroplast DNAs studied show a remarkable uniformity in size, as if the evolutionary flexibility of chloroplast biogenesis is essentially zero.

Nature displays more phantasy when it comes to mtDNAs, as shown in table 2. In addition to the usual contour lengths measurements and renaturation data this table also gives the first results obtained with restriction enzymes. If one cuts up a DNA with restriction enzymes, the added molecular weights of the fragments are equivalent to the genetic complexity of the DNA, whereas the relative amounts of each fragment provide information about gene repetition. Like the renaturation analysis one does not require

TABLE 1

Genome size and complexity of chloroplast DNAs.

Species	Molecular weight ($\times 10^{-6}$) from		Refs
	Contour length	Complexity	
Pea	91	95	(1)
Spinach	89		(2)
Lettuce	90	98	(1, 3)
Bean	90		(1)
Maize	86		(2)
Euglena gracilis	92	90	(4, 5)
Chlamydomonas reinhardii		100	(6)

TABLE 2

Size and complexity of mtDNAs.

Species	Structure	Molecular weight ($\times 10^{-6}$) from			Refs
		EM[a]	Renaturation	Restriction enzymes	
Animals	Circular	9—12	10—11	10[b]	(7—10)
Higher plants	Circular	70	74	-	(11)
Fungi					
Saccharomyces	Circular	41—49	50	40—60[c)] 38—50[d)]	(8, 9, 12—14)
Kluyveromyces	Circular	22	20	18[d]	(15)
Neurospora	Circular	37—39			(16, 17)
Saprolegnia	Circular	28			(18)
Protozoa					
Acanthamoeba	Circular	28			(19)
Tetrahymena	Linear	29	30		(8, 9, 20)
Paramecium	?		35		(21)
Other uni-cellular organisms					
Euglena	?	3?	40		(22—24)
Dictyostelium	?	46[e)]	35—40		(25, 26)

[a)] Recalculated on the basis of internal standards, where possible; [b)] C.A. Hutchison III, S.S. Potter, J.E. Newbold and M.H, Edgell, personal communication, [c)] G. Bernardi, this volume; [d)] J.P.M. Sanders, unpublished results; [e)] from sedimentation analysis.

intact DNA for this analysis. It is potentially more powerful than renaturation analysis in the detection of micro-heterogeneity, if present.

The results in table 2 show that in all mtDNAs the potential genetic information is equivalent to size and that large-scale heterogeneity or gene repetition is absent. Note that there is now general agreement that the size and complexity of *Saccharomyces* mtDNA is around $40—50 \times 10^6$ daltons, but that the mtDNA from *Kluyveromyces* (formerly *Saccharomyces*) *lactis* is only half this value. These two fungal mtDNAs cross-hybridize only to the extent of approximately 20% and the hetero-duplex is poorly matched, melting about 10°C below the homo-duplexes (27).

Note also in table 2 that the earlier reports (22, 23) that the mtDNA of *Euglena* is the smallest in nature, have not been substantiated by a recent renaturation analysis (24), which has yielded a complexity value (40×10^6 daltons) in the same range as obtained with mtDNAs from other uni-cellular organisms. The same value was also inferred from the hybridization plateau of mitochondrial rRNA with mtDNA in *Euglena* (28). It seems likely, therefore, that the smaller DNA obtained previously from *Euglena* mitochondria, represented genome fragments, rather than mini-chromosomes.

With this diversity in size and complexity of mtDNAs, one would expect to find a corresponding diversity in function. Potentially, a comparison of genes on mtDNA in animals, plants and lower organisms could yield insight into mitochondrial evolution and in the cooperation between the mitochondrial and nuclear genetic systems, because one could hope to find genes on mtDNA in protists which are absent in animals, their function being taken over by nuclear genes. Up till now, however, none have been found with the possible exception of some tRNA species (see (9, 10)).

There is then at the moment no good idea what the protist mtDNAs are doing with all their genetic information. This has led to the suggestion that a large fraction of this DNA might have no function, or would act as spacers not required in the more economically designed animal mtDNA. I tend to be more optimistic and think that most of this extra DNA is actually coding for minor proteins. It should be realized that the proteins identified until now as synthesized on mitochondrial ribosomes are major proteins of the mitochondrial membrane. Obviously, there are many proteins in mitochondria present at 10 to 100-fold lower concentrations than cytochrome oxidase and these could not have been found by the methods presently employed, which heavily rely on sodium dodecyl sulphate gels of proteins labelled in the presence of cycloheximide. Identification of these putative genes coding for minor proteins will be a major undertaking in the coming years.

Finally, let me reiterate (see (29)) that we still do not really understand why mtDNA is there at all, notwithstanding the identification of an increasing number of genes on mtDNA. Of course, we know now why mtDNA is indispensable in the present situation — one cannot make functional mitochondria without cytochrome oxidase and ATPase — but we still do not know why these genes should be located on a separate mtDNA rather than on nuclear DNA. Before this question is answered, I consider further speculations on the evolutionary origin of mitochondria futile.

References

1. R. Kolodner and K.K. Tewari, J. Biol. Chem. 247 (1972) 6355.
2. J.E. Manning, D.R. Wolstenholme and O.C. Richards, J. Cell Biol. 53 (1972) 594.
3. R. Wells and M. Birnstiel, Biochem. J. 112 (1969) 777.
4. J.E. Manning and O.C. Richards, Biochim. Biophys. Acta 259 (1972) 285.
5. E. Stutz, FEBS Letters 8 (1970) 25.
6. R. Wells and R. Sager, J. Mol. Biol. 58 (1971) 611.
7. D.R. Wolstenholme, K. Koike and H.C. Renger, Oncology 1 (1970) 627
8. P. Borst and R.A. Flavell, in S.G. Van den Bergh, P. Borst, L.L.M. Van Deenen, J.C. Riemersma, E.C. Slater and J.M. Tager (Eds.), Mitochondria: Biogenesis and Bioener-

getics; Biomembranes: Molecular Arrangements and Transport Mechanisms, Vol. 28, Elsevier, Amsterdam (1972) p. 1.
9. P. Borst, Ann. Rev. Biochem. 41 (1972) 333.
10. P. Nagley and A.W. Linnane, in I.R. Falconer and D. Cove (Eds.), Advances in Molecular Genetics, Elek, London (1976) (in press).
11. R. Koldner and K.K. Tewari, Proc. Nat. Acad. Sci. U.S. 69 (1972) 1830.
12. T.D. Petes, B. Byers and W.L. Fangman, Proc. Nat. Acad. Sci. U.S. 70 (1973) 3072
13. C. Christiansen, G. Christiansen and A. Leth Bak, J. Mol. Biol. 84 (1974) 65.
14. M. Rabinowitz, J. Casey, P. Gordon, J. Locker, H.-J. Hsu and G.S. Getz, in A.M. Kroon and C. Saccone (Eds.), The Biogenesis of Mitochondria, Academic Press, New York (1974) p. 89.
15. J.P.M. Sanders, P.J. Weijers, G.S.P. Groot and P. Borst, Biochim. Biophys. Acta 374 (1974) 136.
16. E. Agsteribbe, A.M. Kroon and E.F.J. Van Bruggen, Biochim. Biophys. Acta 269 (1972) 299.
17. D.A. Clayton and R.M. Brambl, Biochem. Biophys. Res. Commun. 46 (1972) 1477.
18. G.D. Clark-Walker and F.H. Gleason, Arch. Mikrobiol. 92 (1973) 209.
19. H.J. Bohnert, Biochim. Biophys. Acta 324 (1973) 199.
20. R.A. Flavell and I.G. Jones, Biochem. J. 116 (1970) 811.
21. R.A. Flavell and I.G. Jones, Biochim. Biophys. Acta 232 (1971) 255.
22. D.S. Ray and P.C. Hanawalt, J. Mol. Biol. 11 (1965) 760.
23. L. Schori, Y. Ben-Shaul and M. Edelman, Israel J. Chem. 8 (1970) 117.
24. J.L. Talen, J.P.M. Sanders and R.A. Flavell, Biochim. Biophys. Acta 374 (1974) 129.
25. A.T. Khoury and R.A. Deering, J. Mol. Biol. 79 (1973) 267.
26. R.A. Firtel and J. Bonner, J. Mol. Biol. 66 (1972) 339.
27. G.S.P. Groot, R.A. Flavell and J.P.M. Sanders, Biochim. Biophys. Acta 378 (1975) 186.
28. E.J. Crouse, J.P. Vandrey and E. Stutz, FEBS Letters 42 (1974) 262.
29. P. Borst, in N.K. Boardman, A.W. Linnane and R.M. Smillie (Eds.), Autonomy and Biogenesis of Mitochondria and Chloroplasts, North-Holland, Amsterdam (1971) p. 260.

CHAPTER II.3

AN ANALYSIS OF THE MITOCHONDRIAL GENOME OF YEAST WITH RESTRICTION ENZYMES

G. BERNARDI, A. PRUNELL and H. KOPECKA

Laboratoire de Génétique Moléculaire, Institut de Biologie Moléculaire du C.R.N.S. et de l'Université Paris VII, 2 Place Jussieu, 75005 Paris

Several years ago we did some physical and chemical investigations on the mitochondrial DNA from both wild-type yeast (*Saccharomyces cerevisiae*) cells and from cytoplasmic "petite" mutants (1—3). On the basis of the results obtained, we thought that the mitochondrial DNA from wild-type yeast cells was characterized by an interspersion of A+T-rich and G+C-rich stretches, and that the "petite" mutation was essentially the result of deletions originating from internal crossing-overs in the A+T-rich stretches, which were visualized as internally repetitive. Subsequent work on the fragments obtained by enzymatic degradation of wild-type mitochondrial DNA showed that this was very heterogeneous in base composition (4—6) and formed, in roughly equal amounts, by "spacers" having a G+C level lower than 5% and by "genes" having an average G+C level of about 32% (7). According to the model mentioned above, the basis for the high frequency of spontaneous "petite" mutation in yeast is the high level of sequence homology in the AT-rich stretches; in addition, deletions not leading to the petite phenotype could be expected to occur. Both predictions could be put to a direct experimental test by studying the fragments released from wild-type and "petite" mitochondrial DNAs by restriction enzymes.

We would like to report here the results obtained by degrading the mitochondrial DNAs from three *S. cerevisiae* and one *S. carlsbergensis* wild-type strains and three petite *S. cerevisiae* mutants. The *S. cerevisiae* wild-type strains used were strains A, B and DM, already used in previous work (1—3); the *S. carlsbergensis* strain was strain C (NCY 74 S) kindly provided by Dr. C. Christiansen. The *S. cerevisiae* petite mutants were strains a_1^* (D 243-2B-R_1 (A_1)), provided by Dr. J.C. Mounolou, b and DM_1 already described (1—3); they were derived from wild-type strains A, B and DM, respectively, and had G+C contents equal to 17%, 12% and 4%, respectively. Although four different restriction enzymes were used in our work, namely Eco R_1, Hind II + III, Hpa II and Hae III, we will restrict this presentation to the results obtained with Hpa II (all abbreviations of restriction enzymes are according to (8)), except where otherwise stated.

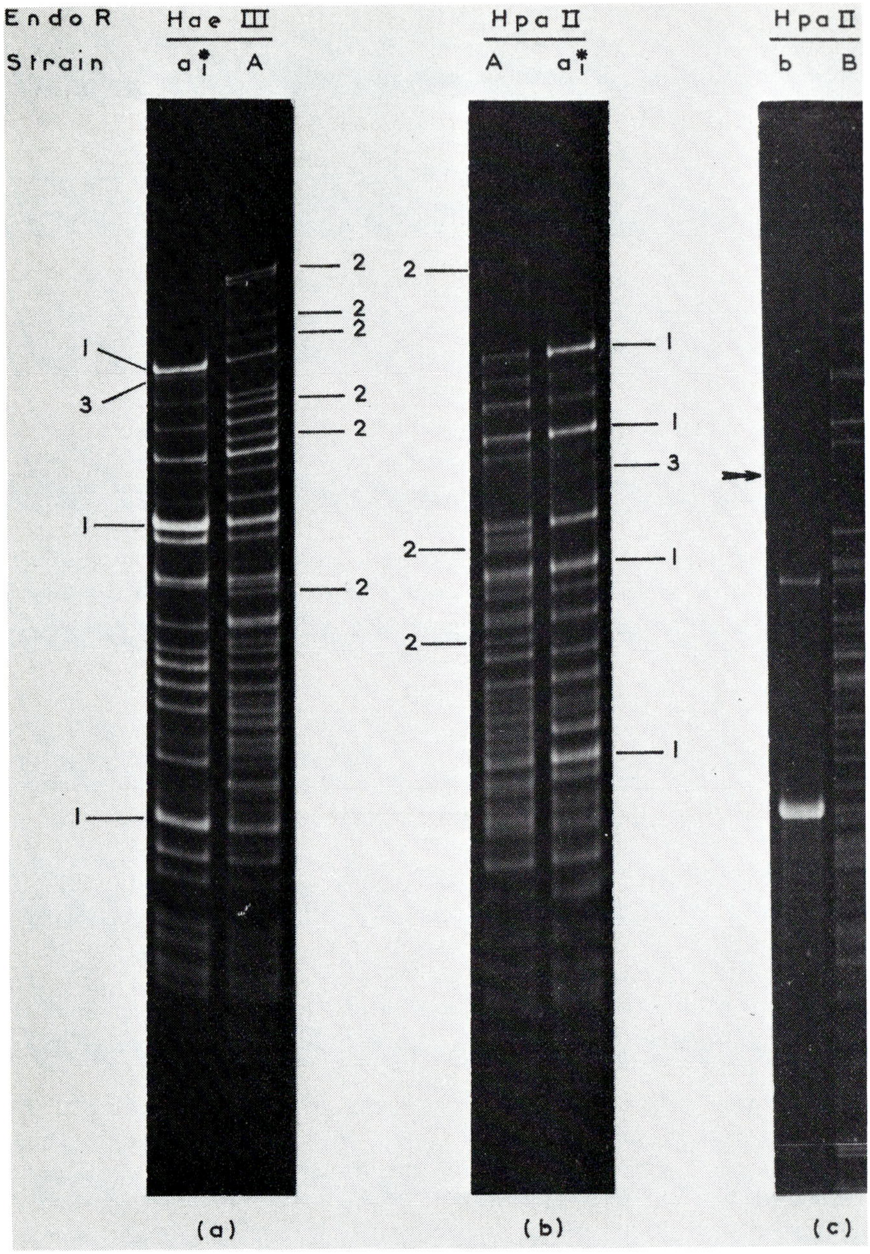

Fig. 1. Band pattern obtained with mitochondrial DNAs as degraded by Hpa II and Hae III. (a) and (b): 1, 2, and 3 indicate bands which are reinforced, missing or novel, respectively, compared to the parent wild-type. (c): the arrow indicates a very faint band. All experiments were performed on 2% polyacrylamide 0.5% agarose gels.

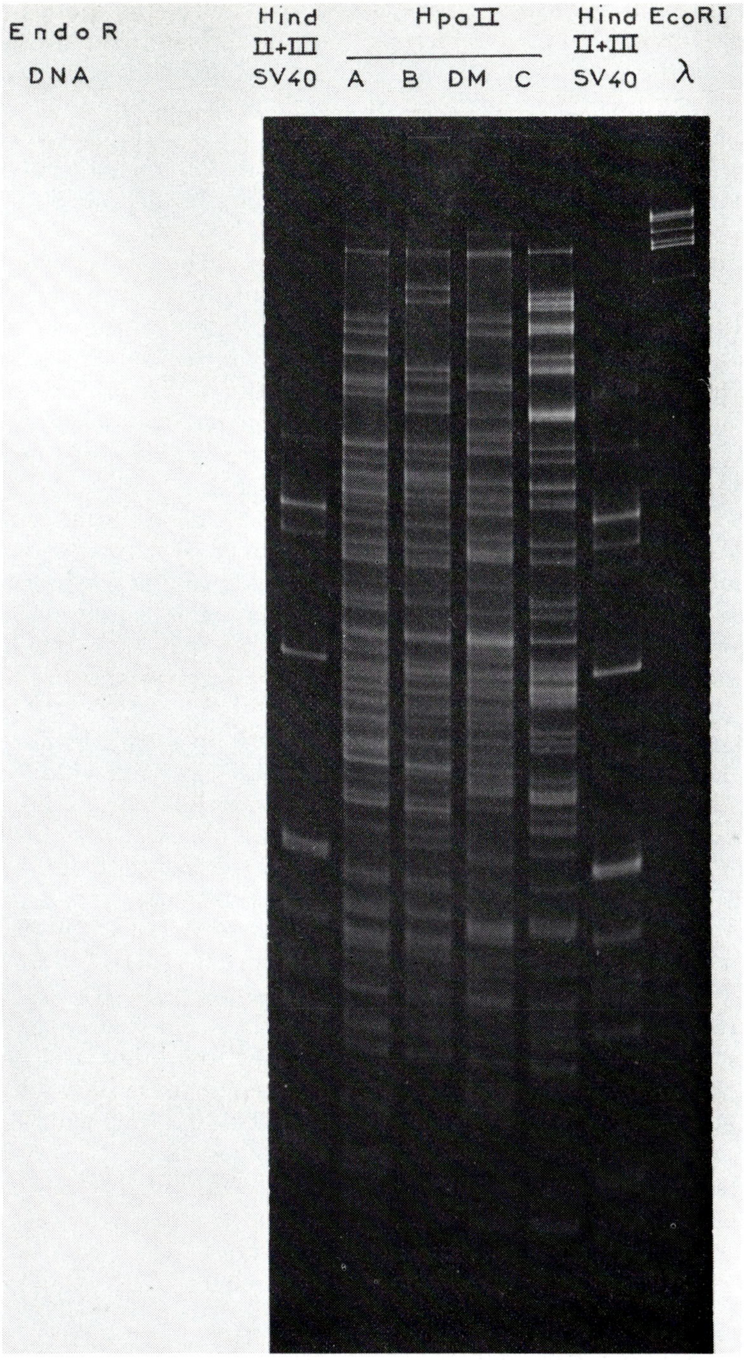

Fig. 2. Band pattern obtained with mitochondrial DNAs as degraded by Hpa II. λ and SV40 DNA digests were used as molecular weight standards. A 2.5% polyacrylamide-0.5% agarose gel was used.

Resolution of the Hpa II fragments was obtained by gel electrophoresis on 0.4 × 16 × 40 cm slabs of 2 to 6% polyacrylamide containing 0.5% agarose. Microdensitometry of the negative pictures of the gels stained with ethidium bromide permitted very precise estimations of the electrophoretic mobility of the fragments. Molecular weights of the fragments were determined by comparing their R_f values with those of DNA markers of known molecular weight, mostly restriction enzyme fragments from lambda phage and SV40 DNAs.

The results obtained with the mitochondrial DNAs from the petite strains can be summarized as follows: a) the fragment pattern of the a_1^* strain DNA was very similar to that of the parent wild-type strain A, but a number of bands were missing or reinforced in the mutant DNA which also showed a few novel bands (fig. 1a, b); b) the fragment pattern of the b strain DNA was characterized by a very small number of bands, compared to the parent wild-type strain B, one of which was amplified a very large number of times (fig. 1c); c) no bands were observed for the DM_1 strain DNA.

The results obtained with the mitochondrial DNAs from the wild-type strains can be outlined as follows: a) a very large number of bands (about 100) could be resolved in all cases; the molecular weights of the fragments ranged from about $4 \cdot 10^6$ to about $4 \cdot 10^4$; b) differences were found in the band patterns shown by the DNAs from the different *S. cerevisiae* strains; such differences were much larger between the *S. carlsbergensis* and the *S. cerevisiae* strains (fig. 2); c) a number of bands showed twice or three times the expected amount of DNA; in several cases this could be shown to arise from a lack of resolution; in other cases duplications could be responsible for this phenomenon; d) the genome size could be evaluated by summing up the molecular weights of all fragments; values ranging from 47 (C; DM) to 49 (B), to 54 million daltons (A) were found for the DNAs originating from the different strains used; these values are in general agreement with electron microscopic estimates of the physical size of yeast mitochondrial DNA (9).

Since the enzyme used specifically splits the sequence CCGG (10) and since the frequency of such sequence is negligible in the spacers (G+C $< 5\%$) and remarkably lower in the ribosomal genes (G+C = 23%; (11)) compared to the other genes, three main sorts of fragments can be expected: a) fragments formed by the G+C-rich genes; these fragments ought to be the smallest ones, since they have the highest frequency of CCGG; b) fragments of intermediate size, formed by enzyme-resistant spacers flanked by G+C-rich genes; c) fragments derived from the ribosomal genes; these might well be the largest fragments, because the number of sites for the enzyme is very low compared to that of the G+C-rich genes. These expectations are borne out by the composition of the fragments as resolved by Sepharose 2B columns (fig. 3).

In conclusion, we have obtained for the first time direct evidence for the deletions and amplifications characterizing the mitochondrial DNAs from

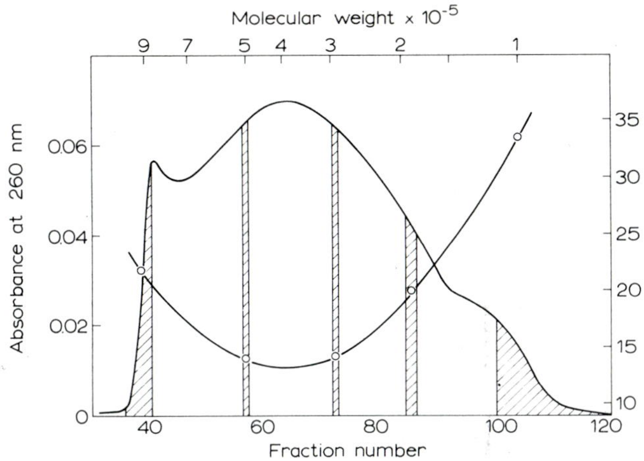

Fig. 3. Sepharose 2B chromatogram of strain B mitochondrial DNA as degraded by Hpa II. Hatched areas indicate the fractions pooled for the base composition.

the petite mutants; novel bands, arising as a result of elimination or insertion of sites recognized by the enzymes, were also found; the results with DM_1 DNA can be explained by the enzyme specificity.

Remarkable differences in nucleotide sequences of the mitochondrial DNAs from different *S. cerevisiae* strains and from the *S. carlsbergensis* strain are reported here for the first time. Fragment patterns are too different from each other to have been originated to more than a minor extent by point mutations at the cleavage sites. We suggest that the differences in nucleotide sequences of DNAs from different wild-type strains have the same molecular basis (deletions, amplifications) as the petite mutation. The significant differences in genome size of the different mitochondrial DNAs support this conclusion.

References

1. G. Bernardi, F. Carnevali, A. Nicolaieff, G. Piperno and G. Tecce, J. Mol. Biol. 37 (1968) 493.
2. G. Bernardi, M. Faurès, G. Piperno and P. Slonimski, J. Mol. Biol. 48 (1970) 23.
3. G. Bernardi and S.N. Timasheff, J. Mol. Biol. 48 (1970) 43.
4. G. Bernardi, G. Piperno and G. Fonty, J. Mol. Biol. 65 (1972) 173.
5. G. Piperno, G. Fonty and G. Bernardi, J. Mol. Biol. 65 (1972) 191.
6. S.D. Ehrlich, J.P. Thiéry and G. Bernardi, J. Mol. Biol. 65 (1972) 207.
7. A. Prunell and G. Bernardi, J. Mol. Biol. 86 (1974) 825.
8. H. Smith and D. Nathans, J. Mol. Biol. 81 (1973) 419.
9. C.P. Hollenberg, P. Borst and E.F.S. Van Bruggen, Biochim. Biophys. Acta 209 (1970) 1.

10. K. Murray and A. Morrison, Personal communication (1974)
D.E. Garfin and H.M. Goodman, Personal communication (1974)
11. L. Reijnders, C.M. Kleisen, L.A. Grivell and P. Borst, Biochim. Biophys. Acta 272 (1972) 396.

Dr. SCHATZ:
What is the effect of glucose repression on the cleavage patterns?

Dr. BERNARDI:
This point has not been tested yet.

Dr. SCHATZ:
Was the digestion by restriction enzymes complete?

Dr. BERNARDI:
Yes. This was tested by studying cleavage patterns at different enzyme concentrations.

Dr. SCHATZ:
Can recombinant molecules be shown by this method?

Dr. BERNARDI:
I think so. We plan to check this point.

CHAPTER II.4

THE NUCLEAR CONTROL OF MITOCHONDRIAL DNA REPLICATION IN *TETRAHYMENA PYRIFORMIS*

P. BORST

Section for Medical Enzymology, Laboratory of Biochemistry, University of Amsterdam, Eerste Constantijn Huygensstraat 20, Amsterdam, The Netherlands

In 1968 Suyama and Miura (1) found that mtDNA from *Tetrahymena pyriformis*, strain ST, consists of a homogeneous population of linear molecules with a molecular weight of about 30×10^6. We have confirmed this (2, 3) and added three additional findings: a) The DNA lacks single-stranded breaks and is stable in alkali (3); b) the gene sequence is unique (4); c) the DNA contains a terminal duplication-inversion (4) like adenovirus DNA. This has increased the likelihood that it is actually linear in vivo, since all linear viral DNAs in nature are able to circularize in one way or another.

Replication of this DNA starts close to the middle of the molecule and proceeds bidirectionally to the ends by way of intermediates with a largely duplex character (5—8).

Previous work by Westergaard and co-workers (9—11) has shown that a number of treatments that affect DNA lead to a 30 to 50-fold induction of a mtDNA polymerase. These treatments include ultraviolet light and ethidium bromide; induction is blocked by cycloheximide, but not by chloramphenicol, suggesting that this DNA polymerase is specified by nuclear genes and made on cell-sap ribosomes. Westergaard interpreted these results to mean that ultraviolet light etc. lead to damage of mtDNA and that the DNA polymerase is sent in by the nucleus to aid in the repair of this damage.

TABLE I

Effects of ethidium bromide on mtDNA synthesis in *Tetrahymena*. See (6) for details.

Time of growth in ethidium (hr)	Inhibition of mtDNA synthesis (% of control)	Replicative intermediates (% of total mtDNA)
0	0	< 3
2—5	> 95	< 10
20—30	< 50	> 50

TABLE 2

Control of mtDNA synthesis.

1) All enzymes required are specified by nuclear genes and made on cell-sap ribosomes
2) A new round of synthesis requires an initiator, acting stoicheiometrically. Initiation inactivates the initiator
3) The basic level of initiator synthesis is under nuclear control. Initiator synthesis may be linked to the cell cycle (Trypanosomes) or not (*Tetrahymena*)
4) The basic level of initiator synthesis is modulated by a mitochondrial repressor

A more detailed analysis of the polymerase induction by ethidium in my laboratory (6) demonstrated, however, that ethidium does not lead to damage of mtDNA. As shown in table 1 it inhibits mtDNA synthesis initially over 95%; if growth in ethidium continues, an increasing fraction of the mtDNA is converted into (essentially normal) replicative intermediates and, hence, the inhibition of mtDNA synthesis is largely overcome. It appears then that the DNA polymerase is not induced to aid in the repair of mtDNA but to overcome a block in mtDNA synthesis.

But how does the nucleus know that mtDNA synthesis is blocked? I think that all our results can be explained with the model put forward by Barath and Küntzel (12), which is a modification of an earlier idea of Williamson (13). In this model mtDNA codes for a repressor protein made on mitochondrial ribosomes, which controls the expression of nuclear genes specifying components of the mitochondrial biosynthetic machinery. On the basis of this scheme we postulate that the treatments used by Westergaard to induce the polymerase all lead to an inhibition of mitochondrial RNA and/or protein synthesis and hence to a block of the synthesis of the mitochondrial repressor that affects the control of polymerase synthesis. This remains to be verified.

These ideas lead to the scheme for the control of mtDNA synthesis presented in table 2. mtDNA is a replicon controlled by the nucleus. Regulation of the number of DNA copies per cell occurs by the production of a defined number of initiator molecules that act stoicheiometrically. The production of initiator molecules may be temporarily altered by nuclear factors; for instance, a nutritional shift-up will lead to preferential replication of mtDNA in *Tetrahymena* (14). The production of initiator molecules is modulated by a mitochondrial repressor. Absence of this repressor leads to over-production of initiator and other proteins specified by nuclear genes. In this last aspect yeast may be an exception, because the meager data available do not provide evidence for over-production of nuclear components of the mitochondrial biosynthetic machinery in petite mutants or cells grown in erythromycin (15).

References

1. Y. Suyama and K. Miura, Proc. Nat. Acad. Sci. U.S. 60 (1968) 235.
2. A.C. Arnberg, E.F.J. Van Bruggen, R.B.H. Schutgens, R.A. Flavell and P. Borst, Biochim. Biophys. Acta 272 (1972) 487.
3. R.B.H. Schutgens, Ph.D. Thesis, De Biogenese van Mitochondriën in *Tetrahymena pyriformis*, Quadriga Press, Amsterdam (1973).
4. A.C. Arnberg, E.F.J. Van Bruggen, P. Borst, R.A. Clegg, R.B.H. Schutgens and P.J. Weijers and R.W. Goldbach, Biochim. Biophys. Acta 383 (1975) 359.
5. R.A. Clegg, P. Borst, P.J. Weijers, A.C. Arnberg and E.F.J. van Bruggen, Biochem. Soc. Trans. 2(1974) 227.
6. W.B. Upholt and P. Borst, J. Cell Biol. 61 (1974) 383.
7. R.A. Clegg, P. Borst and P.J. Weijers, Biochim. Biophys. Acta 361 (1974) 277.
8. A.C. Arnberg. E.F.J. Van Bruggen, R.A. Clegg, W.B. Upholt and P. Borst, Biochim. Biophys. Acta 361 (1974) 266.
9. J. Keiding and O. Westergaard, Exp. Cell Res. 64 (1971) 317.
10. O. Westergaard, K.A. Marcker and J. Keiding, Nature 227 (1970) 708.
11. O. Westergaard and B. Lindberg, Eur. J. Biochem. 28 (1972) 422.
12. Z. Barath and H. Küntzel, Proc. Nat. Acad. Sci. U.S. 69 (1972) 1371.
13. D.H. Williamson, in P.L. Miller (Ed.), Control of Organelle Development, Cambridge University Press, Cambridge (1970) p. 247.
14. J. Engberg, J.R. Nilsson, R.E. Perlman and V. Leick, Proc. Nat. Acad. Sci. U.S. 71 (1974) 894.
15. G.S.P. Groot, in A.M. Kroon and C. Saccone (Eds.), The Biogenesis of Mitochondria, Academic Press, New York (1974) p. 443.

Dr. X:

Does the original replication complex which is inhibited by EtBr treatment continue to replicate on finishing the EtBr?

Dr. BORST:

For technical reasons this experiment has not been done. However, the replicative intermediates accumulating in ethidium-treated cells are not detectably different from those in control cells and there is no detectable degradation of mtDNA following ethidium addition. Moreover, the inhibition of cell growth is fully reversible and since the bulk of mtDNA in ethidium-arrested cells is present in replicative intermediates, it seems likely that these intermediates will go on to yield normal, happy mtDNA molecules, once ethidium is removed.

CHAPTER II.5

EXTENSIVE RECOMBINATION OF YEAST MITOCHONDRIAL DNA DURING VEGETATIVE GROWTH AS REVEALED BY DENSITY LABELLING

D.H. WILLIAMSON

National Institute for Medical Research, Mill Hill, London, U.K.

Log phase cells of the yeast *Saccharomyces cerevisiae*, fully density labelled by growth in minimal medium containing ^{15}N as sole nitrogen source were transferred to medium lacking the density label. Changes in the buoyant density of the nuclear and mitochondrial DNA in samples taken at intervals were monitored by isopycnic banding in CsCl. The nuclear DNA replicated semi-conservatively, but the mitochondrial DNA showed a dispersive pattern, appearing throughout as a single unimodal gradient band, the density of which decreased continuously throughout the period of growth. Analysis of this situation showed that vegetative replication of mitochondrial DNA is accompanied by extensive exchanges of homologous strands between duplexes.

Between 10 and 20% of the DNA of the yeast *Saccharomyces cerevisiae* is mitochondrial in origin and carries genes specifying certain mitochondrial constituents (1,2). These genes are inherited in a non-mendelian fashion, can be eliminated in certain respiratory deficient (cytoplasmic petite) mutants, and comprise what may be termed the mitochondrial genome. The question arises as to the number of mitochondrial genomes in the cell. We know with reasonable certainty that the DNA concerned is distributed amongst the organisms' mitochondria and consists (in a haploid) of about 30 identical molecules each having a molecular weight of $40—50 \times 10^6$ daltons (3, 4). Since mitochondria are known to be equipped to transcribe and translate the information in mitochondrial DNA (mitDNA), it might be supposed that all 30 mitDNA molecules would be capable of expression as independent genetic entities. In fact however, this naive supposition is not fulfilled. Analyses of the kinetics of induction of the cytoplasmic petite mutant by heat (5), acriflavine (6), UV-light (7, 8) or ethidium bromide (9) all suggest the presence of rather less than 10 mitochondrial genomes, and this view is supported, if rather weakly, by analyses of the inheritance patterns of mitochondrial markers in crosses (10). None of these estimates is free from criticism, if only because of the various assumptions that have to be made in interpreting the data. However, taken together, they reveal a disparity between the cell content of mitDNA molecules and of genetic determinants that is too striking to be ignored.

Various explanations of this situation can be devised. One possibility is that only one (or a few) of the cells' complement of mitDNA molecules is capable of replication, the rest merely being "sterile" copies capable only of transcription.

We explored this possibility using density label transfer experiments of the type first used by Meselson and Stahl (11) to show that replication of bacterial DNA is semi-conservative. In this type of experiment, cells growing in a density labelling environment are transferred to a medium lacking the density label, and changes in the buoyant density of the DNA in samples taken at intervals are monitored by isopycnic banding in CsCl.

The prediction for semi-conservative replication is summarised in fig. 1a. Following one round of replication in the "light" medium, all the duplexes are of hybrid density, while by two doublings, equal numbers of hybrid and fully light duplexes are present. This situation is clearly distinguishable from that predictable in the "master replicating copy" situation discussed above. In this case, as illustrated in fig. 1b, fully light molecules would be observed within one generation, and a large proportion of the original heavy duplexes present at the start should still be present at one doubling, in unaltered form. At two doublings, fully light and fully heavy molecules should be present in proportions of approximately 3 : 1, and hybrid molecules would probably not be detected at any stage.

The details of our experiments with yeast have been reported elsewhere (12). A diploid strain of *Saccharomyces cerevisiae* was fully density labelled by growth for two subcultures in minimal medium in which the sole source of nitrogen was $(^{15}NH_4)_2SO_4$. These fully heavy (HH) cells contained about as much mitochondrial DNA as their counterparts in "light" (LL) medium,

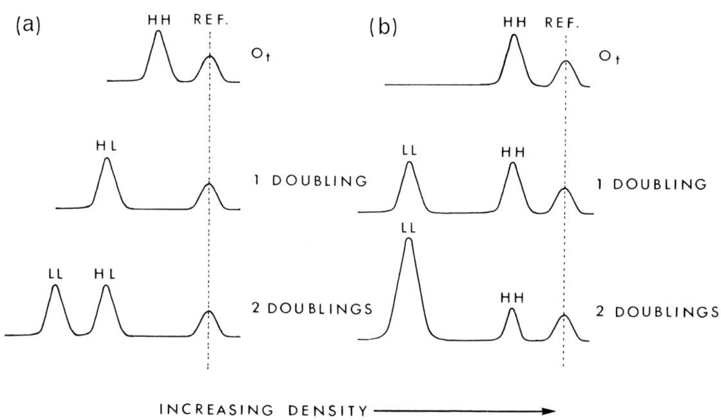

Fig. 1. Schematic representation of the densitometer tracings from analytical isopycnic gradients following density transfer experiments; a) for semi-conservative replication, b) for the "master replicating copy" situation described in the text.

but both the nuclear and mitochondrial DNAs were about 13 mg/cc denser than in the LL cells.

Logarithmically growing HH cells were rapidly harvested by filtration and transferred to prewarmed minimal medium containing only $(^{14}NH_4)_2SO_4$ as nitrogen source. Cell division and the synthesis of both nuclear and mitochondrial DNA continued exponentially throughout the period of the experiment, all parameters having a doubling time of about 3 hr.

Analytical isopycnic gradients of the DNA of samples at 0 hr (the time of the density transfer), 3 hr (one doubling), and 6 hr (2 doublings) are shown in fig. 2. The nuclear DNA behaved consistently with semiconservative repli-

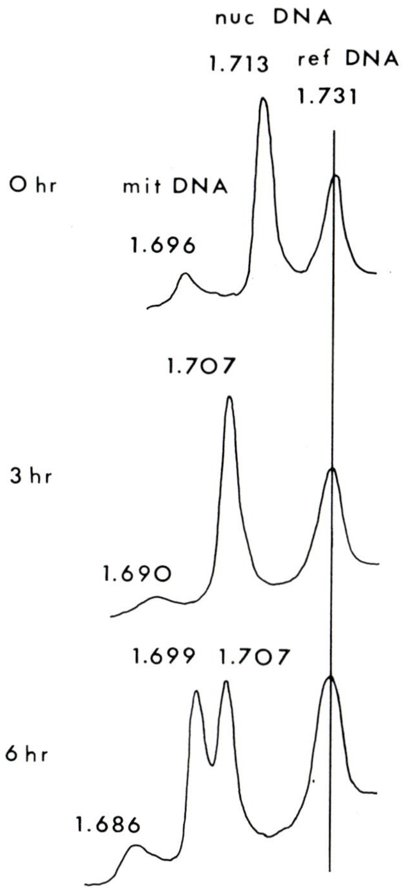

Fig. 2. Analytical CsCl gradient patterns of nuclear DNA (the major component) and mitochondrial DNA at the time of the density transfer (0_t), one doubling and two doublings. The reference DNA was from *M. lysodeikticus*. The figures are the buoyant densities in g/cc.

cation; at one doubling time all the duplex molecules were of hybrid density, while by two doublings, equal amounts of hybrid and fully light duplexes were observed.

In contrast, mitDNA replicated in an apparently dispersive manner. The unimodal band of fully labelled molecules present at the start underwent a continuous decrease in density throughout the experiment, and at no stage was more than one band observed. At one doubling the modal density of the mitDNA duplexes corresponded to that of hybrid molecules, but at two doublings the mitDNA density was intermediate between hybrid and fully light. Denaturation and isopycnic banding of the mitDNA showed that by two doublings all the single strands were of a uniform density compatible with that of their parent duplexes. Thus by this time the ^{15}N atoms in these duplexes had been distributed equally between old and new strands. The operation of this redistribution process was also apparent at one doubling, for denaturation showed that even by this time, pure heavy and light single strand stretches were in the minority. This observation was confirmed by experiments involving the use of double radioactive labels in combination with the density transfer.

The question arises as to the molecular mechanism of the redistribution

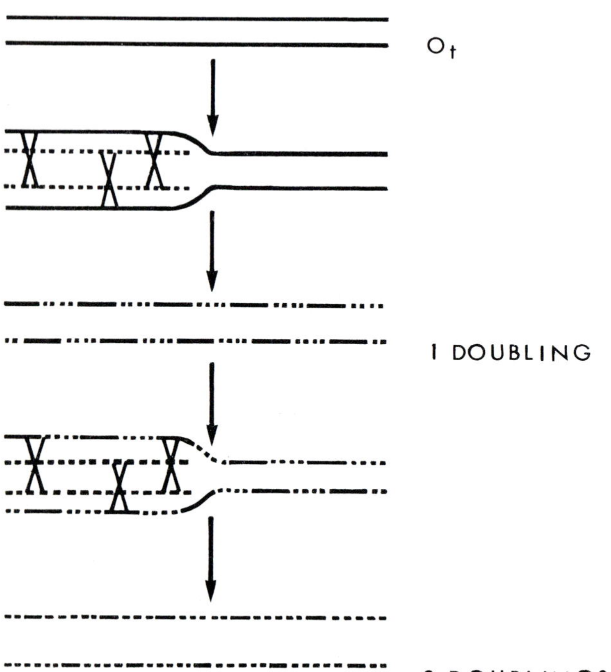

Fig. 3. Model of recombination process of mitDNA molecules: ———, heavy "old" strands; . . . . . ., light "new" strands. Only a few exchanges are shown.

process. Two basic possibilities come to mind. One would involve continuous degradation of the duplexes and re-incorporation of small fragments randomly into old or new strands, i.e. a form of metabolic turnover. The alternative would be based on the exchange of homologous single strand stretches between duplexes. Analysis of the time course of density loss of the mitochondrial component following the density transfer showed that essentially all the previously incorporated ^{15}N was conserved in the population of duplexes. This argues strongly against the first of the possibilities referred to above, and the most likely explanation of our findings is that the replication of mitDNA is accompanied by extensive exchanges of homologous strands between duplexes.

The model we favour is illustrated in fig. 3, which depicts the most likely possibility, namely that the exchanges occur between true sister duplexes during, or soon after replication. A corollary of this model is that probably all the mitDNA molecules of the cell normally undergo replication, and the hypothesis of a "master replicating copy" referred to above is therefore unlikely to be valid. On the other hand our observations suggest that the explanation for the apparently abberant genetic behaviour of the mitochondrial system may be sought along entirely different lines. It is conceivable that the observations we interpret in the simple model shown in fig. 3 are really only the end result of a more complex process involving interactions between all the duplexes in the cell, designed to stabilise and unify the mitochondrial genome. In this case it may be that we cannot determine the number of mitochondrial genomes in conventional genetic ways. Rather, the mitochondrial system may comprise a flexible pool of genes whose behaviour in genetic tests at any given time may vary according to the spatial distribution of the mitDNA molecules within the cell, the opportunities these have for interaction with each other, as well as the nature of the process which governs their inheritance.

References

1. D.H. Williamson, Symp. Soc. Exp. Biol., Control of Organelle Development: University Press, Cambridge (1970) p. 247.
2. P. Borst, Ann. Rev. Biochem. 41 (1972) 333.
3. C.P. Hollenberg and E.F.J. van Bruggen. Biochim. Biophys. Acta 209 (1970) 1.
4. T.D. Petes, B. Byers and W.L. Fangman, Proc. Nat. Acad. Sci. U.S. 70 (1973) 3072.
5. F. Sherman, J. Cell Comp. Physiol. 54 (1959) 37.
6. T. Sugimura, K. Okabe and A. Imamura, Nature 212 (1966) 304.
7. D. Wilkie, J. Mol. Biol. 7 (1963) 527.
8. N.G. Maroudas and D. Wilkie, Biochim. Biophys. Acta 166 (1968) 681.
9. P.P. Slonimski, G. Perrodin and J.H. Croft, Biochem. Biophys. Res. Commun. 30 (1968) 232.

10. D. Wilkie and D.Y. Thomas, Genetics 73 (1973) 367.
11. M. Meselson and F.W. Stahl, Proc. Nat. Acad. Sci. U.S. 44 (1958) 671.
12. D.H. Williamson and D.J. Fennell, Mol. Gen. Genetics 131 (1974) 193.

Dr. SCHATZ:

Could you comment on Luck's earlier experiments?

Dr. WILLIAMSON:

In Luck's earlier experiments with *Neurospora*, there were considerable problems with pool; also the exact doubling time of mitochondrial DNA was not determined. The combination of these factors makes it very difficult to interpret his results, and although Luck concluded that replication was semi-conservative, I think any conclusion should only be very tentative. At the same time, it does not follow that all fungi will be found to behave in the same way in this regard.

Dr. MORPURGO:

If I remember well in some bacteriophages there is also an apparent dispersive replication and it is also explained by crossing-over. Secondly I would remember that a few years ago we have crossed 2 petite colonies with different density obtaining a new strain with intermediate density and this is in favor of your hypothesis?

Dr. WILLIAMSON:

I think the type of recombination to give the results you mentioned on crossing 2 haploid petites must be of a quite different nature from the sister strand exchange phenomenon we observed in vegetative growth, since we interpret our observations in terms of exchanges of homologous strands between identical duplexes. Moreover, we really do not know enough about the behaviour of mitochondrial DNA in zygotes, particularly in crosses involving petites.

CHAPTER II.6

SENSITIVITY OF CHLOROPLAST AND NUCLEAR DNA OF *CHLAMYDOMONAS* TO MYXIN

R.M. BEHKI* and P.F. LURQUIN

Radiobiology Department, Centre d'Etude de l'Energie Nucléaire, CEN/SCK, 2400 Mol, Belgium

Introduction

In *Chlamydomonas* the replication of chloroplast DNA is independent of, and temporarily separated from the replication of chromosomal DNA (1, 2). The two replication cycles, however, are tightly coupled (3). Since the chloroplast possesses DNA molecules distinct from the nuclear DNA (4, 5) it has been feasible to study the effect of certain antibiotics on nuclear and chloroplast functional systems in *Chlamydomonas*. Streptomycin, for example, allows about one round of chloroplast DNA replication in the absence of nuclear DNA synthesis (6). Rifampin inhibits only chloroplast — located DNA — dependent RNA synthesis (7) in *C. reinhardi*. Similarly ethidium bromide inhibits the replication of chloroplast DNA only and causes the preferential but reversible degradation of pre-existing chloroplast DNA in these cells (6).

We have investigated the effect of the wide-spectrum antibiotic (8) myxin (1-hydroxy-6-methoxyphenazine-5,10 dioxide) on the DNA metabolism in *C. reinhardi* cells. The primary effect of this antibiotic on *Escherichia coli* cells has been shown to be the inhibition of DNA biosynthesis, followed by an extensive degradation of intracellular DNA (9, 10).

Materials and methods

Wild-type *Chlamydomonas reinhardi* (mt +) strain 137 C (from Prof. R.P. Levine) was grown in high salt medium (HSM) in suspension culture (11) under phototrophic or mixotrophic conditions in the light or under heterotrophic conditions in the dark. The cells were used at a density of about 2×10^6/ml. In vivo labelling of the nucleic acids was carried out by the addition of 8-^{3}H-adenine at a concentration of 5—10 μCi/ml. The incorpora-

*Visiting scientist from the Chemistry and Biology Research Institute, Research Branch, Agriculture Canada, Ottawa, Ontario, Canada.

tion of the label into nucleic acids was determined by precipitation of the cell aliquots with cold 10% TCA and filtration on 0.45 μ Millipore filters. The incorporation into DNA was similarly measured after hydrolysis of RNA in the samples (0.5 M KOH for 24 hr at 39° C). Degradation of intracellular DNA was followed in cells labeled with ^{3}H-adenine for at least 3 generations. The decrease in the amount of radioactivity in alkali-resistant, acid-precipitable fraction of the cells was taken as a measure of DNA degradation.

The extraction and purification of *Chlamydomonas* DNA was as follows: lysis of the cells with 1.5% sodium sarkosylate, a one hour pronase treatment at 37° C and subsequent molecular sieving of the lysate on Sepharose 4B (Pharmacia) in the presence of 2 M NaCl (12). The excluded fractions which contained DNA free of RNA and protein, were pooled and dialysed against 0.15 M NaCl. Cesium chloride (Merck), 1.25 g per ml, was then added for neutral CsCl density gradient centrifugation in a Spinco 40-rotor at 33,000 rpm for 63 hr at 20° C. The tubes were then punctured and about 100 2-drop fractions were collected from the bottom. Radioactivity determinations were made in a Packard—Tricarb Scintillation Spectrometer in the presence of Packard Instagel.

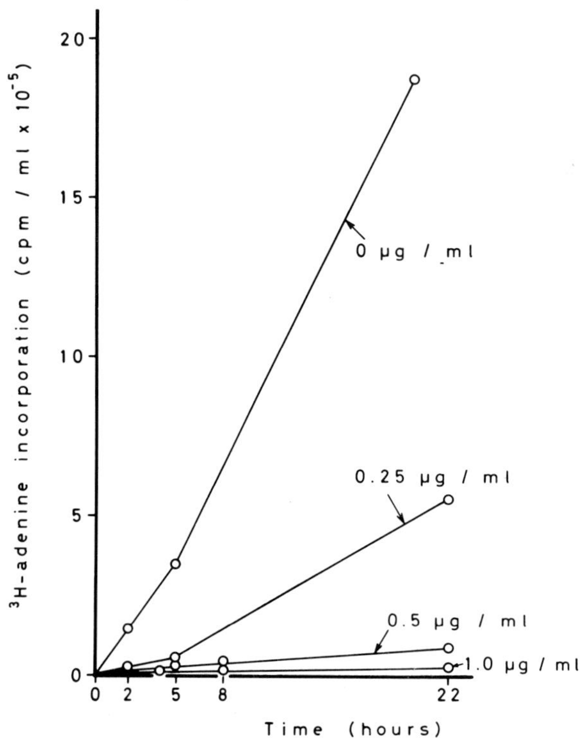

Fig. 1. Effect of different myxin concentrations on ^{3}H-adenine incorporation into TCA-precipitable fraction of *C. reinhardi* cells with time.

Results

Effect on ^{3}H-adenine incorporation

The rates of ^{3}H-adenine incorporation into the cold TCA-precipitable fraction of *C. reinhardi* cells at different myxin concentrations is shown in fig. 1. Following a severe initial inhibition of adenine incorporation at all three levels of myxin examined, the inhibitory effect at 0.25 μg/ml myxin was about 70% compared to the untreated control by 22 hr. The inhibition of the label incorporation into DNA, however, exceeded 85% at this concentration of myxin. Bleaching of the cultures was not observed up to 0.5 μg/ml myxin under these conditions.

Analysis of the DNA isolated from the control and from 0.25 μg/ml myxin-treated cultures by CsCl density gradient centrifugation revealed that in the control culture, without myxin treatment radioactivity in chloroplast DNA (d = 1.695 g/cm^3) separable from nuclear DNA (d = 1.724 g/cm^3) represented about 11% of the total DNA counts. In myxin treated cells chloroplast DNA accounted for about 30% of the total DNA radioactivity. The amount of the label incorporated into both DNA species was, however, severely inhibited.

Effect on DNA degradation

Table 1 gives the results on the degradation of prelabelled intracellular DNA in *C. reinhardi* cells in the presence of different myxin concentrations. The amount of DNA degraded increases with increasing myxin concentration. More than 50% of the labeled DNA was degraded by 18 hr at 1.0 μg/ml myxin. The decrease in the alkali-resistant, TCA-insoluble radioactivity with time in a myxin treated culture was fully accounted for by an identical increase in the amount of acid-soluble radioactivity in the cell-free superna-

TABLE 1

Effect of different myxin concentrations on DNA degradation in *C. reinhardi* cells. ^{3}H-adenine labeled cells were centrifuged, washed with HSM (room temperature) and resuspended in HSM containing appropriate amounts of myxin, all under sterile conditions. The alkali-resistent, cold 10% TCA precipitable radioactivity (^{3}H-DNA) was determined in the samples at 0 and 18 hr.

Myxin μg/ml	^{3}H-DNA at		% degradation
	0 hr	18 hr	
0	86,700	89,600	—
0.25	88,900	84,300	5.2
0.5	86,900	62,600	28.0
1.0	85,400	39,800	53

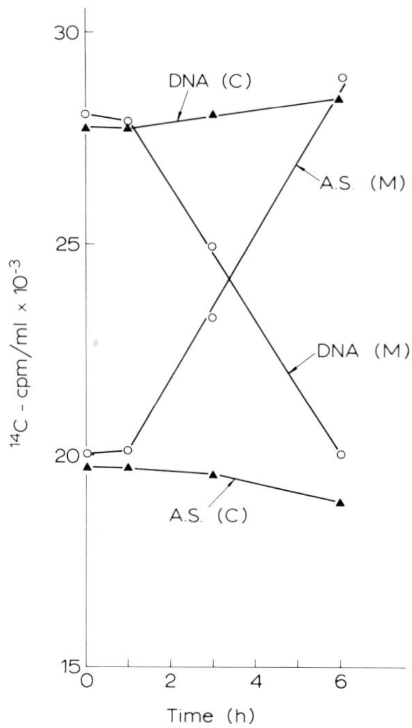

Fig. 2. Degradation of ^{14}C-labeled *C. reinhardi* DNA with time at 1 μg/ml myxin. The conditions were the same as in table 1 except that ^{14}C-adenine was used for in vivo labeling of nucleic acids. The acid-soluble (A.S.) radioactivity was determined in the supernatant fraction following centrifugation of the cells. C and M refer to control and myxin-treated cultures respectively.

tant fraction (fig.2) where ^{14}C-adenine labeled *C. reinhardi* cells were treated with 1 μg/ml myxin.

The neutral CsCl density gradient centrifugation profiles of ^{3}H-DNA isolated from the control culture and from cultures treated with different myxin concentrations for 22 hr is shown in fig. 3. At a myxin level of 0.25 μg/ml, there is almost a complete disappearance of the chloroplast DNA peak while there is little effect on the amount of radioactivity in the nuclear DNA fraction. At higher myxin concentrations the amount of radioactivity in the nuclear DNA fractions is also progressively lowered. There is no trace of radioactivity in the chloroplast DNA fractions, however. It was subsequently determined that the preferential degradation of *C. reinhardi* chloroplast DNA (9 hr treatment with 0.5 μg/ml myxin) is also accompanied by the presence of single-strand breaks in a certain proportion of the nuclear DNA molecules (13). The solubilization of the nuclear DNA apparently occurs only after longer exposure (fig. 3c) or at high myxin levels (fig. 3d).

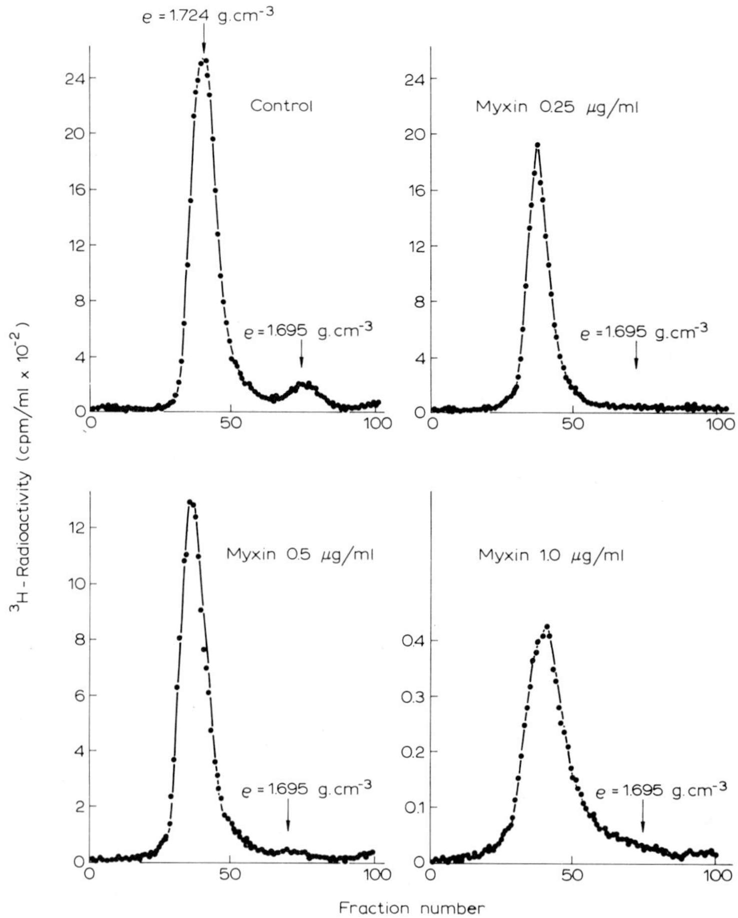

Fig. 3. Neutral CsCl density gradient centrifugation profiles of DNA extracted from myxin-treated and untreated *C. reinhardi* cells. DNA was isolated after 22 hr treatment with 0.25, 0.5 and 1.0 μg/ml myxin.

Effect on DNA degradation following a short exposure to myxin

When *C. reinhardi* cells are exposed to myxin (0.5 μg/ml or more) under phototrophic or mixotrophic conditions for 15 min and the antibiotic is removed by centrifugation and resuspension of the cells in fresh medium, they continue to lose intracellular DNA during subsequent incubation in light. As shown in fig. 4 the amount of DNA degradation in cells exposed to myxin was quantitatively very similar to that found in cells where myxin was added but was not removed. When myxin is added to a *Chlamydomonas* culture growing under heterotrophic conditions the degradation or intracellular DNA is greatly reduced (13). This also happens when the cells ex-

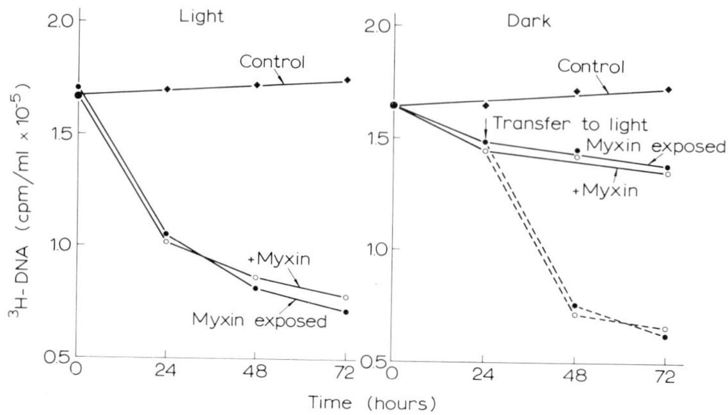

Fig. 4. Left: Degradation of ^{3}H-adenine labeled *C. reinhardi* DNA under mixotrophic growth conditions. Myxin (0.5 μg/ml) was either added to the cells or the cells were exposed to this level of myxin for 15 min and the antibiotic was then removed. Right: The same as above except, that the growth was under heterotrophic conditions. Two duplicate cultures (dotted lines) were transferred to mixotrophic growth after 24 hr.

posed to myxin for 15 min as above are subsequently incubated under heterotrophic conditions. In both cases the amount of DNA degradation is small but almost identical (fig. 4). The limited DNA degradation under heterotrophic conditions may represent either the repair of the cellular damage to the DNA or the arrest of some other cellular activity connected with the solubilization of DNA. This was tested by exposing the cells prelabeled with ^{3}H-adenine in HSM-acetate medium to 0.5 μg/ml myxin for 15 min in light. The cells were then centrifuged and resuspended (sterile conditions) in an equal volume of fresh HSM acetate medium. Two aliquots of the culture were incubated in the dark for the following 24 hr. One culture was then transferred to light (mixotrophic). The degradation of DNA was followed in both with time. As shown in fig. 4, the transfer of the treated culture to light triggers the DNA degradation process which was arrested under heterotrophic conditions. In addition the extent of DNA degradation approaches the values found in cells grown under mixotrophic conditions throughout.

Discussion and summary

Myxin, similar to the reported effects of this antibiotic in *E. coli* (9, 10) inhibits DNA biosynthesis, causes extensive degradation of intracellular DNA and exerts a lethal effect in *C. reinhardi* cells (13). However, myxin degrades chloroplast DNA preferentially in these cells. Nuclear DNA is degraded either after prolonged exposure of the cells to myxin or when myxin is used at higher concentrations.

Both the rate and the extent of DNA degradation are significantly enhanced when protein synthesis is simultaneously inhibited during the initial 10—12 min exposure of *E. coli* cells to myxin (10). In *Chlamydomonas* on the other hand, inhibition of organelle and cytoplasmic protein synthesis has no additional effect on myxin-induced DNA degradation (13). In addition, DNA degradation continues in *C. reinhardi* cells exposed to myxin for only a short time. In contrast, removal of myxin arrests further DNA degradation in *E. coli*. To obtain continued degradation in *E. coli* it is essential not only to inhibit protein synthesis during exposure of the cells to myxin but also to maintain inhibition of protein synthesis subsequently (10). Based on these results and on the examination of the cellular damage to DNA under these conditions it was suggested that the lethal and metabolic damage in myxin treated *E. coli* represents the competition between damage to, and the repair of, cellular DNA (10).

The presence or the localization of the DNA-repair enzymes (14, 15, 16) in *Chlamydomonas* is not known. According to Swinton and Hanawalt (17) there is no repair-replication of DNA in *C. reinhardi* cells irradiated with U.V. Our results support this conclusion since *C. reinhardi* cells, in contrast to the *E. coli* system (18) fail to show metabolic recovery following only a short exposure to myxin. In fact the degradation of DNA continue unrestricted in these cells.

In addition to the differential sensitivity of chloroplast and nuclear DNA species to myxin the antibiotic affects the DNA degradative process differently in the light incubated and the dark incubated cultures of *Chlamydomonas*. The degradation of DNA in myxin-treated culture is minimal under heterotrophic conditions. If the heterotrophic condition is followed by returning the treated cultures to mixotrophic state DNA degradation resumes at a higher rate indicating that the damage to the cellular DNA occurs under both these conditions but remains unexpressed in the dark. The reasons for the arrest of myxin-induced DNA degradation in *Chlamydomonas* in the dark and the subsequent manifestation of this breakdown in the light is under investigation.

References

1. K.S. Chiang and N. Sueoka, Proc. Nat. Acad. Sci. U.S. 57 (1967) 1506.
2. K.S. Chiang and N. Sueoka, J. Cell. Physiol. 70, suppl. 1 (1967) 89.
3. R. Sager, Cytoplasmic Genes and Organelles, Academic Press, N.Y. (1972).
4. R. Sager and M.A. Ishida, Proc. Nat. Acad. Sci. U.S. 50 (1963) 725.
5. E.H.L. Chun, M.H. Vaughan and A. Rich, J. Mol. Biol. 7 (1963) 130.
6. V.R. Flechtner and R. Sager, Nature New Biol. 241 (1973) 277.
7. S.J. Surzycki, Proc. Nat. Acad. Sci. U.S. 63 (1969) 1327.
8. E.A. Peterson, D.C. Gillespie and F.D. Cook, Can. J. Microbiol. 12 (1966) 221.
9. S.M. Lesley and R.M. Behki, J. Bacteriol. 94 (1967) 1837.

10. R.M. Behki and S.M. Lesley, J. Bacteriol. 109 (1972) 250.
11. N. Sueoka, K.S. Chiang and J.M. Kates, J. Mol. Biol. 25 (1967) 47.
12. P.F. Lurquin et al., Anal. Biochem. (in press).
13. R.M. Behki and P.F. Lurquin, Plant Sci. Letters 3 (1974) 419.
14. P. de Lucia and J. Cairns, Nature 224 (1969) 1164.
15. L. Kanner and P.C. Hawawalt, Biochem. Biophys. Res. Commun. 39 (1970) 149.
16. M. Gellert, J.W. Little, C.K. Oskinsky and S.B. Zimmerman, Cold Spring Harb. Symp. Quant. Biol. 33 (1968) 21.
17. D.C. Swinton and P.C. Hanawalt, Biochim. Biophys. Acta 294 (1973) 385.
18. S.M. Lesley and R.M. Behki, Can. J. Microbiol. 17 (1971) 1327.

Dr. BORST:

The effects you find for myxin (introduction of SS breaks into DNA) resemble that of streptonigrin, which is now known to lead to O_2^- production, the O_2^- giving SS breaks in DNA. Could myxin act in the same way?

Dr. LURQUIN:

An important difference between myxin and streptonigrin is that the latter is able to interact with DNA in vitro (H.L. White and J.A. White, Biochim. Biophys. Acta 123 (1966) 648), whereas myxin appears to be unable to do so. However, both drugs inhibit RNA and DNA syntheses and cause SS breaks in DNA. Another analogy lies in the fact that myxin is reduced in vivo and that its reduced form is biologically inactive. It would certainly be interesting to study the effects of streptonigrin on *Chlamydomonas*.

SECTION III

INTERGENOMIC COOPERATION AND THE TRANSCRIPTIONAL EQUIPMENT OF THE CELL

CHAPTER III.1

GENES FOR PROTEINS OF CHLOROPLAST RIBOSOMES AND THE EVOLUTION OF EUKARYOTIC GENOMES*

L. BOGORAD, J.N. DAVIDSON, M.R. HANSON and L.J. METS**

Harvard University, The Biological Laboratories, 16 Divinity Ave., Cambridge, Mass. 02138, U.S.A.

Introduction

Attempts to locate structural genes for proteins of organelles lead to examinations of irreducible fundamental interactions among nuclei, plastids and mitochondria. The results of these investigations compel us to re-examine and question currently popular ideas about the origin of eukaryotism and to consider mechanisms by which modern eukaryotic cells evolved.

When RNA is the final gene product, as in the case of rRNAs or tRNAs, the site of the gene can be sought in the DNAs of the nucleus or of the organelle using RNA—DNA hybridization techniques. In every case examined to date the rRNA of the organelle has been found to hydridize with its own DNA. However, in some plants chloroplast rRNAs have hybridized with nuclear as well as plastid DNA; there are several plausible explanations of this observation but they do lead to some uncertainty about the location of these genes (e.g. (1)).

How can genes for organelle-localized proteins be traced to the nuclear, plastid or mitochondrial genome of the eukaryotic cell? Until the present work was done, the gene for *iso*-1-cytochrome *c* in the yeast was the only determinant for an organelle-localized protein which had been mapped (2).

We have located 3 genes which specify chloroplast ribosomal proteins involved in erythromycin sensitivity in the flagellated green alga *Chlamydomonas reinhardi.* We selected this problem and this organism because:

1) Easy-to-score genetic markers are needed to trace a gene to a genome and ultimately to a locus within a genome. The usefulness of mutations to antibiotic resistance as genetic markers has been demonstrated repeatedly in studies of the genetics of micro-organisms. At the time this work was con-

* Supported in part by a research grant from the National Institute of General Medical Science and in part by the Maria Moors Cabot Foundation of Harvard University.

** Present address: National Institute of Arthritis, Metabolism and Digestive Diseases, Room 107, Building 6, NIH, Bethesda, Md. 20014, U.S.A.

templated, *C. reinhardi* was known to be sensitive to erythromycin and mutant strains resistant to this antibiotic were also known (3). Erythromycin binds to bacterial ribosomes (4) and blocks protein synthesis; resistance to erythromycin by *Escherichia coli* has been correlated with an alteration in a single protein of the 50S ribosomal subunit (5).

Thus, published work on *Chlamydomonas* made it seem reasonable that we could obtain resistant mutants and the bacterial work suggested that these mutants might have altered ribosomal proteins. We were not certain, however, that erythromycin affects *Chlamydomonas* chloroplast ribosomes solely or at all.

2) One of the objects of obtaining erythromycin-resistant mutants would be to determine, at the first level of genetic analysis, whether the genetic determinants were located in the nuclear or organellar genomes. Many characters are transmitted in a normal Mendelian manner in *Chlamydomonas*; the genes for these characters are in the nuclear genome. In addition, a uniparentally transmitted genome, thought to be in the chloroplast, has been studied in great detail. Markers on a number of linkage groups of the nuclear genome are known and a number of genes in the uniparental genome have been ordered on a single chromosome (6). Relatively simple genetic testing can reveal whether a gene is located in the uniparental (chloroplast) or in the Mendelian (nuclear) genome in *Chlamydomonas*. Furthermore, the availability of many genetic markers makes it possible to assign a gene to a position on a particular linkage group. In all of these respects then, *Chlamydomonas* seemed a suitable organism for our investigation.

3) Even if the antibiotic-resistance trait could be mapped to and within a particular linkage group it would still remain to be shown that it is a structural gene. In order to establish this for a protein it is necessary to demonstrate that a mutation of the gene results in an alteration of the primary amino acid sequence and not in a secondary modification. This kind of question can be resolved only by isolating the protein in large enough quantities to be able to study some of its properties. Highly purified ribosomal subunits had been obtained in reasonable quantities from a number of organisms and techniques for analysis of ribosomal proteins by polyacrylamide gel electrophoresis in urea were quite advanced. Also, about 30% of the *Chlamydomonas* ribosomes are in the chloroplast.

To these reasons had to be added the attractiveness of studying the distribution of genes for an integrated multi-molecular structure as a way of exploring intergenomic relationships within a eukaryotic cell.

We believed that we could find the site of at least one structural gene for a chloroplast ribosomal protein in *C. reinhardi* by the following steps: a) determining whether erythromycin interacts exclusively with chloroplast ribosomes; b) isolating resistant mutants in which the character could be used as a genetic marker; c) determining whether resistance resulted from alterations in the ribosomes themselves; d) establishing by genetic analyses first whether the resistance is transmitted in a Mendelian manner (and therefore that the gene is in the nuclear genome) or uniparentally (indicating that the gene is located in the extra-nuclear genome of the chloroplast) and then to attempt to map the gene to a place on a known linkage group by cross-over frequency techniques, and e) examining ribosomal proteins of mutant and wild-type cells in order to establish the molecular nature of the alteration in resistant strains — i.e. to determine if the gene specifies the primary structure of the protein.

Chloroplast ribosomes bind erythromycin

Wild-type *Chlamydomonas* does not grow in 1.5×10^{-4} M erythromycin, an inhibitor of protein synthesis in bacteria. Does this antibiotic interact with *Chlamydomonas* ribosomes? If so, which ribosomes? Which subunits?

^{14}C-erythromycin was mixed with a preparation of ribosomes from *C. reinhardi* and the mixture was centrifuged through a sucrose density gradient. The positions of ribosomes and dissociated subunits were determined by their absorption at 254 nm. The distribution of the radioactive erythromycin was determined by analyzing samples of the gradient for radioactivity.

Erythromycin binds only to intact 68S chloroplast ribosomes and their 52S subunits, not to the small subunit of the chloroplast ribosomes, not to any component of the 81S ribosomes of the cytoplasm. There is on the average one binding site per 52S subunit. The equilibrium constant for the binding is about 8×10^{-4} mole^{-1} (7).

The isolation of erythromycin resistant mutants

The next step was to isolate mutants which are resistant to the antibiotic. From among about 75 resistant isolates, 9 were selected for further analyses (7).

Ribosomes of resistant mutants do not bind erythromycin

Why are the cells resistant? Because they no longer take up erythromycin? Because they destroy it? Or, because mutant 52S subunits no longer bind the antibiotic?

Ribosomes from none of these 9 resistant strains bind erythromycin under the conditions established for the wild-type. This satisfied us that some property of the 52S subunits was altered genetically and that this warranted further investigation (7).

Genes for erythromycin resistance are in both the Mendelian and the uniparental genome

By manipulation of culture conditions vegetative *Chlamydomonas* cells can be converted to gametes. A mating-type minus ("—") gamete can fuse with one of mating-type plus ("+") to form a zygote. The zygote undergoes meosis and germinates to give rise to 4 haploid zoospores (or 8 in some strains). These vegetative zoospores can, in turn, be converted to gametes to continue the cycle (3).

"Mating-type" is transmitted as a Mendelian character. Thus, one gamete is of the "+" type and the other of the "—" type; 2 of the 4 zoospores which are released on germination of the zygote are mating type "+" and the other two are mating type "—". Other genetic determinants carried on chromosomes of the nuclear genome of *Chlamydomonas* are also transmitted in this Mendelian manner.

Some other characters are transmitted uniparentally, i.e. only from the "+" parent. Thus, if the trait is carried by the "+" mating type gamete, all 4 offspring will carry it; if, on the other hand, the same trait is carried by the "—" mating type parent, none of the zoospores will carry it. Uniparentally transmitted characters introduced into the cross by the "—" parent are not recovered in subsequent crosses; therefore, it can be assumed that the genetic material carried by this parent is destroyed between the time of formation of the zygote and its germination.

This simple difference, 2 to 2 segregation if the character is transmitted in the Mendelian manner and is in the nuclear genome and either 4 : 0 or 0 : 4 uniparental transmission when the character is in the extra-nuclear plastid genome, provides a relatively easy-to-use tool for determining which genome carries the character. Crossing erythromycin-resistant mutants with wild-type, i.e. sensitive cells, of either mating type should yield two resistant and two sensitive offspring if erythromycin resistance is carried in the nuclear genome. The "0 : 4—4 : 0" pattern of uniparental transmission of resistance would show the gene to be in the extra-nuclear genome.

Crosses between wild-type cells and our mutants showed that erythromycin resistance of 8 strains was transmitted according to the Mendelian pattern. These 8 nuclear mutants were assigned the class designation "*ery*-M", to indicate Mendelian transmission of the character. These strains were further subdivided into two groups on the basis of their behavior when crossed with one another. When some erythromycin strains were crossed with others, all offspring were resistant. In some other crosses between two erythromycin resistant strains some offspring were sensitive, others were resistant, and still others failed to survive. "Compatible" strains — i.e. those which when crossed with one another yielded only resistant strains — were grouped. Two groups, *ery*-M1 and *ery*-M2, were established. Four of our mutant strains fit into each group. It was assumed tentatively that these represented alterations at two different loci in the nuclear genome.

Resistance in the 9th mutant was transmitted uniparentally; this locus was designated *ery*-U1; the mutant, *ery*-U1a. Three loci, one in the chloroplast and 2 in the nuclear genome thus appeared to be involved in erythromycin resistance among the strains which we had isolated (7,8).

Why does erythromycin fail to bind to ribosomes from resistant strains? What is the molecular change in the ribosomes?

Comparatively large amounts of purified ribosomal subunits were needed for the next step: to separate and study the proteins. Procedures were worked out to grow 13-l cultures of *Chlamydomonas*, to recover the ribosomes, and to separate them into their subunits in a large sucrose gradient using a zonal centrifuge. Fig. 1 shows the separation of the 4 subunits of chloroplast 68S and cytoplasmic 81S ribosomes (8). We collected purified 52S chloroplast ribosomal subunits of wild-type and *ery*-M2b cells from sucrose density gradients and suspended them in LiCl-urea. Under these circumstances the proteins are dissociated and solubilized but the rRNA is precipitated. The supernatant solution was dialyzed to remove LiCl and lypholized. The lypholized proteins were solubilized in 8 M urea and analyzed by polyacrylamide gel electrophoresis.

Fig. 2 shows gels prepared from proteins of wild-type and *ery*-M2b cells. The arrow shows the altered position of one of the polypeptides. There is no protein at the normal position in the gel of the *ery*-M2b strain but a new band with a slightly lower electrophoretic mobility is present. The mutation in *ery*-M2b results in a change in behavior of the protein designated by the arrow in fig. 2.

Ery-U1a: Although it was easy to detect an alteration in the migration of the 52S *ery*-M2 protein we were unable to detect a change in a protein of the 52S ribosomal subunit of *ery*-U1a by the urea polyacrylamide gel electrophoresis method. Consequently, a 2-dimensional gel electrophoresis method which could give much better resolution was developed (9). In this procedure, ribosomal proteins are prepared in the manner already described and

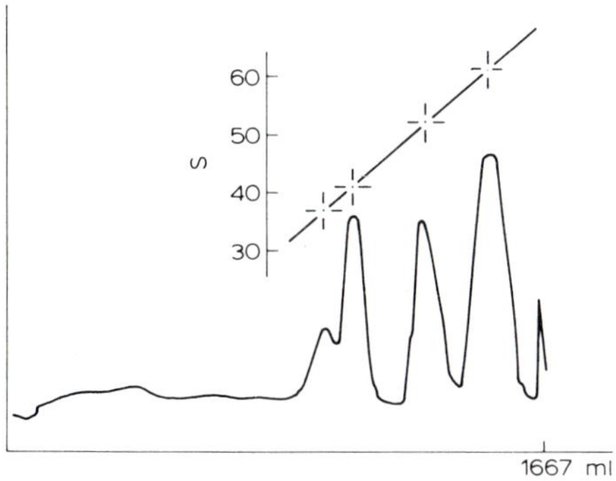

Fig. 1. Separation of the subunits of 68S chloroplast ribosomes and 81S cytoplasmic ribosomes of *Chlamydomonas reinhardi* from one another on a gradient of about 7.25—40% sucrose (w/v) in a B-15Ti zonal rotor. Conditions were selected to optimize the separation of the 52S subunit of the chloroplast ribosomes. (From Mets and Bogorad (8).)

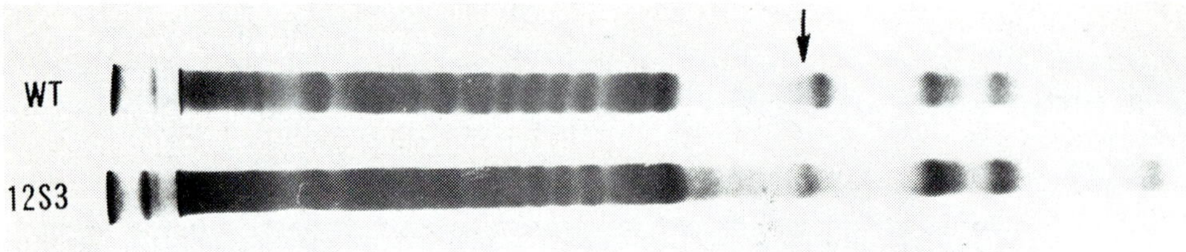

Fig. 2. Electrophoresis in urea-containing polyacrylamide gels of proteins from the 52S subunits of chloroplast ribosomes from wild-type *C. reinhardi* (above) and from mutant *ery*-M2d. The position of the altered band in *ery*-M2d is shown with an arrow above the upper gel. (After Mets and Bogorad (8).)

applied to a cylinder of 4% polyacrylamide gel in buffered 8 M urea. After proteins separate on this gel, it is cemented to the upper edge of a rectangular slab of 10% polyacrylamide gel in a specially-designed apparatus. A buffer

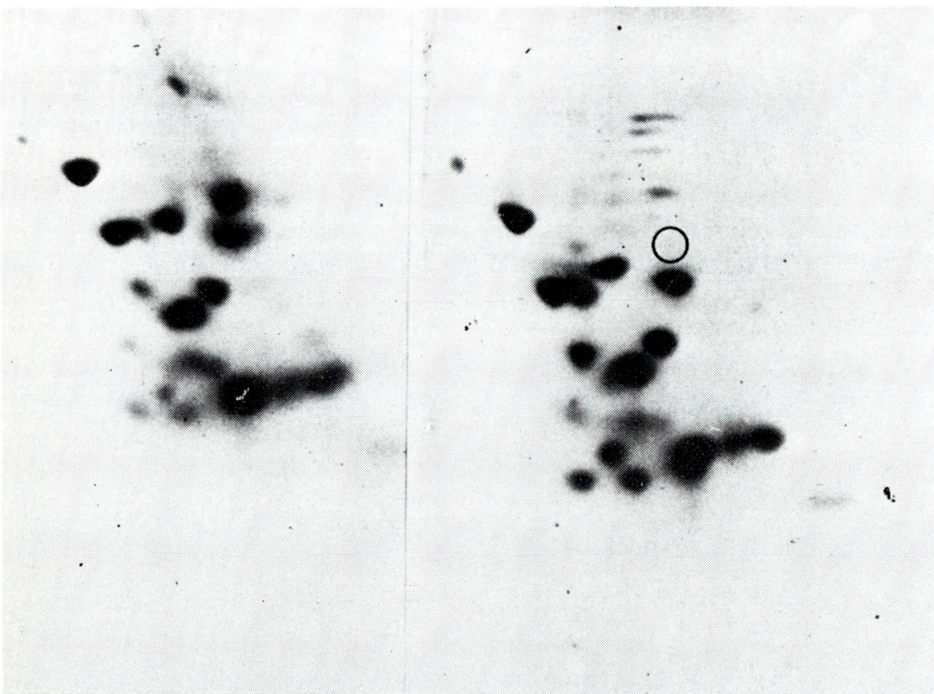

Fig. 3. Polyacrylamide gel electrophoresis in 2 dimensions (in urea, left to right); in sodium dodecyl sulfate (top to bottom). Comparable portions of two gel slabs are shown. a) Proteins of 52S subunit of chloroplast ribosomes of wild-type *C. reinhardi*. b) Similar preparation from the uniparental mutant *ery*-U1a. The position of the missing protein LC4 is shown by a circle. Aggregates of altered LC4 molecules appear as a series of horizontal streaks above its normal position (Mets and Bogorad (8).)

solution containing SDS is placed in the chamber over the cemented-on gel cylinder. SDS migrates electrophoretically down through the gel cylinder and through the slab. During this electrophoresis, the ribosomal proteins present in the gel cylinder are distributed in the gel slab according to molecular weight. The ribosomal proteins, in this way, are spread out on a 2-dimensional grid where they are separated horizontally according to charge in the presence of 8 M urea (using a very porous gel which hardly impedes their progress) and are separated in the second dimension of the grid in proportion to the log of their molecular weights (9,10).

Fig. 3a shows a portion of a gel on which proteins of the 52S subunits of the wild-type cells have been separated and 3b shows the separation of proteins from the same subunit of *ery*-U1a. The circle in fig. 3b shows the position of the protein which is lacking from *ery*-U1a but which is present in the wild-type. This approximately 30,000 dalton protein is designated LC4 on a 2-dimensional gel map of the ribosomal proteins of the 52S subunit (10) (fig. 4). Several horizontal streaks are seen in fig. 3b just above the circle on the gel of the proteins from *ery*-U1a. These occur at discreet intervals and appear to be aggregated forms of the "missing protein". Thus the locus *ery*-U1 appears to be associated with a greater tendency of LC4 to aggregate (8). This was the first demonstration of the molecular consequence of an alteration in a chloroplast gene. Later the form of the small subunit of ribulose-diphosphate carboxylase was shown to be inherited maternally (11).

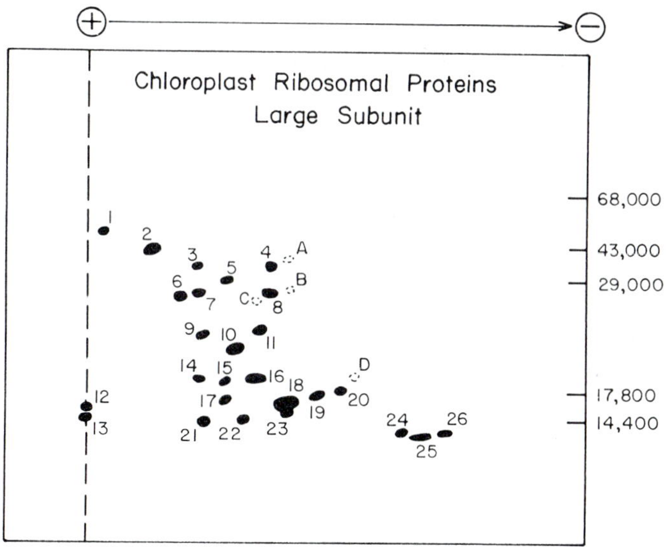

Fig. 4. Schematic diagram showing the positions of proteins from the large (52S) subunit of the chloroplast ribosome from wild-type cells of *C. reinhardi*. Unreproducible components are designated by letters. (Hanson et al. (10).)

Ery-M1: The evidence presented above indicates that perhaps two different 52S proteins are associated with mutation to erythromycin resistance; one of these proteins may be specified by the Mendelian locus, *ery*-M2, and the other, identified as LC4, may be specified by the uniparental locus *ery*-U1. However, in neither case is it beyond possibility that a secondary alteration in the protein has occurred rather than a change in the primary amino acid sequence. For example, if *ery*-M2 affected an enzyme responsible for the phosphorylation, methylation or some other alteration in a polypeptide chain, change in the specificity of such an enzyme could result in its altering a polypeptide which it had not attacked previously. If such a change fortuitously reduced the affinity of the ribosome for erythromycin, the effect observed would be explained. All of this is possible but is not any more likely than that a change in the primary amino acid sequence has occurred. In both *ery*-M2 and *ery*-U1, a single protein in the ribosome is altered whereas one might expect to see more proteins altered and the cell in great difficulty if a protein-modifying enzyme had been changed. But still, the evidence available does not exclude the possibility of a secondary alteration.

The most direct way to determine whether a primary or secondary change has occurred is to compare the amino acid sequences of wild-type and mutant proteins. Or, to compare fingerprints of the two polypeptides, to isolate an altered polypeptide, and to determine that the amino acid composition of the two polypeptides of the digest differ in amino acid composition. (Differences in migration of peptides after digestion could result from secondary alterations of a primary sequence. The detection of such differences without knowledge of the amino acid compositions of the peptides does not reveal any more than difference in electrophoretic behavior of undigested proteins.) We have approached this problem more indirectly by taking advantage of having several mutants.

First, the locus *ery*-M1 has been located by conventional mapping techniques. It is about 13 map units to the right of the marker "paralyzed flagella 2" on nuclear linkage group XI (12). We had guessed that 4 of the mutants which we had are at a single locus. These strains were designated *ery*-M1a, b, c, and d. Extensive genetic analyses revealed that within the limits of the error of the number of crosses analyzed all 4 of these mutations map to the same position. In other words, all of these 4 mutations are of a single gene; they are alleles. If this is so, we should expect the same protein to be altered in all 4 mutants. This proved to be the case.

By modifying the one-dimensional urea polyacrylamide gel electrophoresis system we had used before, we were able to detect an altered protein in the 52S ribosomal subunit of each of the mutant strains. The protein marked with an arrow in the wild-type (left gel, fig. 5) is not present at that position in gels of 52S proteins of any of the 4 *ery*-M1 mutants. *Ery*-M1a and 1c 52S subunits contain a protein which moves more slowly than the wild-type. The

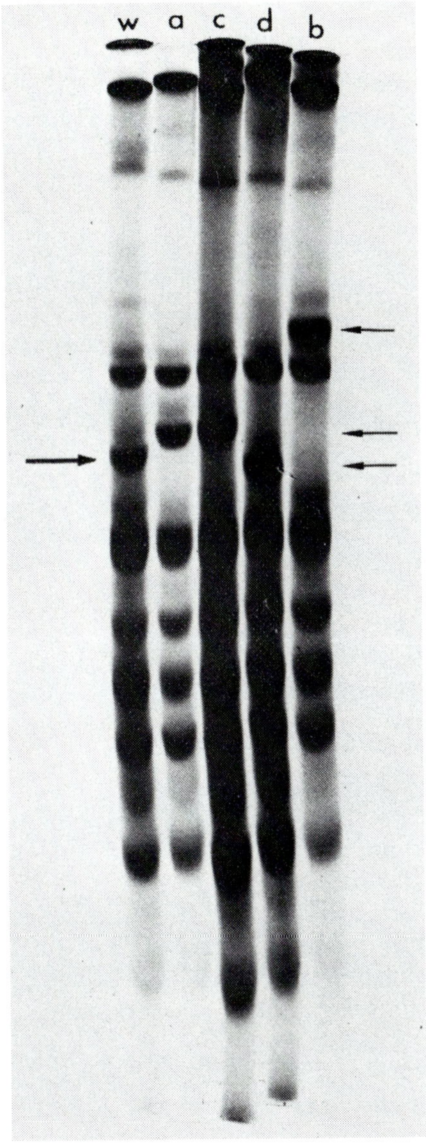

Fig. 5. Electrophoresis in urea-containing polyacrylamide gels of proteins from the 52S subunits of chloroplast ribosomes from wild-type, and of *ery*-M1a, b, c, d *C. reinhardi*. Arrows indicate the positions of normal and mutant forms of protein LC6. (Davidson et al. (12).)

protein missing from *ery*-M1d is replaced by another slightly more rapidly moving one. In *ery*-M1b a new protein appears which moves much less rapidly than the wild-type protein. In all, we can distinguish three alterations in this same protein among the 4 mutant strains. The protein in *ery*-M1a and 1c may be different but this is not revealed by our analytical methods. The protein, which has been identified as LC6 from the 2-dimensional gel map, is clearly specified by the locus *ery*-M1 (10,12). Two-dimensional gel analyses have shown that the mutant forms of LC6 in *ery*-M1a, c and d are not distinguishably different from the wild-type in molecular weight but LC6 in *ery*-M1b is about 28% smaller than the wild-type protein (12).

These data on *ery*-M1 and LC6 establish for the first time the site of a structural gene for a eukaryotic ribosomal protein. *Ery*-M1 is also the first structural gene located for a chloroplast protein.

Three different 52S ribosomal proteins have been identified as being specified by genes at 3 different loci in 2 genomes. At least one, *ery*-U1, is located in the uniparental, presumably chloroplast, genome. The locus *ery*-M2 is not linked to *ery*-M1; thus, genes for chloroplast ribosomal proteins are scattered even within the nuclear genome.

Intracellular gene dispersal. Problems in the origin and evolution of eukaryotism

Ribosomes and all the genes for their rRNAs and proteins were together when eukaryotic cells were first formed. Now the products are all in the plastid compartment, but genes for parts of chloroplast ribosomes in *Chlamydomonas* are distributed in 2 different compartments. How did this separation come about?

First, we should consider how the eukaryotic habit might have originated. Two conspicuous possibilities are: a) by the combination of two kinds of cells, i.e. by endosymbiosis; or b) the creation of compartments within a previously uncompartmentalized cell, i.e. in the cluster-clone way (13).

According to the endosymbiont hypothesis today's plastid is a descendant of an invading prokaryotic alga or photosynthetic bacterium which was accepted into an anaerobic eukaryotic cell. A mitochondrion is a descendant of an aerobic prokaryote or modified alga or photosynthetic bacterium (e.g. 14).

How might the separation of ribosomal proteins and ribosomal genes have come about if eukaryotes originated by endosymbiosis? We can consider the two possibilities: a) gene transfer; and/or b) protein and gene substitution.

Fig. 6 illustrates gene transfer. We do not know the rules for intracellular evolution or for the migration of genes among genomes in a single cell but the latter is not excluded by other experiences in molecular biology. For example, the present distribution of genes for *Chlamydomonas* chloroplast

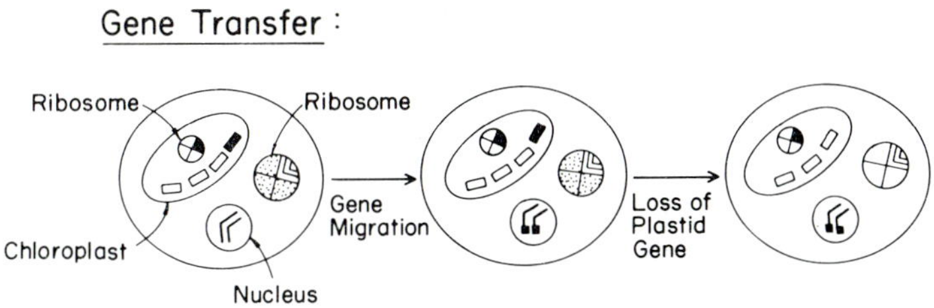

Fig. 6. Diagramatic representation of gene dispersal by transfer of a gene from the chloroplast to the nuclear genome. The case of a structural gene for a protein of the chloroplast ribosome is used as an illustration. (Also see ref. (13).)

ribosomal proteins might have come about if the gene, or a copy of the gene, had become integrated into the nuclear genome and was subsequently lost from the chloroplast genome.

Gene and protein substitution is illustrated in fig. 7. If the gene for a chloroplast ribosomal protein became altered in a way that the protein was no longer useful, the ribosome would not function unless another protein could substitute. If a protein, coded for by a nuclear gene present throughout the cell including the organelle could substitute for the defective one, the new ribosomal protein of the organelle would be coded by a nuclear gene. The new protein would probably function more poorly or have a lower affinity for the ribosomal site than the original. (If there were two proteins

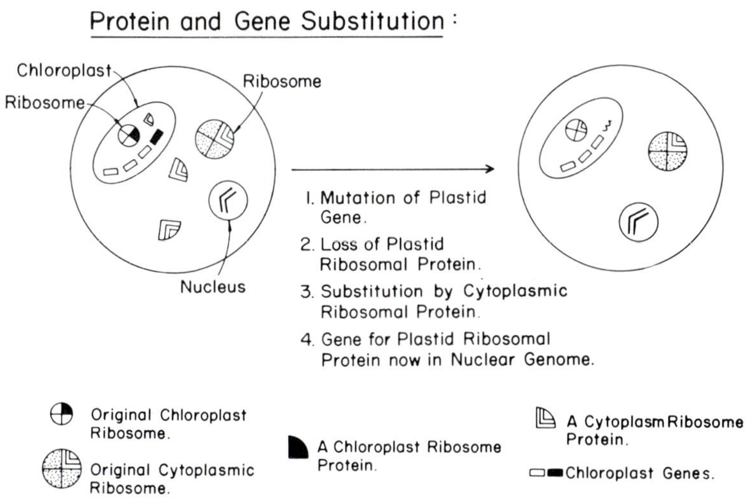

Fig. 7. Diagrammatic representation of the protein and gene substitution mechanism of intracellular gene dispersal. The example is of a cytoplasmic ribosome protein and its nuclear gene which come to substitute for a chloroplast ribosome protein and its structural gene within the organelle. (Also see ref. (13).)

with equal affinity and effectiveness for a (e.g.) ribosomal site in any cell, one gene would probably be lost. In a multi-genomic cell either the nuclear or the organellar gene for the proteins would probably be lost; we do not know the rules, if any, for intracellular gene selection.) Fig. 7 suggests that the substituting protein in the chloroplast ribosome example is a cytoplasmic ribosomal protein but this, of course, need not necessarily be the case. Some other protein might work better. (If the substituting protein were from a cytoplasmic ribosome, a subsequent mutation might make it unsuitable for its cytoplasmic site without altering its usefulness for the chloroplast ribosome. By gene duplication and a few alterations two separate proteins, one for the chloroplast and one for the cytoplasmic ribosome, might easily be imagined to evolve.)

The suggested sequence of events for the cluster-clone formation of a eukaryotic cell are: 1) The grouping of genes. Each group or cluster would become a genome; 2) Each cluster would become separated from the others by a membrane. One membrane-limited cluster would become the nucleus and another a mitochondrion, a third a plastid; 3) Each cluster would reproduce. In this way, a clone of nuclei, a clone of chloroplasts, and a clone of mitochondria would be established within a single cell.

Many crucial problems are disregarded in this abbreviated chronology but the subsequent evolution of genomes in the cell might depend most on whether there was one copy or more of each gene. But what could that single, original cell have been like? Fig. 8 shows one of the possibilities if

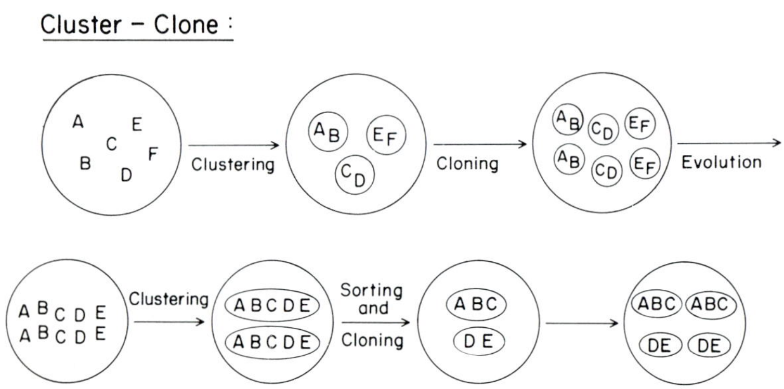

Fig. 8. Diagrams showing steps in the origin and evolution of eukaryotism by the cluster-clone mechanism. a) If the progenitor cell had one copy of each gene, each cluster would contain a unique complement of genes. Genes for complex structures could have become separated at the time of origin of the eukaryotic cell in this way. b) If the progenitor cell had, e.g., two copies of each gene, one extreme type of separation would be of a copy of every gene in each cluster. Sorting and elimination of genes to yield unique clones would follow during evolution. Evolution of (e.g.) two or three ribosome types in the modern eukaryotic cell would have had to occur later in either case.

only a single copy of each gene was present: each compartment would contain an assortment, perhaps at random, of genes. In some clones many of the gene products might remain within the membrane — as in the plastid and mitochondrial clones while expression of genes in the nuclear clones is carried on by equipment outside. Reassortment of genes might occur in the course of evolution. An example at the other extreme, fig. 8b shows two copies of each gene in the cell and each clone initially contains one copy of every gene. Much like a modern multinucleate organism. Then modern eukaryotic cells might have developed by elimination of functionally redundant genes, according to the rules which we do not understand.

Gene transfer and substitution mechanisms could have played roles in evolution of eukaryotic cells regardless of how they originated but gene dispersal without such devices could have occurred only if eukaryotism arose in the cluster-clone way. But why aren't all the genes in the nucleus? What limits gene dispersal?

Intergenomic cooperation: A principle of organelle biology?

Soon after organelles were found to have genetic information storage and processing capacities the notion of autonomous organelles was favored widely. Now, several situations are known in which the formation of a single organellar structure requires the cooperation of two translation systems or two information storage-transcription-translation systems of the cell. At the level where the site of information storage is known there are the following cases: 1) Genes for *Chlamydomonas* chloroplast ribosomal proteins have been located and been identified as discussed above (7,8,10,12); 2) Apel and Schweiger (15) have shown by nuclear transplantation that some chloroplast thylakoid membrane proteins of *Acetabularia* are specified by genes in the nucleus; 3) The electrophoretic behavior of some *Acetabularia* chloroplast ribosomal proteins is determined by nuclear genes (16); 4) The enzyme ribulose disphosphate carboxylase is composed of large and small subunits. In *Nicotiana*, information for one of the subunits is transmitted in a Mendelian way with suppression of one of the parental types (17). The other subunit is specified by a uniparentally transmitted gene (11).

At the level of the site of synthesis of proteins there are the following cases: 1) Based on sequential antibiotic experiments it has been judged that some proteins of *Chlamydomonas* chloroplast thylakoids are formed on 81S, cytoplasmic ribosomes while others are produced on 68S plastid ribosomes (18); 2) One subunit of ribulose diphosphate carboxylase is formed by chloroplasts in vitro (19). The other subunit is not, it is presumably formed on the cytoplasmic, 80S, ribosomes; 3) Experiments (20) with antibiotics which differentially affect the ribosomes of mitochondria and cytoplasm to show that some subunits of yeast mitochondrial ATPase are formed on one class of ribosomes and other subunits on the other; 4) Schatz (21) and

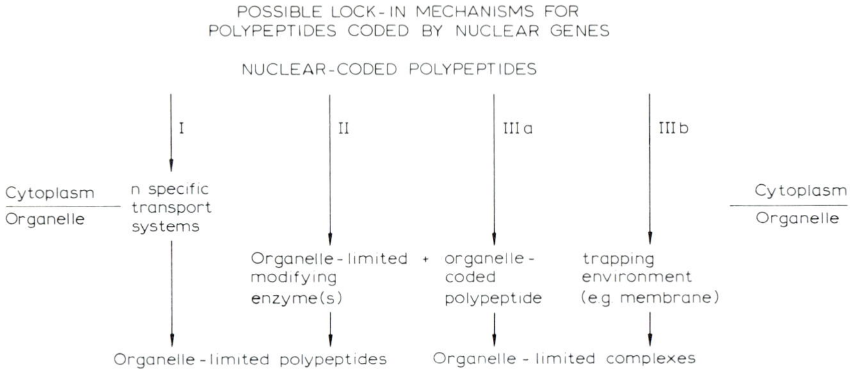

Fig. 9. An enumeration of some devices which may serve to lock in organelle-limited proteins coded by nuclear genes and synthesized on cytoplasmic ribosomes.

Tzagoloff (e.g. 22) have shown by sequential antibiotic techniques that some polypeptides of yeast mitochondrial cytochrome oxidase are produced on cytoplasmic and others on mitochondrial ribosomes.

If we assume, with no evidence, that the site of synthesis of a protein reflects the location of the structural gene and indeed that all of the cases mentioned deal with locations of structural genes, we have a large number of cases in which the genes for proteins of multi-subunit structures are separated in at least 2 genomes. Again, what are the limits of gene separation?

What mechanisms are used to lock nuclear-specified, cytoplasmically-synthesized proteins into the organelle? Fig. 9 describes some possible lock-in devices. In some cases examples exist: 1) Nothing is known about specific transport systems although this might be the system of choice for small, soluble proteins; 2) A cytoplasmic leucyl tRNA synthetase in *Neurospora crassa* is apparently altered for use in mitochondria (23): 3a) Examples are components of ribosomes of *Chlamydomonas* and *Acetabularia* as well as ribulose diphosphate carboxylase; 3b) This is a pattern seen in membrane assembly in *Acetabularia* and *Chlamydomonas* as well as in cytochrome oxidase and ATPase of yeast.

Now to try to bring together the evolutionary mechanisms and possible lock-in devices: Devices I and II describe secondary adjustments of the organelle or the protein and would be compatible with the protein and gene substitution mechanism for gene dispersal. Class 3 devices are probably not so limited. They might be employed in connection with either gene transfer or gene substitution mechanisms.

This enumeration of some evolutionary possibilities is designed to provide a framework for the study of intergenomic cooperation. A more extended discussion of some of these problems appears elsewhere (24).

References

1. R.M. Smillie, N.S. Scott and D.G. Bishop, in, J.K. Pollak and J.W. Lee (Eds.) Biochemistry of Gene Expression in Higher Organisms, Australian and New Zealand Book Co., Sydney (1973) p. 479.
2. F. Sherman, J.W. Steward, E. Margoliash, J. Parker and W. Campbell, Proc. Nat. Acad. Sci. U.S. 55 (1966) 1498.
3. R. Sager and Z. Ramanis, Proc. Nat. Acad. Sci. U.S. 65 (1970) 593.
4. H. Teraoka, J. Mol. Biol. 48 (1970) 511.
5. E. Otaka, H. Teraoka, M. Tamaki, K. Tanaka and S. Osawa, J. Mol. Biol. 48 (1970) 499.
6. R. Sager, Cytoplasmic Genes and Organelles, Academic Press, New York (1972) p. 405.
7. L.J. Mets and L. Bogorad, Science 174 (1971) 707.
8. L.J. Mets and L. Bogorad, Proc. Nat. Acad. Sci. U.S. 69 (1972) 3779.
9. L.J. Mets and L. Bogorad, Anal. Biochem. 57 (1974) 200.
10. M.R. Hanson, J.N. Davidson, L.J. Mets and L. Bogorad, Mol. Gen. Genet. (in press).
11. P. Chan and S.G. Wildman, Biochim. Biophys. Acta 277 (1973) 677.
12. J.N. Davidson, M.R. Hanson and L. Bogorad, Mol. Gen. Genet. (in press).
13. L. Bogorad, L.J. Mets, K.P. Mullinix, H.J. Smith and G.C. Strain, Biochem. Soc. Symp. 38 (1973) 17.
14. L. Margulis, Origin of Eukaryotic Cells, Yale Univ. Press, New Haven (1970).
15. K. Apel and H.G. Schweiger, Eur. J. Biochem. 25 (1972) 229.
16. K. Kloppstech and H.G. Schweiger, Exp. Cell Res. 80 (1973) 69.
17. N. Kawashima and S.G. Wildman, Biochim. Biophys. Acta 262 (1972) 42.
18. I. Ohad, this volume (1974).
19. G.E. Blair and R.J. Ellis, Biochem. J. 127 (1972) 42.
20. A. Tzagoloff and P. Meagher, J. Biol. Chem. 247 (1972) 594.
21. G. Schatz, this volume (1974).
22. M.S. Rubin and A. Tzagoloff, J. Biol. Chem. 248 (1973) 4275.
23. O. Weeks and S.K. Gross, Biochem. Genet. 5 (1971) 505.
24. L. Bogorad in, A. Tzagoloff (Ed.) Organelle Development (1975).

Dr. BONOTTO:

Would it be possible to perform transformation experiments by supplying nuclear or chloroplastic DNA?

Dr. BOGORAD:

Perhaps.

Dr. SCHWEIGER:

Can it be completely excluded that the primary site of attack in the case of the *ery*-U1a mutant is the rDNA of the organelle?

Dr. BOGORAD:

No. In fact, this was one of our first thoughts when we began to understand this mutation, but it is not so likely. First, we do have evidence of an alteration in ribosomal protein LC4. Second, mutations in rDNA are to my knowledge unknown. There is a case in bacteria (C. Lai and B. Weisblum, Proc. Natl Acad. Sci. U.S. 68 (1971) 856), where cells become resistant to erythromycin in the presence of the antibiotic; the resistance is correlated with more than normal methylation of the rRNA. This appears to be not a heritable change but an effect on the methylating enzyme for methylation returns to normal in progeny freed of erythromycin. To sum it up, *ery*-U1a could have altered rDNA but that would have to be in addition to a change in the protein and there aren't good precedents. Unfortunately, detection of small alterations in rDNA or rRNA is very difficult.

Dr. MARGULIES:

Any evidence on the site of synthesis of LC4?

Dr. BOGORAD:

No. We have studied only the locations of the genes.

Dr. WEIL:

If the hypothesis of gene (or protein) substitution is correct, we would expect to find the same protein in the cytoplasm and the chloroplasts. In the case of the chloroplast ribosomal proteins coded by nuclear genes (mutants *ery* Mendelian 1 and 2), do you find the same ribosomal protein in the chloroplasts and the cytoplasm?

Dr. BOGORAD:

Four chloroplast ribosomal proteins migrate in the urea-SDS two-dimensional system identically to four cytoplasmic ribosomal proteins. This does not prove that they are the same but gives the maximum number which might be identical. LC6, the protein coded for by the gene at locus *ery*-M1 is not among these. However, the protein and the gene substitution hypothesis does not require that ribosomal proteins can be substituted for only by other ribosomal proteins. This was shown only to limit the oral presentation but even if cytoplasmic ribosomal protein substituted for one in the plastid, subsequent gene duplication and mutations of one of the genes could result in changes which might make the gene product unusable by the chloroplast ribosomes but not unsatisfactory for the cytoplasmic ribosomes.

Dr. LURQUIN:

When one speaks in terms of uniparental heredity, how can one be sure that one deals with the chloroplast genome since nothing is known about mitochondrial DNA in *Chlamydomonas?*

Dr. BOGORAD:

The question even extends to whether or not the extra-nuclear genes are in the chloroplast. The information presently available shows that a number of genes are in a single linkage group which is transmitted uniparentally. Because transmission is distinctly different from Mendelian transmission of the nuclear genome, it is assumed that these are physically outside of the nucleus. It seems reasonable to assume for the present, at least, that these genes are in the chloroplast. In the case of our mutant, *ery*-U1a, we see for the first time the molecular consequence, i.e., a change in LC4, of an alteration of a gene in that extra-nuclear genome.

Dr. WILLIAMSON:

Ruth Sager has presented some data which suggest that all the "70S" ribosomes in streptomycine resistant mutants were affected by a mutation in the chloroplast genome. This implies the mitochondrial ribosomes were also at least partly coded by chloroplast genes. Would you care to comment on this possibility?

Dr. BOGORAD:

We have been concerned with the question of mitochondrial ribosomes as possible contaminants in our analyses of chloroplast ribosomal proteins in *Chlamydomonas*. About 30% of the total cell ribosomes are 70S in *Chlamydomonas*. Judging from mitochondrial profiles in electron micrographs of *Chlamydomonas*, we estimate that only a very small fraction of this 30% could be mitochondrial ribosomes. This fraction is probably too small to assess separately by the analytical methods used by us or anyone else working with ribosomes prepared directly from whole *Chlamydomonas* cells.

CHAPTER III.2

XENOPUS OOCYTES: AN IN VIVO ASSAY TO TEST MITOCHONDRIAL FUNCTIONS

H. PINON, M. BARAT, C. DUFRESNE and J.C. MOUNOLOU

Laboratoire de Biologie Générale, Université Paris-Sud, 91405 Orsay, France

Yeast mitochondria injected into an immature oocyte were readily dispersed and kept their overall integrity for several hours. Purified yeast mitochondrial DNA was actively used as a template by the cytoplasmic DNA polymerase of the immature oocyte.

The transfer of organelles or of informative molecules from their mother cell into a different one has been a very successful method in elucidating some aspects of the control of their functions. Injections of frog brain nuclei into unfertilized eggs of *Xenopus* enabled Gurdon (1) to demonstrate the influence of the cytoplasm on the differentiation state of the nucleus. Later, the occurrence of DNA polymerase activities in the egg or oocyte cytoplasm was reported (2—5), these activities exhibiting various preferences for templates of different forms and origins.

At last the translation machinery of *Xenopus* oocytes has been used to identify RNA as messenger RNA and to study translation mechanisms and controls (6—8). We chose to utilize such an in vivo system to study the activities of injected mitochondria and the replication of injected mitochondrial DNA. Yeast mitochondria and yeast mitochondrial DNA were used since they differ in many respects from those of *Xenopus*.

The first obvious question relates to the possibility of injecting and monitoring mitochondria in the oocyte while the second question relates to the fate of the population of exogenous organelles. To check whether the foreign material was accepted by the oocyte, yeast mitochondria were isolated, labelled with ^{125}I (glucose-oxidase + lactoperoxidase method of C. Jacquemin and B. Lambert) and injected. The localization of the labelled mitochondria was analyzed by autoradiographs of sections of the injected oocytes. It was seen that the radioactive material did not aggregate (fig. 1) but became progressively dispersed 2 hr after injection.

Low concentrations of rotenone (2×10^{-6} M) completely inhibit the NADH oxidase activity of vertebrate mitochondria (9), and especially of *Xenopus* mitochondria, but yeast mitochondrial NADH oxidase is insensitive to this drug (10). This difference was used to follow the activity of yeast mitochondria in the oocyte. A mitochondrial fraction was isolated from

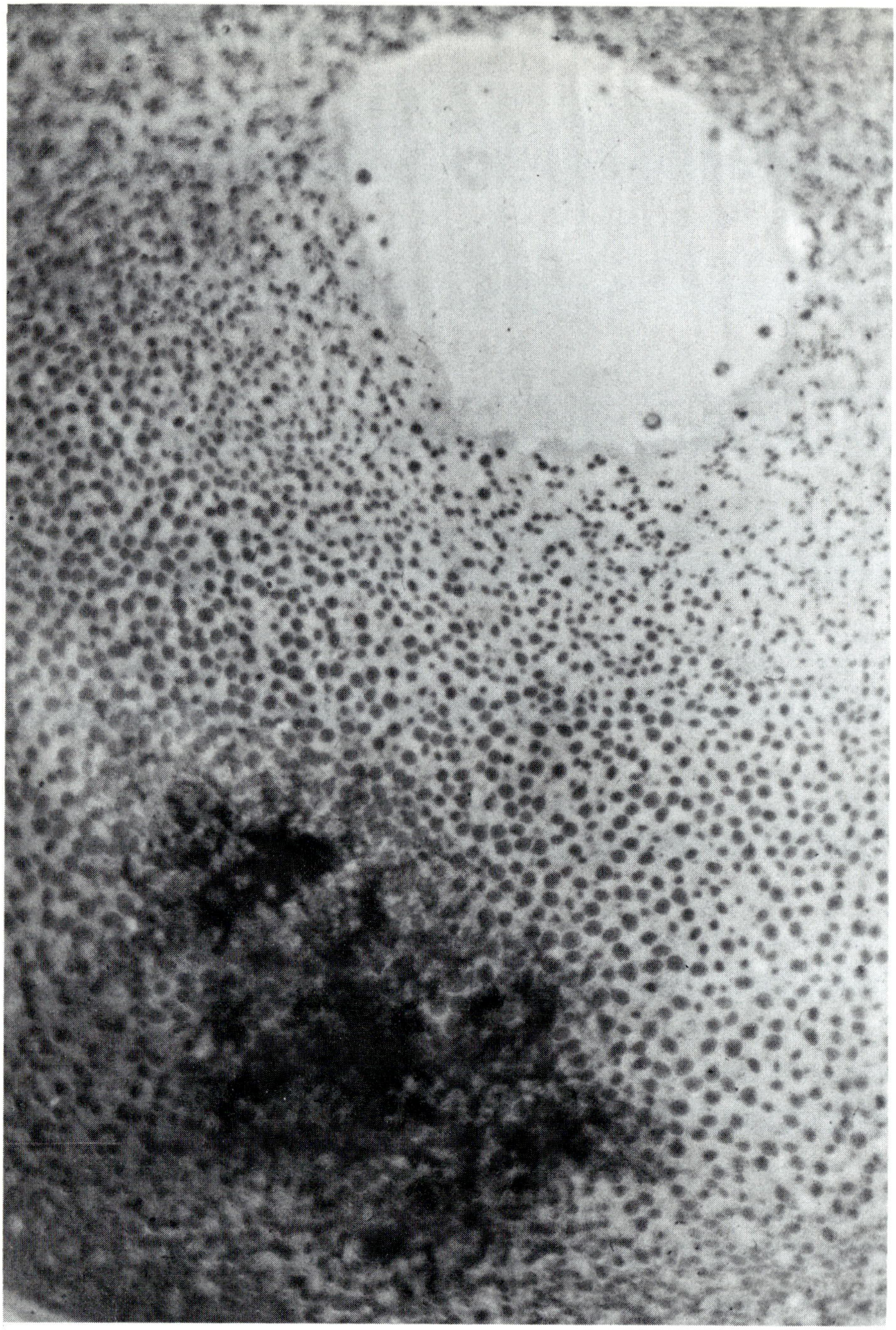

Fig. 1. Autoradiograph of a section of an oocyte injected with ^{125}I-labelled yeast mitochondria (0.2 μg yeast proteins and 2×10^4 dpm into the oocyte) and incubated for 2 hr. The germinal vesicle appears as a clear area. The vitellus platelets are scattered all over the cytoplasm and the labelled material is dispersed among them.

oocytes that were injected with a known quantity of yeast mitochondria and incubated for 5 hr. Both the total NADH oxidase activity and the rotenone insensitive activity were determined in this fraction. We observed in these conditions that 70—80% of the injected yeast NADH oxidase activity was recovered in the mitochondrial fraction.

From these two types of experiments we conclude that yeast mitochondria are dispersed in the oocyte cytoplasm and that they roughly keep their integrity (sedimentation properties and NADH oxidase activity).

Our second aim was to determine whether yeast mitochondria keep on synthesizing their own DNA in the oocyte. ^{3}H-deoxyadenosine was added in the cold to the yeast preparation just before injection (0.2 μg yeast mitochondrial proteins + 50 mμCi ^{3}H-deoxyadenosine/oocyte). Oocytes were incubated and homogenized 6 hr after injection. Follicular cells, nuclei and

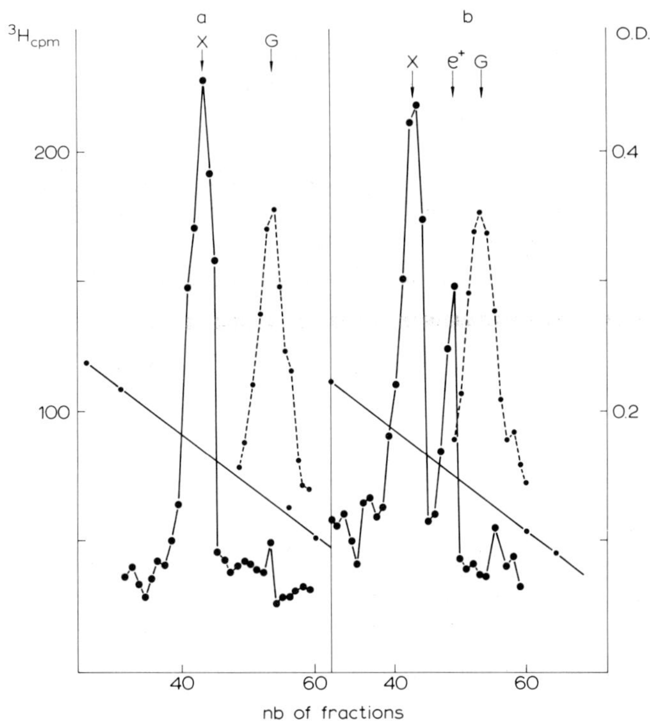

Fig. 2. a) 150 oocytes were injected with ^{3}H-deoxyadenosine (50 mμCi each), incubated for 6 hr and homogenized in 1 × SSC. Follicular cells, nuclei and debris were spun down and DNA was isolated from the supernate and analyzed on a CsCl density gradient. b) 150 oocytes were injected with ^{3}H-deoxyadenosine (50 mμCi each) and yeast mitochondria (0.2 μg proteins) and treated as in (a). In the collected fractions, radioactivity traces *Xenopus* DNA (X) and yeast mitochondrial DNA (ρ^+), optical density is due to glycogen (G).

debris were spun down at a low speed and DNA was rapidly isolated from the supernate and analyzed in a CsCl density gradient. In oocytes injected with ^{3}H-deoxyadenosine alone a peak of radioactivity appeared at the density of *Xenopus* DNA (fig. 2). In oocytes injected with the ^{3}H-deoxyadenosine and yeast mitochondria a second peak of radioactivity was seen at the density of yeast mitochondrial DNA. Although this peak was remarkably homogeneous and banded exactly as native double-stranded yeast mitochondrial DNA, it is not clear at present if this reveals a true mitochondrial activity.

Finally, since the DNA polymerase activity of immature oocytes has a peculiar affinity for denatured DNA or DNase treated DNA (4,5), we injected native and denatured mitochondrial DNA. It appears that native mitochondrial DNA is a very good template for the DNA polymerase of the oocyte cytoplasm. We compared the relative efficiency of DNA synthesis when yeast mitochondria or yeast mitochondrial DNA were injected in the oocyte (*Xenopus* DNA synthesis was the reference and all the results are given for equal amounts of template injected). Purified yeast mitochondrial DNA was about 5 times more efficient than yeast mitochondria. The activity observed with yeast mitochondria could be the consequence of a leakage of DNA out of the organelle. However, when purified DNA was injected in the oocyte the peak of radioactivity was skew on the heavy side (4), indicating partial synthesis and denaturation; this was not observed with yeast mitochondria. We are currently investigating DNA replication by the oocyte polymerase of mitochondrial DNA of grandes and cytoplasmic petites mutants of various base composition, sequence and length.

References

1. J.B. Gurdon, Proc. Nat. Acad. Sci. U.S. 58 (1967) 545.
2. J.B. Gurdon, M.L. Birnstiel and V.A. Speight, Biochim. Biophys. Acta 17 (1969) 614.
3 P. Grippo and A. Lo Scavo, Biochem. Biophys. Res. Commun. 48 (1972) 280.
4. H.R. Woodland, C.C. Ford, J.B. Gurdon and C.D. Lane, in M. Susman (Ed.) Genetics and Developmental Biology, Prentice-Hall (1972) 393.
5 B.L. Brandt and J.S. Watchel, Biochem. Biophys. Res. Commun. 53 (1973) 1017.
6 J.B. Gurdon, C.D. Lane, H.R. Woodland and G. Marbaix, Nature 233 (1971) 177.
7. R.H. Stevens and A.R. Williamson, Nature 239 (1972) 143.
8. B. Giglioni, A.H. Gianni, S. Ottolenghi and D. Runger, Nature New Biol. 246 (1973) 99.
9. D.J. Horgan, H. Ohno, T.P. Singer and J.E. Casida, J. Biol. Chem. 243 (1968) 5967.
10 G. Schatz and E. Racker, Biochem. Biophys. Res. Commun. 22 (1966) 579.

Dr. BONOTTO:

Did you inject circular DNA?

Dr. PINON:

Yeast circular mitochondrial DNA is not actually available.

Dr. BONOTTO:

Do you have any data on the further fate of the injected DNA into the *Xenopus* oocytes?

Dr. PINON:

Oocytes have been incubated for 6 hr. Anyway large oocytes cannot be fertilized.

Dr. MOUNOLOU:

The difference between Pinon's experiments and Gurdon's one is 1) that Gurdon used eggs and not oocytes; 2) that the cytoplasmic polymerase of the egg acceptes native DNA as a template while the enzyme of the oocyte does not.

CHAPTER III.3

LIGHT-INDUCED CHANGES IN THE PLASTID RNA POLYMERASE ACTIVITY OF GREENING MAIZE

K. APEL and L. BOGORAD

The Biological Laboratories, Harvard University, 16 Divinity Avenue, Cambridge, Massachusetts 02138, U.S.A.

Illumination of dark-grown maize plants induces a rapid increase of the RNA polymerase activity of isolated plastids. Within 16 hr of illumination the activity increases 5 to 6-fold as compared to dark-grown controls. This increase in enzymatic activity cannot be explained by either a change in the specific activity of the enzyme or a difference in the amount of RNA polymerase protein present in the plastids. This suggests that the light-induced increase in enzymatic activity might be caused by a modification of the DNA template.

It has been well established that in plant cells chloroplasts as well as nuclei contain DNA from which RNA is transcribed (1). The morphogenesis of the chloroplast depends on the interaction of both the nuclear and chloroplast genomes (2—5). It is not known, however, how this interaction is regulated. One possible means of regulation may be exerted at the level of transcription involving both the nuclear (6,7) and chloroplast (8) DNA-dependent RNA polymerases.

In order to investigate the possible involvement of the RNA polymerases in the regulation of chloroplast development, we have begun to study the RNA polymerases of the greening maize plant. It is well documented that after illumination of dark-grown plants a number of biochemical and morphological changes occur in the plastids (9). One of these changes is a rapid and specific increase in RNA synthesis (9—11). The data presented in this paper suggest that the chloroplast RNA polymerase might be involved in the regulation of this light dependent RNA synthesis.

Etiolated maize plants (5—7 days old) were illuminated for different lengths of time. The plastids were isolated and osmotically shocked. The RNA polymerase activities of the broken plastids were determined as described by Bottomley et al. (8). For the first 6 hr of illumination, there is no difference in activity between the broken plastids from illuminated plants and from dark-grown controls. Within the next 6 to 10 hr of illumination, the activity increases rapidly until a point (about 16 hr) at which the activity is increased 5 to 6-fold as compared to the dark-grown controls (fig. 1).

We have eliminated the possibilities that this increase is caused by differ-

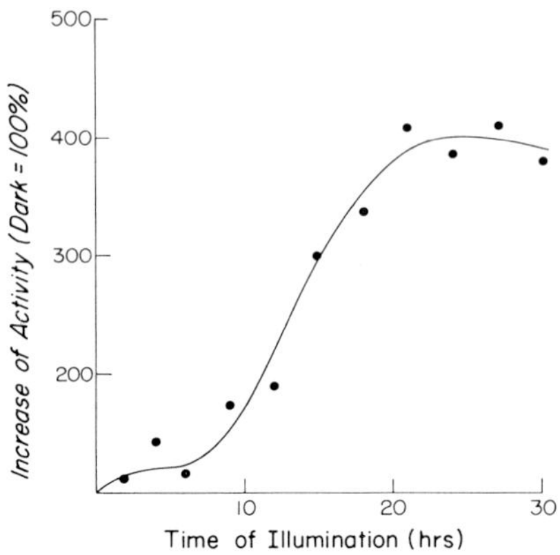

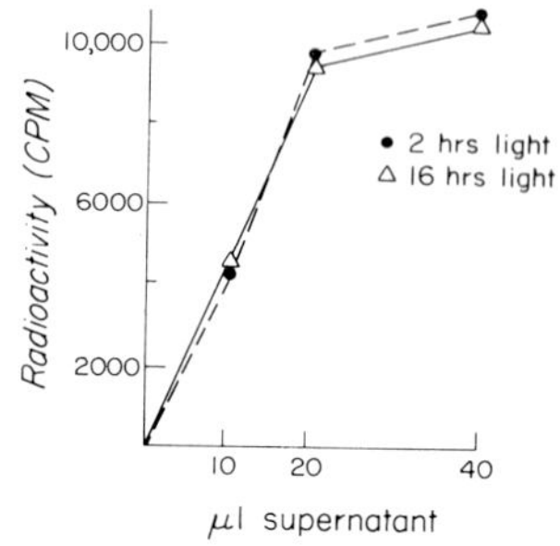

Fig. 1. RNA polymerase activity in suspensions of maize plastids isolated from etiolated plants illuminated for various periods of time.

Fig. 2. Solubilized plastid RNA polymerase activity from etiolated maize plants illuminated for 2 or 16 hr. The RNA polymerase was separated from the plastid membranes and assayed with calf thymus DNA as described by Bottomley et al. (8).

ences in optimal assay conditions, or differences in concentration of nucleases and proteases in samples from illuminated and dark-grown plants (Apel and Bogorad, in preparation). Also, because osmotically shocked plastids were used, the differences in activity cannot be accounted for by changes in the permeability of the plastid membrane. The increase in RNA polymerase activity in the plastid following exposure to light might be explained by the following mechanisms. 1) The specific activity of the enzyme may change upon illumination. 2) Light may induce a rapid de novo synthesis of RNA polymerase molecules in the plastids. 3) Illumination may modify the template.

In order to test these possibilities, the enzyme was solubilized and purified according to Bottomley et al. (8). The specific enzymatic activity and the amount of enzyme per g of leaf tissue were determined for plants which were illuminated for 2 and 16 hr.

After the enzyme of the two samples was solubilized and separated from the bulk of plastid membranes, the enzymatic activities in the supernatant were assayed by adding calf thymus DNA or chloroplast DNA as template. Surprisingly, the "2 hr" and "16 hr" samples no longer exhibited the differences in activity, which were present before the enzyme was solubilized (fig. 2).

The solubilized enzyme was further purified by glycerol density gradient centrifugation and phosphocellulose chromatography (P.C.). SDS polyacryl-

TABLE 1

Specific activity and amount per g of leaf tissue of chloroplast RNA polymerase after 2 and 16 hr illumination.

Fraction	%* of solubilized enzyme activity	%** recovery of the solub. enzymatic activity after P.C. purification	Specific activity*** μmole AMP incorporated for 20 min at 48°C per mg of protein	Amount of enzyme per g of leaf tissue nmole AMP incorporated for 20 min at 48°C
2 hr	93	61	0.207	2.6
2 hr	89	56	0.168	3.0
2 hr	87	52	0.195	1.82
16 hr	76	68	0.155	2.11
16 hr	85	53	0.158	1.39
16 hr	91	55	0.149	1.27

* After EDTA treatment, the plastid suspension was centrifuged and the activities of the supernatant and the pellet were assayed.
** The solubilized enzyme was purified by glycerol density gradient centrifugation and phosphocellulose chromatography. 100% is the activity of the solubilized enzyme before glycerol gradient centrifugation.
*** Specific activity of the P.C.-purified enzyme.

amide gel electrophoresis of the P.C. — purified plastid enzyme of "2 hr" and "16 hr" samples revealed no detectable differences in the polypeptide composition of the two samples. In addition, the specific activity of the P.C. — purified enzyme was approximately the same for both the "2 hr" and the "16 hr" sample (table 1). Thus, the observed difference in RNA polymerase activity between the "2 hr" and "16 hr" plastids cannot be explained by a change in specific activity of the solubilized enzyme.

As for the second possibility mentioned above, the amount of enzyme per g of leaf tissue for both the "2 hr" and "16 hr" plants was approximately the same (table 1). This result suggests that the difference in enzymatic activity in the plastids after 2 and 16 hr illumination was not due to an increased number of RNA polymerase molecules in the "16 hr" sample.

The fact that the differences in activity seen in the two plastid preparations disappeared after the enzyme was solubilized and its activity assayed with an exogenous template and the results shown in table 1 argue against the first two hypotheses, thus suggesting that the third may be correct. Light — induced modifications in the templates of the two different plastid preparations might have caused the difference in RNA polymerase activity. This possibility is currently under investigation.

References

1. J.T.O. Kirk and R.A.E. Tilney-Bassett, The Plastids, W.H. Freeman and Co, San Francisco (1967).
2. K. Apel and H.G. Schweiger, Eur. J. Biochem. 25 (1972) 229.
3. K. Apel and H.G. Schweiger, Eur. J. Biochem. 38 (1973) 373.
4. L.J. Mets and L. Bogorad, Science 174 (1971) 707.
5. L.J. Mets and L. Bogorad, Proc. Nat. Acad. Sci. U.S. 69 (1972) 3779.
6. G.C. Strain, K.P. Mullinix and L. Bogorad, Proc. Nat. Acad. Sci. U.S. 68 (1971) 2647.
7. K.P. Mullinix, G.C. Strain and L. Bogorad, Proc. Nat. Acad. Sci. U.S. 70 (1973) 2386.
8. W. Bottomley, H.J. Smith and L. Bogorad, Proc. Nat. Acad. Sci. U.S. 68 (1971) 2412.
9. L. Bogorad, in T.W. Goodwin (Ed.) Biochemistry of Chloroplasts, Vol. 2, Academic Press, London and New York (1967) p. 615.
10. L. Bogorad, Devel. Biol. Supplement 1 (1967) 1.
11. J. Ingle, Plant Physiol. 43 (1968) 1850.

CHAPTER III.4

REPAIRS OF NUCLEAR AND MITOCHONDRIAL DNAS IN *TETRAHYMENA PYRIFORMIS*

M.S. NETRAWALI, K. PASUPATHY, D.S. PRADHAN and A. SREENIVASAN

Biochemistry and Food Technology Division, Bhabha Atomic Research Centre, Bombay 400 085, India

Repairs of nuclear and mitochondrial DNAs can occur in *Tetrahymena pyriformis* cells exposed to γ-radiation. Both nuclei and mitochondria have been found to be autonomous in DNA repair as evidenced by their ability to rejoin single-strand scissions in their DNAs when incubated in isotonic buffered medium. 2,4-Dinitrophenol and procaine hydrochloride suppressed mitochondrial DNA repair completely but not nuclear DNA repair. These agents seem to exert their effects by the action on mitochondrial membranes rather than by uncoupling of oxidative phosphorylation.

Introduction

As in prokaryotes, DNA is the important and perhaps the principal target responsible for radiation-induced loss of viability in unicellular eukaryotic systems. Radiation injury and repair mechanisms in eukaryotic cells can, however, be expected to exhibit several distinguishing features. For, in eukaryotic cells, the DNA is mainly confined to the discrete nucleus being cloistered there with proteins such as histones. Besides, certain cytoplasmic organelles, especially mitochondria, also contain DNA. We have been studying radiation injury and repair mechanisms in eukaryotic cells using the ciliate *Tetrahymena pyriformis* as a representative organism. We present evidence to show that in this organism both nuclei and mitochondria are capable of rejoining single-strand breaks caused in their DNAs by ionising radiations.

Methods

T. pyriformis (W) was grown in peptone medium containing [^{3}H] thymidine (1 μc/ml; specific activity 6.2 Ci/mmole) to label nuclear and mitochondrial DNAs for 4 days at 25°C. To examine DNA repair in vivo, the cells were harvested, washed in 0.1 M phosphate buffer, pH 7.4, suspended in the same buffer to the density of 6×10^{6} cells/ml and the suspension was exposed to 200 krad (in a Gamma cell-220, Atomic Energy of Canada Ltd.; dose rate 8.04 krad/min) at 0°C. Post-irradiation incubation was carried out by resuspending the cells in fresh growth medium at the density of 6×10^6/ml. Strand breaks in nuclear and mitochondrial DNAs in the cells (with and without post-irradiation incubation) were studied by isolating the organelles and treating them essentially according to the procedure of McGrath and Williams as described for bacterial DNA (1). Nuclei and mitochondria were isolated by the procedures of Mita et al. (2) and Suyama and Preer (3), respectively. The organelles were finally suspended in buffered isotonic solution (0.25 M sucrose, 0.001 M $MgCl_2$, 0.01 M Tris—HCl, pH 7.4) at concentrations corresponding to 50 to 75 μg DNA/ml. Aliquots of suspensions (0.2 ml) were directly layered and lysed with SDS on 5—20% linear sucrose gradient in 10-ml centrifuge tubes and subjected to centrifugation at 125,000 *g* for 3 hr. After centrifugation, 35- 40 equal fractions per tube were collected for determination of radioactivity.

To study DNA repair in vitro, nuclei and mitochondria were isolated from unirradiated cells (pre-labelled with [^{3}H]thymidine), suspended in the buffered isotonic sucrose solution to the concentration corresponding to 50—75 μg DNA/ml, irradiated at 200 krad and resuspended in the fresh sucrose solution for incubation at 25° C for various periods. Sedimentation profiles of DNAs of irradiated organelles (with and without post-irradiation incubation) were obtained as described above.

Results and discussion

Results on radiation effects on nuclear and mitochondrial DNAs and their repairs during post-irradiation incubation are illustrated in fig. 1. The procedure for assessment of radiation-induced damage to nuclear and mitochondrial DNAs involving lysing of the respective organelles on alkaline sucrose density gradient in the manner similar to that adopted for bacterial DNA seems to work quite well. The average sedimentation constant of the nuclear DNA (denatured in alkaline gradient) of unirradiated cells has been found to have a value of 110S corresponding to molecular weight of about 2×10^8 daltons. Nuclear DNA seems to be considerably broken down in this procedure since studies on kinetic complexity indicate that this DNA may have a molecular weight of about 4×10^{10} daltons (4). Sedimentation values of the nuclear DNA fragments nevertheless fall in the range quite suitable for assessing the extent of strand scissions caused by γ-irradiation. It will be apparent that mitochondrial DNA has the average sedimentation value of 45S in this procedure. The S value corresponds well with the molecular weight of 3×10^7 dalton — as calculated from the reported size of this DNA of 17 μ (5,6). This suggests that mitochondrial DNA remains undegraded during sedimentation analysis. As seen from the figure, 200 krad resulted in shifting of average sedimentation value of nuclear DNA from 110 to 54S and of mitochondrial DNA from 45 to 25S. When irradiated cells were incubated in growth promoting medium for 1 hr at 25° C, the S values of both nuclear and

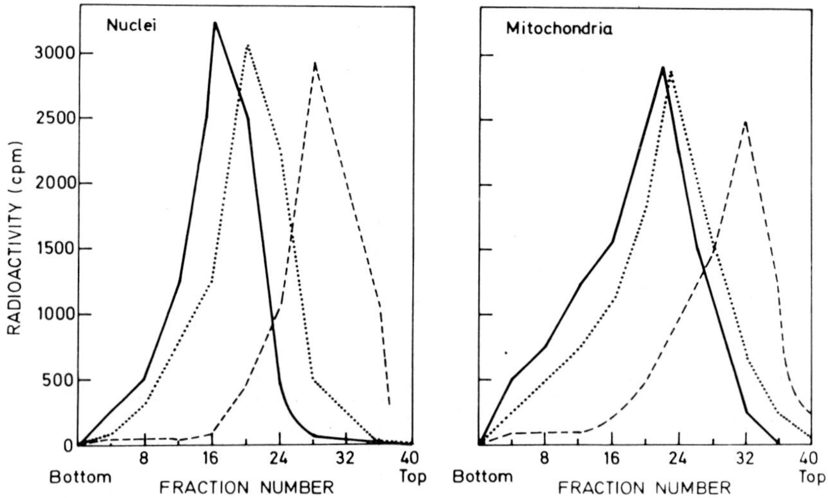

Fig. 1. Repairs of nuclear and mitochondrial DNAs in irradiated cells of *Tetrahymena pyriformis.* Control, ———; irradiated (200 krad), - - - - - -; irradiated (200 krad) and incubated (60 min), ...

mitochondrial DNAs were substantially shifted to those of unirradiated cells indicating restitution of single-strand breaks.

The repairs could conceivably be brought about by independent machine-

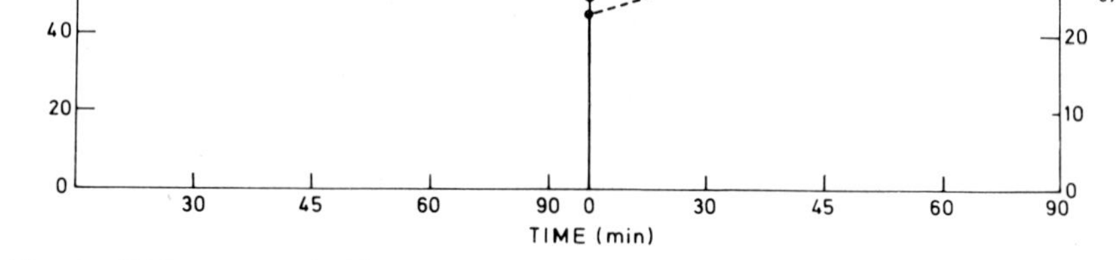

Fig. 2. Efficiency of DNA repair in vivo and in vitro in nuclei and mitochondria of *Tetrahymena pyriformis.* Immediately after irradiation at 200 krad, cells (in vivo experiments) and nuclei and mitochondria (in vitro experiments) were suspended in appropriate media and incubated at 25°C. At the times indicated in the figure, aliquots of respective suspensions were removed for analysing sedimentation behaviours of nuclear and mitochondrial DNAs in alkaline sucrose gradients (3). Average S values were calculated from the DNA sedimentation profiles thus obtained according to the method of Burgi and Hershey (13). In vivo, ○———○; in vitro, ●- - - - - -●.

ries but it is quite likely that some enzymes presumably localised in the cytoplasm could be common to both mitochondrial and nuclear repairs. To clarify some aspects of these repairs as well as their interrelationships, it was important to know whether these organelles could exhibit repair in vitro. Indeed it was observed that nuclei and mitochondria irradiated in vitro and then incubated in isotonic sucrose at 25° C for a period upto 1 hr could rejoin single-strand breaks in their DNAs.

It will be apparent from fig. 2 that repair efficiencies (rate of increase in S values of DNAs in alkaline sucrose gradients) exhibited by either of the organelles are the same in vitro as in vivo further confirming that the repair functions are free of influence from cellular environment external to the organelles.

It would seem that structural integrity of the organelles is absolutely essential since suspension of these in hypotonic conditions failed to bring about repair. 2,4-Dinitrophenol (DNP), which has been suggested to uncouple oxidative phosphorylation in mitochondria through its action on membrane (7), was tried in our repair systems. This agent affects DNA repair in bacteria presumably by interfering with the production of ATP known to be necessary for polynucleotide ligase action (8). At a very low concentration (5×10^{-5} M), DNP was found to inhibit effectively mitochondrial DNA repair but not nuclear DNA repair. Oxidative phosphorylation was supressed only partially at this concentration which would imply that DNP may block mitochondrial DNA repair by a mechanism other than inhibiting production of ATP. Perhaps DNP may exert its action by structurally altering mitochondrial membrane. Procaine hydrochloride, the local anaesthetic believed to act on cellular membrane (9—11), likewise inhibited mitochondrial repair but not nuclear repair (at concentration of 20 mM).

The observed ineffectiveness of the membrane acting agents on nuclear DNA repair could be interpreted as due to non-involvement of membranes in the repair of nuclear DNA. However, it may be mentioned in this connection that the macronucleus of *Tetrahymena* is quite peculiar. The organization of the DNA in the macronucleus has been thought to be either consisting of a population of at least 100 copies of DNA molecules or a collection of 50 diploid sub-nuclei (12). There is no definite information on whether inside the macronucleus, sub-nuclei, i.e. the entities surrounded by membrane, exist. It would therefore seem more likely that the double membrane envelope of the macronucleus being quite distinct from that of mitochondria is either not affected by these agents or that it does not allow the agents to reach the sub-nuclei where membrane-associated DNA repair may be occurring. The indication that nuclear DNA repair may also be associated with nuclear membrane comes from studies with EDTA which seems to be membrane-acting. At the concentration of 1×10^{-3} M, EDTA effectively inhibits both nuclear and mitochondrial repairs in vitro. With the in vitro systems now at hand, it should be possible to uncover similarities and differences between biochemical processes dependent on mitochondrial DNA and those dependent on nuclear DNA.

References

1. R.A. McGrath and R.W. Williams, Nature 212 (1966) 534.
2. J. Mita, M. Shiomi and K. Iwai, Exp. Cell Res. 43 (1966) 696.
3. Y. Suyama and J.R. Preer, Jr., Genetics 52 (1965) 1051.
4. R.A. Flavell and I.G. Jones, Biochem. J. 116 (1970) 155.
5. R.A. Flavell and I.G. Jones, Biochem. J. 116 (1970) 811.
6. Y. Suyama and K. Miura, Proc. Nat. Acad. Sci. U.S. 60 (1968) 235.
7. H.C. Hemker, Biochim. Biophys. Acta 8 (1964) 1.
8. E. Van der Schueren, K.C. Smith and H.S. Kaplan, Radiation Res. 55 (1973) 346.
9. D. Papahadjopoulous, Biochim. Biophys. Acta 265 (1972) 169.
10. H. Hawser, S.A. Penket and D. Chapman, Biochim. Biophys. Acta 183 (1969) 466.
11. P. Seeman, T. Sauks, N. Argent and W.O. Kwant, Biochim. Biophys. Acta 183 (1969) 476.
12. A.M. Elliott, in L. Levine (Ed.) The Cell in Mitosis, Academic Press, New York (1963) p. 107.
13. E. Burgi and A.D. Hershey, Biophys. J. 3 (1963) 309.

Dr. BORST:

In our hands mtDNA from *Tetrahymena* always gives a narrow unimodal band in sucrose gradients, with a variable amount of DNA at the top. We have never observed the broad bands presented here. I wonder therefore if the mtDNA preparations studied by Dr. Netrawali could be heavily contaminated by nuclear DNA. Is there any direct evidence to exclude this possibility?

Dr. NETRAWALI:

It is unlikely that the mitochondrial preparation could be contaminated with nuclear DNA, since this in addition to repeated washings with sucrose, was also treated with DNase according to Gross et al. (J. Biol. Chem., 244 (1969) 1552), prior to sucrose density centrifugation.

Dr. BORST:

Have you measured any changes in the amounts of base damage following γ-irradiation as there is increasing evidence that this base damage is responsible for a high % of biological inactivity?

Dr. NETRAWALI:

No. We have only concerned ourselves with γ-radiation-induced strand breaks in the context of possible repair in *Tetrahymena.*

CHAPTER III.5

STUDIES ON THE DNA DEPENDENT DNA POLYMERASE ACTIVITIES OF *PARAMECIUM*

A. TAIT* and D.J. CUMMINGS

University of Colorado Medical Center, Department of Microbiology, Denver, Colorado 80220, U.S.A.

Three DNA dependent DNA polymerase activities (A, B and C) have been isolated from the macronucleus of *Paramecium*. The enzyme activities can be distinguished by their chromatographic behaviour and the effect of KCl, araCTP and DNA templates on their activities. All three activities have the same molecular weight (90—100,000 daltons).

When the polymerases are isolated from cells at different growth stages, two changes are observed, firstly in the specific activity and secondly in the relative proportions of the three activities. These observations have then been used to deduce the possible functional roles of these activities. Two further distinct polymerase activities have been detected, one associated with the micronuclei and the other with the mitochondria. In the light of these studies, the use of this ciliate as a model for the study of the function of various macro-molecules in gene replication and expression is considered.

Introduction

In the last few years DNA polymerases have been isolated and characterised from a wide range of eukaryotic cells (e.g. 1,2,3), but in many cases investigation of the function of these enzymes has been limited by the properties of the organisms studied. In this context, the ciliate protozoa could be a possibly useful group of organisms to use as a model system in view of their nuclear dualism, the ease with which the physiological conditions can be altered and the lack of differentiation except at the sub-cellular level. DNA polymerase activities have been isolated from *Tetrahymena* (4,5) where a single enzyme species (apart from the mitochondria) was observed on gel filtration of whole cell extracts.

Paramecium contains one macronucleus and two micronuclei per cell; the macronucleus is largely concerned with gene expression and the control of the vegetative cell while the micronuclei are concerned with the maintenance of the genotype and can be considered as analogous to the germ line (6,7). The gross functional differences between the nuclei and the existence of

*Present address: Institute of Animal Genetics, University of Edinburgh, West Mains Road, Edinburgh, Scotland.

temperature sensitive DNA synthesis mutants (8) make this organism a useful system for the study of the function of DNA polymerases in eukaryotic cells. The work reported here constitutes a preliminary study on the feasibility of using this organism, together with some data on the possible role of the DNA polymerases of the macronucleus.

Isolation and purification

Macro and micronuclei were isolated by modifications (9) of the methods of Cummings (10). DNA polymerase activity was extracted from purified macronuclei by extraction with 4 M NaCl, followed by dialysis and centrifugation to remove the precipitated nucleohistone; this method resulted in the extraction of 96% of the activity of the nuclear extract and so is thought to be a good representive sample of macronuclear DNA polymerases. The supernatant of this macronuclear extract was chromatographed on DEAE-cellulose to remove DNA and was eluted as a single peak. This peak was then further purified by phosphocellulose chromatography and was eluted from the column by a salt gradient. In preparations from late log phase cells two peaks of activity A and B were observed (Fig. 1a) which elute at 0.25 M and

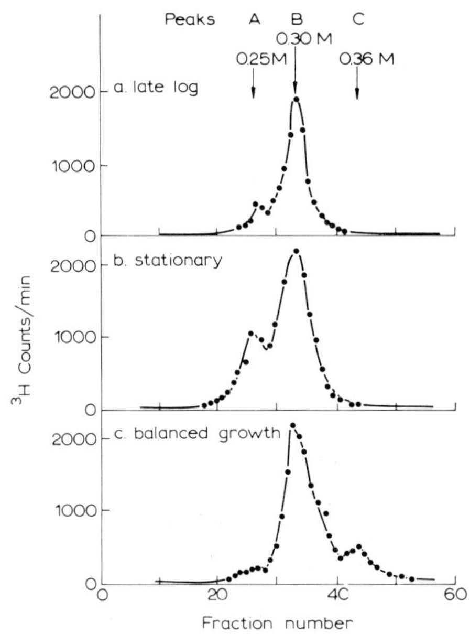

Fig. 1. Phosphocellulose chromatography elution profiles of DNA polymerase prepared from cells at three different growth stages. The column was equilibrated with 0.05 M K phosphate pH = 6.8, 10% glycerol, 1 mM EDTA, 0.5 mM dithiothreitol, 0.15 M KCl and eluted with a linear gradient of KCl (0.15—0.65 M) in the same buffer. 2.4-ml fractions were collected and assayed for DNA polymerase activity as previously described (11).

0.3 M salt respectively. The purification, relative to the nuclear extract was 22-fold for polymerase A and 156-fold for polymerase B (11).

Effect of growth stage

Macronuclear extracts were prepared from cells in three different growth stages namely, balanced growth, late log phase and stationary phase; the specific activities of DNA polymerase were measured and shown to increase 3.1-fold in late log and 9.5-fold in balanced growth relative to stationary phase. The extracts were then purified as described and the elution profiles on phosphocellulose chromatography are shown in fig. 1.

In stationary phase polymerase A increases relative to polymerase B while in balanced growth a third peak of activity (C) is observed and polymerase A is reduced. These three polymerase activities have been shown to be dependent on template, Mg^{2+} and dithiothreitol for activity and to differ in their response to different KCl concentrations, different araCTP concentrations and the physical state of the DNA template. It can be tentatively concluded that polymerase A is associated with a repair function while polymerase C is closely associated with replication. Several mechanisms for these changes can be envisaged, but it must first be established whether these three activities are distinct proteins or are modified forms of a single polymerase. It has been shown (12) in mouse tissues that different polymerases can be interconverted; this involves changes in molecular weight and so with *Paramecium* this possibility can be excluded as the three activities all show identical molecular weights by gel filtration.

Preliminary results show that two further distinct polymerase activities can be isolated from *Paramecium*, one associated with the micronucleus and the other with the mitochondria. Neither of these activities has been purified, but it would clearly be of interest to compare them with the macronuclear enzymes, and to investigate their functional roles.

Discussion

It is clear from these results that this type of study is feasible with *Paramecium* and could be used to elucidate the control and function of DNA polymerases as well as being extended to other macromolecules involved with gene replication and expression. The exact roles of polymerases A, B and C as well as their relationship to each other can be further investigated by examining the effect of altering the physiological conditions (e.g. UV irradiation, age, inhibitors, etc.) as well as examining the temperature sensitive mutants. One important question concerns the relationship of the macronuclear polymerases to those isolated from other eukaryotic cells; it is clear that they are distinct from the mammalian polymerases but show some similarities to both bacterial and yeast DNA polymerases. Although this could be seen as an objection to using *Paramecium* as a eukaryotic model system it is thought that many of the mechanisms for gene replication and expression may well be common to other eukaryotes.

Acknowledgement

This work was supported by National Science Foundation Grant GB 35383.

References

1. L.M.S. Chang and F.J. Bollum, Biochemistry 11 (1972) 1264.
2. N.B. Hecht, Biochim. Biophys. Acta 312 (1973) 471.
3. U. Wintersberger and E. Wintersberger, Eur. J. Biochem. 13 (1970) 11.
4. M. Crerar and R.E. Pearlman, FEBS Letters 18 (1971) 231.
5. O. Westergaard, Biochim. Biophys. Acta 213 (1970) 36.
6. I. Gibson, Adv. Morphog. 8 (1970) 159.
7. T.M. Sonneborn, Methods Cell Physiol. 4 (1970) 241.
8. P.J. Baumann, E.L. Peterson and J.D. Berger, J. Protozool. 20 (1973) 507.
9. D.J. Cummings and A. Tait, Methods Cell Physiol. (in press).
10. D.J. Cummings, J. Cell. Biol. 53 (1972) 105.
11. A. Tait and D.J. Cummings, Biochim. Biophys. Acta 378 (1975) 282.
12. N.B. Hecht, Nature New Biol. 245 (1973) 199.

CHAPTER III.6

DOUBLE STRANDED RNA CONTAINING PARTICLES FROM THE YEAST *SACCHAROMYCES CEREVISIAE* AND THEIR RELATIONSHIP TO THE 'KILLER' CHARACTER

A.J. HERRING and E.A. BEVAN

Department of Plant Biology and Microbiology, Queen Mary College, Mile End Road, London E1 4NS, U.K.

The cytoplasmically inherited killer character in yeast has been found in genetic and mutational studies to be completely correlated with the possession by killer cells of two species of double stranded RNA of molecular weights 2.5×10^6 and 1.4×10^6 daltons. These dsRNAs are contained in isometric virus-like particles and the two species are separately encapsidated. Sensitive strains of yeast may possess double stranded RNA but, with a very few exceptions, they possess only the larger of the two molecules.

Killer yeast cells release a protein toxin which, under stringent conditions of temperature and pH, is able to kill cells of sensitive strains (1). It has been found that the toxin is very heterogeneous in molecular weight, presumably due to a sub-unit structure; the range being from 2.0×10^5 to 2.0×10^6 daltons (2). In addition to the killer and sensitive phenotypes some strains are of a third type, called neutral; these strains do not show killing but are not themselves killed.

Both killer and neutral phenotypes are cytoplasmically inherited and give 4 : 0 killer or neutral to sensitive segregation ratios in appropriate crosses (3). However, three nuclear genes affect the maintenance of the killer and neutral determinants: the *M/m* gene (3) cells must possess the *M* allele to be killers; the *pet.s* allele, a segregational petite allele with the additional property of rendering the cell a sensitive, and the ur_3 allele which has the effect of making the killer character unstable. Killer by neutral crosses may yield weak killer diploids which may segregate neutral colonies mitotically (4).

The *pet.s* phenotype suggests a connection between the killer character and mitochondrial genome but no other evidence has been found to substantiate such a relationship. Indeed, it has been shown (5,6) that petite cells, which entirely lack any detectable mitochondrial DNA and which show zero suppressiveness in crosses with grandes, fully retain the killer character.

The realisation that the killer character could not be mitochondrially coded led to the search for an extrachromosomal nucleic acid which might code for the character. This search failed to reveal any difference in the DNA constituents of killer and sensitive cells but led to the discovery of high molecular weight double stranded RNA (dsRNA) in killer yeast cells (7).

The isolation of dsRNAs from a variety of strains showed that both killer and neutral strains possess two species of dsRNA, of molecular weight 2.5×10^6 and 1.4×10^6 daltons but that sensitive strains either lack dsRNA entirely or possess only the larger molecule. A few sensitive strains have been found which have two dsRNA species but in all cases the molecular weight of the smaller molecule has been found to be significantly smaller than the 1.4×10^6 daltons species found in the killer strains. In all strains possessing dsRNA the larger dsRNA species appears to be identical in molecular weight (8).

This finding naturally led to the hypothesis that the dsRNAs constitute the genetic determinant in killer cells. Consequently, an investigation was made of the dsRNA of colonies formed under conditions where the killer and sensitive phenotypes were segregating from one another. The situations examined were meiotic segregation under the influence of the *m* allele, mitotic segregation under the influence of the ur_3 gene and spontaneous sensitive mutation which occurs at very low frequency in growing killer strains. Additionally, situations under which high proportions of sensitive cells are segregated from killer strains were investigated. These conditions are growth in the presence of partially inhibitory concentrations of 5′-fluorouracil or cycloheximide or growth at high temperature. In all cases sensitive clones segregated from killer parents were found to lack the smaller dsRNA species or, exceptionally, to lack dsRNA altogether (3,8,9). This finding leads us to conclude that the killer character is indeed determined by the two dsRNA species which appear to constitute a novel cytoplasmic genetic system. Clearly, the finding that the smaller dsRNA is coordinately lost with killer function implicates this molecule as possibly coding for the killer toxin or for some factor which causes the cell to release the toxin and also, as loss of this molecule confers sensitivity, it is also implicated in conferring immunity to killing.

High molecular weight dsRNA has been found in a variety of species of fungi (10—12) and in all cases has been associated with the presence of isometric "virus" particles. These particles are termed "virus" mainly by virtue of their structure and it is controversial whether any of them fulfill all of Koch's postulates; the criterion of infection being the major stumbling block. All of these fungal dsRNA viruses so far described have been found to be of the multi-component type; that is at least two and as many as five different size dsRNA molecules are found to be separately encapsidated in one type of capsid as judged by the size, electrophoretic mobility and serological properties of the capsid (12).

Yeast dsRNA proved to be no exception. It has been shown (13), by sucrose density gradient analysis, that the dsRNA present in a post-mitochondrial supernatant from disrupted yeast cells was associated with a particle sedimenting at approximately 150S. Examination of the dsRNA containing fractions from the gradient revealed virus particles of isometric shape with a diameter of 39 nm. These particles, isolated from a killer strain, may be seen in fig. 1.

Recently, in this laboratory, we have developed techniques for the bulk purification of these virus-like particles (VLPs) and these will be reported in detail elsewhere. Using these purified preparations, we have been able to answer some of the questions which arise naturally from the findings with other dsRNA VLPs. While the situation in the sensitive strains, which have only a single dsRNA component, is unique, the killer strains, with two dsRNA components, resemble the other fungal VLP systems and, as the data in fig. 2 show, these two dsRNA species are also separately encapsidated. Fig. 2 shows the profile of a sucrose gradient loaded with a crude VLP preparation from a killer strain; three peaks may be seen on the gradient, labelled a, b and c. Polyacrylamide gel electrophoresis of the dsRNA content of the individual fractions has shown that peak *a* contains no dsRNA and consists of empty capsids, peak *b* consists of VLPs containing the smaller dsRNA species and peak *c* consists of VLPs containing the larger dsRNA species.

Only one size class of particle is found in both killer and sensitive strains. However, in other fungi, notably *Aspergillus foetidus* and *Penicillium stoloniferum*, two types of particle are found in the same strain which are recognised by their differing electrophoretic and serological properties but are indistinguishable by the criterion of size (12). Thus, it was possible that the particles found in killer strains were of two different types. To investigate this possibility we have analysed the constituent polypeptides of the two types of VLP capsid by SDS-polyacrylamide gel electrophoresis. The results may be seen in fig. 3, which shows the absorbance profile of two Coomassie blue stained gels loaded with SDS dissociated protein from VLPs containing the smaller dsRNA molecule labelled P2, and the larger dsRNA molecule labelled P1. The two types of particle were prepared by sucrose gradient centrifugation as seen in fig. 2. The gels are both dominated by a single band, the VLP capsid polypeptide which has an identical mobility, and hence molecular weight, for each particle type. There are several minor bands on the gels but these, too, show an identical pattern.

This analysis thus reveals no difference between the two types of particle and this conclusion has been borne out by a recent series of serological experiments in which it was found that antibody raised against capsid protein isolated from a sensitive strain with only one type of particle reacted to both types of particle from the killer strain (unpublished results).

These findings lead us to conclude that both types of particle capsid and, in all probability, the enzyme(s) necessary for the replication of the dsRNA are coded by the larger dsRNA molecule. The smaller dsRNA molecule is replicated as a "satellite" of the larger species and its gene product(s) confers

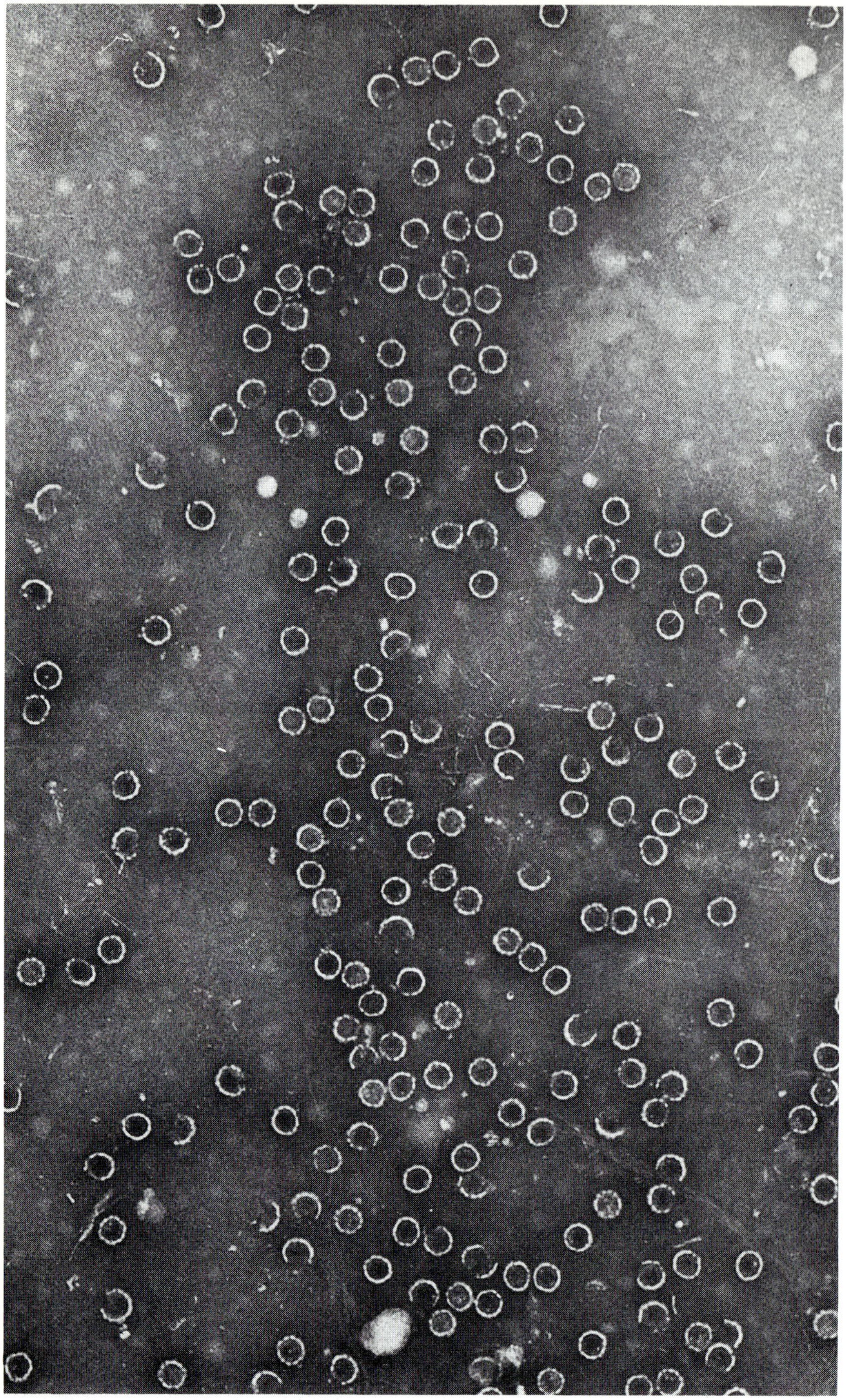

Fig. 1. Yeast virus like particles isolated from the killer strain 4/4D.

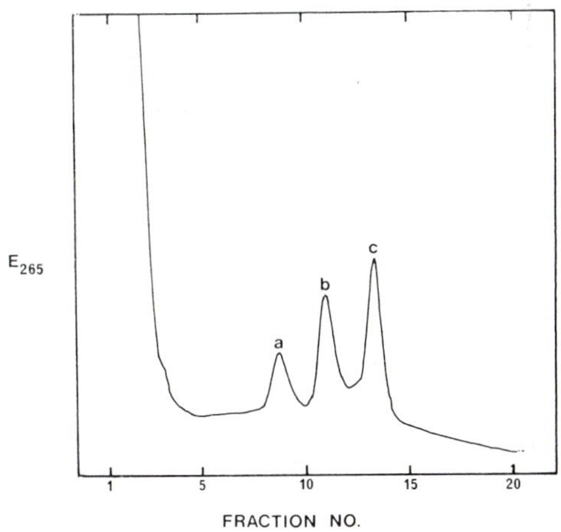

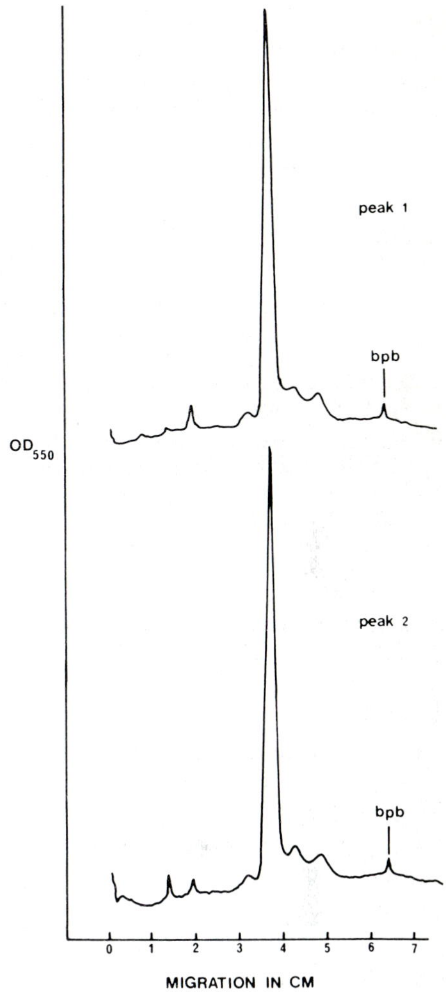

Fig. 2. The OD_{265} profile of a 10—40% sucrose gradient loaded with a crude preparation of virus like particles from the killer strain 4/4D. See text for explanation of peaks labelled a, b and c.

Fig. 3. The OD_{550} profile of two SDS-polyacrylamide gels stained with Coomassie blue and loaded with protein from virus like particles containing the larger dsRNA species (labelled peak 1) and the smaller dsRNA species (labelled peak 2).

the properties of killing and immunity on the cell. Two lines of thought give credence to the idea that the smaller dsRNA molecule may code directly for the killer toxin. Firstly, many of the bacteriocins and also the killer toxins released by *Paramecium* are thought to be the result of "defective" bacteriophage infection of the bacterial cells or, in the case of *Paramecium*, of the bacterial endosymbionts, hence other toxins are thought to be viral products. Secondly, the ability of the protein toxin to polymerise to give such a large active molecule with, it must be assumed, some affinity for the yeast cell wall is a property which would be expected of the capsid protein of an infective yeast virus, ancestral to the present day VLP.

References

1. D.R. Woods and E.A. Bevan, J. Gen. Microbiol. 51 (1968) 115.
2. H. Bussey, Nature New Biol. 235 (1972) 73.
3. J.M. Somers and E.A. Bevan, Genet. Res. 13 (1969) 71.
4. E.A. Bevan and J.M. Somers, Genet. Res. 14 (1969) 71.
5. G.R. Fink and C.A. Styles, Proc. Nat. Acad. Sci. U.S. 69 (1972) 2846.
6. K. Al-Aidroos, H. Bussey and J.M. Somers, Mol. Gen. Genet. 122 (1973) 323.
7. E.A. Berry and E.A. Bevan, Nature 239 (1972) 279.
8. E.A. Bevan, A.J. Herring and D.J. Mitchell, Nature 245 (1973) 81.
9. D.J. Mitchell, E.A. Bevan and A.J. Herring, Heredity 31 (1973) 133.
10. R.F. Bozarth, Environ. Health Perspectives 128 (1972) 23.
11. P.A. Lemke and C.H. Nash, Bacteriol. Rev. 38 (1974) 29.
12. H.A. Woods, J. Gen. Virol. 20 (1973) 61.
13. A.J. Herring and E.A. Bevan, J. Gen. Virol. 22 (1974) 387.

Dr. WILKIE:

What is the *M* gene doing in the replication system of the killer particle?

Dr. HERRING:

The *M* allele is dominant in diploids. It thus appears that the *M* gene product is necessary for the continuence of the satellite relationship between the two types of dsRNA genome; that is for the encapsidation, expression and replication of the smaller dsRNA molecule by the gene products of the larger dsRNA molecule.

Dr. WILLIAMSON:

Have you any ideas about how cells carrying the killer virus manage to survive?

Dr. HERRING:

With regard to the second question I think that it is reasonably clear from the text that the "virus" has not got a cycle of vegetative growth or infection that has yet been demonstrated but is maintained at controlled levels by the cells.

SECTION IV

INTERGENOMIC COOPERATION IN SPECIFYING MEMBRANE CHARACTERISTICS

CHAPTER IV.1

THE BIOSYNTHESIS OF CYTOCHROME *c* OXIDASE IN BAKER'S YEAST

G. SCHATZ

Biocenter, University of Basel, Switzerland

Cytochrome *c* oxidase from *Saccharomyces cerevisiae* consists of seven polypeptides ranging in molecular weight from 40,000 to less than 5000. The three large polypeptides are synthesized on mitochondrial ribosomes and the four small ones on cytoplasmic ribosomes. This paper describes the chemical properties, the three-dimensional arrangement and the genetic control of these subunits.

Introduction

Until a few years ago most studies on the biogenesis of mitochondria centered around the properties and the synthesis of mitochondrial nucleic acids (1). This probably reflected the fact that adequate methods were available to deal with these molecules. More recently, however, increasing attention is being devoted to the question of how the various proteins of the mitochondrial inner membrane are combined with each other to form a functional unit (2). This question is intriguing since the mitochondrial inner membrane is made up of two types of proteins: those made on extramitochondrial (cytoplasmic) ribosomes and those made on intramitochondrial ribosomes. Although the latter class of proteins constitutes no more than 10 to 15% of the total proteins of the mitochondrial inner membrane it is indispensable for the functioning of that membrane (2).

Results from several different laboratories indicate that mitochondria synthesize roughly one dozen hydrophobic polypeptides which are tightly associated with the mitochondrial inner membrane (3—6). Three of these proteins are associated with cytochrome *c* oxidase (7—9), four with the oligomycin-sensitive ATPase complex (8) and one with cytochrome *b* (10). The remaining few polypeptides have not yet been identified.

The studies outlined below are based on the premise that the biosynthesis of cytochrome *c* oxidase reflects in many ways the biosynthesis of the inner membrane as a whole. For example, both assembly processes require a concerted action of mitochondrial and cytoplasmic protein synthesis and are dependent on the presence of oxygen. It is also relevant that cytochrome *c* oxidase is not merely *associated* with the mitochondrial inner membrane but is actually an intrinsic component of that membrane; the enzyme can only be solubilized under conditions which disrupt membrane structure.

Polypeptide composition of cytochrome *c* oxidase from baker's yeast

Cytochrome *c* oxidase can be solubilized from isolated yeast mitochondria with bile salts and purified by repeated ammonium sulfate fractionation and DEAE-cellulose chromatography to a heme content of 10 to 14 nmoles per mg of protein (7, 8). Upon SDS-acrylamide gel electrophoresis the purified enzyme can be resolved into seven polypeptide bands ranging in molecular weight from 40,000 to less than 5000 (7, 11, 12) (table 1). Labeling studies in the presence of specific inhibitors of either cytoplasmic or mitochondrial protein synthesis have established that the three large polypeptides are synthesized on mitochondrial ribosomes whereas the four small polypeptides are synthesized on cytoplasmic ribosomes (7, 8). Moreover, the synthesis (or the integration) of the two largest polypeptides is dependent on the presence of oxygen (7).

Properties and function of the individual cytochrome *c* oxidase polypeptides

R. Poyton in our laboratory has recently isolated each of the seven polypeptides by methods which exploit differences in polypeptide charge, solubility and size (11). The critical step in this procedure was the use of guanidine HCl to partially dissociate the enzyme. If yeast cytochrome *c* oxidase is exposed to 6 M guanidine HCl and reducing agent at room temperature,

TABLE 1

Properties of polypeptides associated with cytochrome *c* oxidase from *Saccharomyces cerevisiae*.

Polypeptide No.	Mol. weight (daltons)	Polarity index*	Site of synthesis
I	40,000	35	mitochondria
II	33,000	42	mitochondria
III	22,000	not determined	mitochondria
IV	14,500	48	cytoplasm
V	12,700	50	cytoplasm
VI	12,000	51	cytoplasm
VII	4,500	41	cytoplasm

* Defined as mole percent of hydrophilic amino acids Lys, His, Arg. Asp, Thr, Ser, Glu (11a).

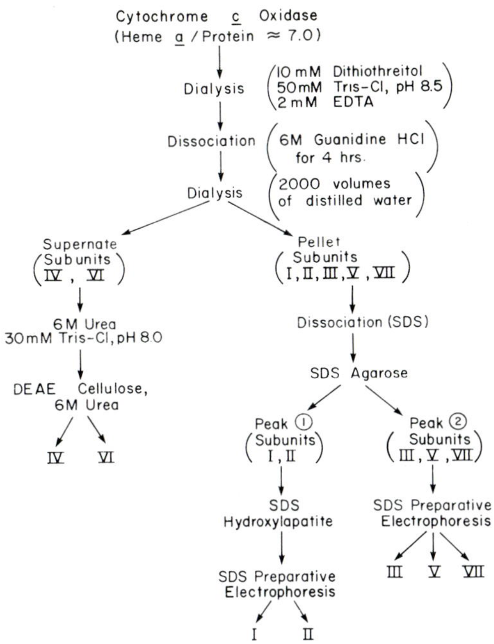

Fig. 1. Purification scheme for the subunits of yeast cytochrome *c* oxidase. The enzyme used as starting material is only partially purified.

subunits IV and VI (cf. table 1) are selectively removed from an aggregate of the other five subunits and can be further purified by conventional methods (fig. 1). Exposure of the remaining enzyme "core" to higher concentrations of guanidine HCl or to formic acid completely dissociates all subunits, permitting their isolation in large amounts. On a smaller scale, the individual subunits can also be separated from each other by SDS-polyacrylamide gel electrophoresis and subsequent elution from the polyacrylamide gels.

The amino acid composition of six of the seven polypeptides has been determined (table 1). The results indicate that the two large subunits are quite hydrophobic whereas the cytoplasmically-synthesized subunits IV, V and VI exhibit polarity indices of around 50% which is typical of most water-soluble proteins. Indeed, subunits IV and VI are water-soluble in the absence of any detergent (11). Subunit VII has a rather low polarity index but the significance of this parameter may be questioned because of the extremely small size of this polypeptide. In collaboration with A. Tsugita we are now characterizing each of these polypeptides in more detail. Work on the amino acid sequence on subunit VI is nearly completed.

In an attempt to gain further insight into the function of each of these seven subunits we prepared antibodies against them and tested the effects of these antibodies on the enzymic activity of solubilized and membrane-bound

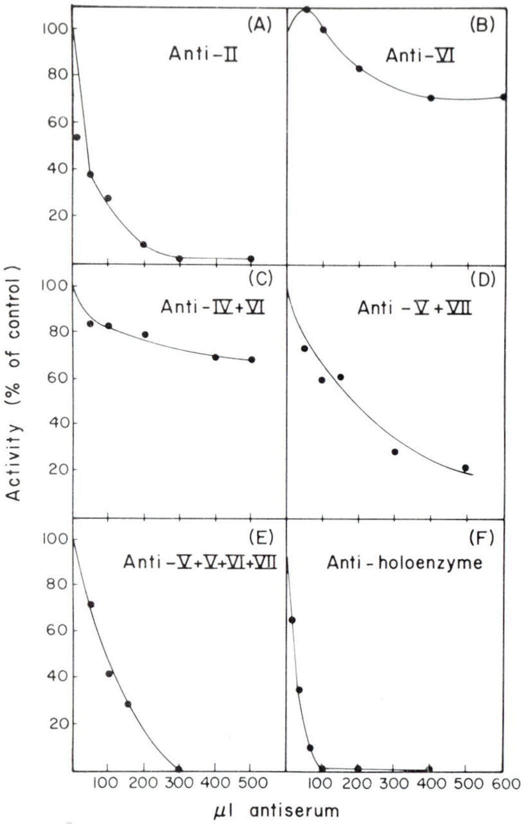

Fig. 2. Inhibition of purified cytochrome *c* oxidase by subunit-specific antisera. Purified cytochrome *c* oxidase (8.5 nmoles of heme *a* per mg protein) was incubated for 10 min at 23°C with the antisera listed in the various subpanels and assayed for cytochrome *c* oxidase activity.

cytochrome *c* oxidase (fig. 2). We found that each of these subunit-specific antibodies could inhibit the enzyme (13). It is particularly noteworthy that an antibody against the mitochondrially-made polypeptide II inhibits the enzyme almost completely. Although by no means conclusive, this result strongly implies that this polypeptide is not merely a tightly-bound reproducible contaminant, but a bona fide component of cytochrome *c* oxidase. The combined results presented in fig. 2 suggest, therefore, that the activity of cytochrome *c* oxidase involves both cytoplasmically and mitochondrially-made polypeptides.

Three-dimensional arrangement of the cytochrome *c* oxidase polypeptides

In order to study the arrangement of the seven polypeptides in the active holoenzyme the reactivity of each polypeptide with a variety of "surface

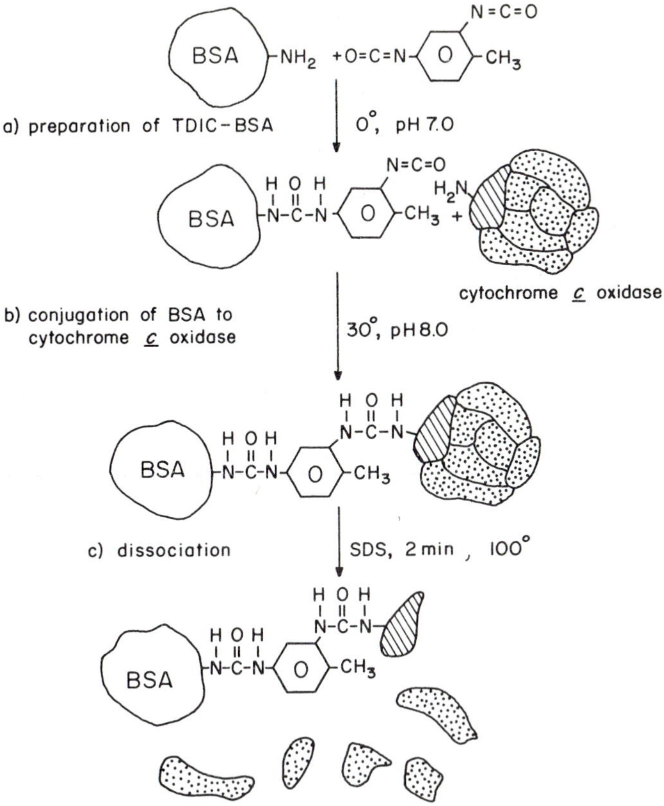

Fig. 3. Tolene-diisocyanate-bovine serum albumin (TDIC-BSA) as a surface probe for cytochrome *c* oxidase. The arrangement of the cytochrome *c* oxidase subunits in this schematic drawing is not intended to reflect the actual arrangement.

probes" was tested with isolated cytochrome *c* oxidase, with cytochrome *c* oxidase incorporated into liposomes and with mitochondrially-bound cytochrome *c* oxidase (14). The surface probes included iodination with lactoperoxidase and coupling with p-diazonium benzene sulfonate. In addition, externally located subunits were identified by linking them to bovine serum albumin which carried a covalently bound isocyanate group. Since this last-mentioned method has not yet been employed for this purpose it is briefly outlined in fig. 3.

The results of the various methods are summarized in fig. 4. In this figure, the height of each bar denotes the accessibility of a given subunit to the probe employed. It is evident that subunit I is almost completely inaccessible to all surface probes. Subunit II is also largely inaccessible. All other subunits are accessible. (We do not yet know why subunit IV is available to enzymic iodination in the membrane-bound enzyme but not in the solubilized one.

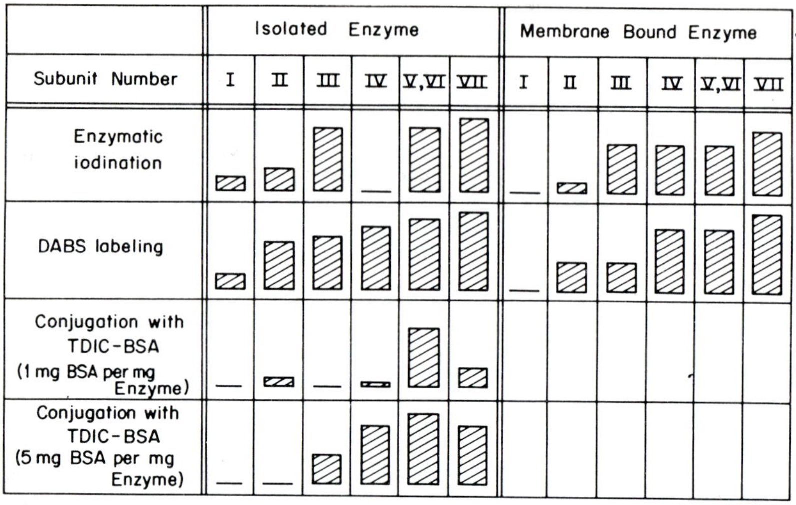

Fig. 4. Reactivity of the different cytochrome *c* oxidase polypeptides towards TDIC-BSA (cf. fig. 3), enzymic iodination and ^{35}S-diazonium benzene sulfonate (DABS).

Perhaps removal of the enzyme from the membrane causes subtle conformational changes which do not abolish enzymic activity but render the tyrosyl residues of subunit IV inaccessible to lactoperoxidase.)

In summary, the seven subunits of cytochrome *c* oxidase from baker's yeast appear to be arranged so that the two largest mitochondrially-made subunits are internal whereas most or all of the cytoplasmically-made subunits are external. The information on the arrangement of the individual polypeptides should be useful for understanding the function and the assembly of the enzyme.

Effect of oxygen on the accumulation of mitochondrially-made cytochrome *c* oxidase polypeptides

If aerobically grown yeast cells are suspended in phosphate buffer, poisoned with cycloheximide and labeled with radioactive amino acids in the absence of oxygen, the two largest polypeptides associated with the cytochrome oxidase are not labeled (7). However, the relevance of this experiment may be questioned since it involves short-term labeling under "shift-down" conditions during which substantial pools of individual cytochrome *c* oxidase subunits may preexist in the cells. What is the effect of oxygen during long-term experiments in which yeast cells are grown for many generations in the absence of oxygen? An experiment designed to answer this question is outlined in fig. 5. Yeast cells were grown anaerobically for 24 hr, suspended in buffered glucose containing cycloheximide and ^{3}H-leucine and

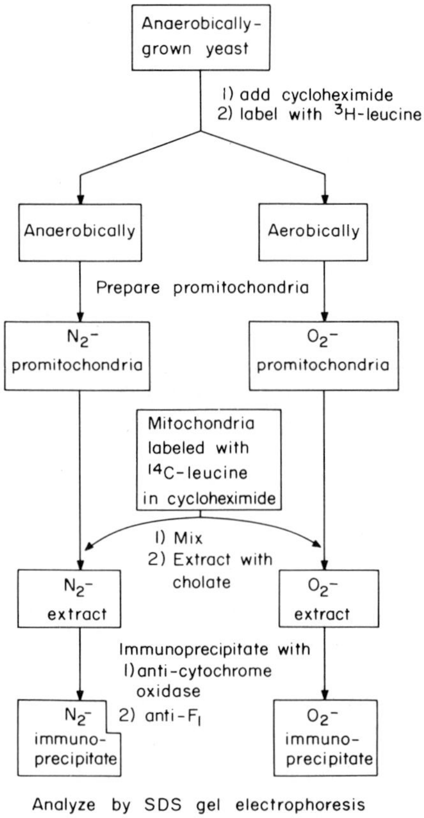

Fig. 5. Experimental approach for studying the effect of oxygen on the synthesis of cytochrome *c* oxidase subunits.

divided into two aliquots. One aliquot was incubated anaerobically and the other aerobically. The promitochondria were then isolated and mixed with mitochondria from yeast cells which had been grown aerobically in the presence of ^{14}C-leucine. The mixtures were extracted with cholate and KCl (15) and subjected to immunoprecipitation with an antiserum against cytochrome oxidase or an antiserum against mitochondrial ATPase. The immunoprecipitates were then analyzed for ^{14}C and ^{3}H (fig. 6). It can be seen that the presence of oxygen during labeling of the anaerobically-grown cells dramatically increased the percentage of promitochondrially-made polypeptides recovered in the immunoprecipitated cytochrome oxidase. In contrast, oxygen had little, if any effect on the recovery of promitochondrially-made polypeptides associated with the oligomycin-sensitive ATPase complex. This agreed with the well-known fact that the synthesis of functional cytochrome

Antiserum against	% of promitochondrial radioactivity recovered by immunoprecipitation		Ratio
	Anaerobic labeling	Aerobic labeling	
Cytochrome oxidase	0.56	7.67	13.7
F_1 - ATPase	2.39	3.34	1.4

Fig. 6. Effect of oxygen on the labeling of promitochondrially-made polypeptides associated with cytochrome *c* oxidase and the oligomycin-sensitive ATPase complex. See fig. 5 for details.

c oxidase is absolutely dependent on oxygen whereas the synthesis of oligomycin-sensitive ATPase is only little affected by oxygen.

When the immunoprecipitate obtained with anti-cytochrome *c* oxidase from anaerobically-labeled promitochondria was analyzed by SDS-acrylamide gel electrophoresis (fig. 7), none of the three mitochondrially-made cytochrome *c* oxidase polypeptides contained any significant radioactivity.

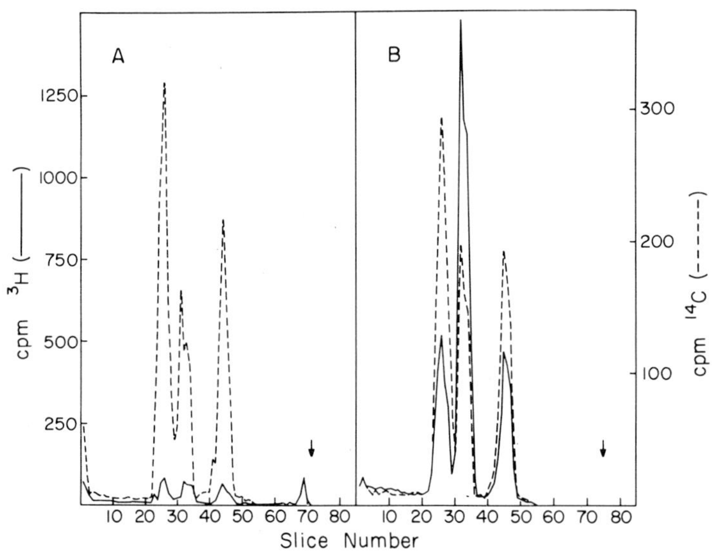

Fig. 7. Analysis of immunoprecipitated cytochrome *c* oxidase polypeptides (obtained as described in fig. 5). A) Anaerobic labeling; B) Aerobic labeling. ———: promitochondrially-made polypeptides (^{3}H); - - - - - -: polypeptides made by fully developed mitochondria in the presence of oxygen (^{14}C).

In contrast, the corresponding immunoprecipitate from the aerobic incubation revealed significant labeling of all three large subunits with subunit II being the one most highly labeled. This "shift-up" experiment may allow us to delineate the sequence of events in the assembly of the functional enzyme.

Analysis of cytochrome *c* oxidase-less yeast mutants

Several years ago E. Ebner isolated many cytochrome *c* oxidase-less yeast mutants (16) and analyzed three of these mutants for residual cytochrome *c* oxidase subunits by immunodiffusion and electrophoretic analysis of radioactively labeled immunoprecipitates (15). All of these mutants (pet 494-1, pet 1030-1 and pet E11-1) were nuclear mutants which lacked only cytochrome *c* oxidase. Mutants pet 1030-1 and pet E11-1 lacked all three mitochondrially-made polypeptides whereas pet 494-1 lacked only the smallest of these three polypeptides. All three mutants appeared to contain virtually normal levels of the cytoplasmically-made cytochrome *c* oxidase subunits, although the analytical techniques employed would not have detected minor modifications of these subunits.

How can the mutation of a single nuclear gene cause the loss of one or more polypeptides made on mitochondrial ribosomes? One explanation would be that the messenger RNAs for the missing subunits are transcribed from a single nuclear gene and translated on mitochondrial ribosomes. On the other hand, the three mutants could be regulatory mutants in which the loss of the mitochondrially-made cytochrome *c* oxidase subunit(s) is the indirect result of the loss or the alteration of a nuclearly-coded cytoplasmically-made protein.

Recent data by B. Ono (17) strongly favor the second alternative. These studies were prompted by the finding that the pet 494-1 phenotype could be suppressed by nuclear amber suppressors. This result not only identified pet *494-1* as an amber marker, but also provided a means for identifying the protein which was the primary target of the pet *494-1* mutation. There is now good evidence that nonsense suppression in bacteria (18) and very probably also in yeast (19) involves altered tRNAs. Also, the mitochondrial genetic system appears to use tRNAs which are different from those used in cytoplasm (1). Thus, any suppressor acting on one of the two genetic systems should not affect the other. In other words, if a mutation in a eukaryote can be suppressed by nuclearly-coded nonsense suppressors one may conclude with reasonable certainty that the protein primarily affected by the mutation is translated in the cytoplasm.

These suppression studies were done in two different ways. a) By constructing pet 494-1 strains containing various amber and ochre markers, isolating respiring revertants from these strains, and demonstrating coreversion of the amber (but not of the ochre) markers. This approach is outlined

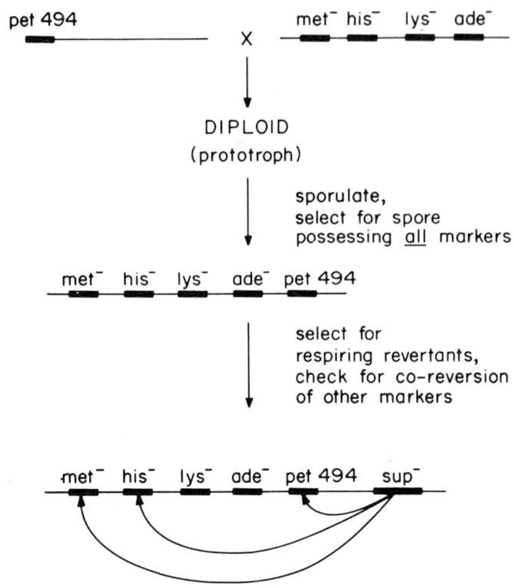

Fig. 8. Experimental protocol for identifying the pet *494-1* allele as an amber marker. *Met 8-1* and *his 4-580* are amber markers and *lys 1-1* and *ade 2-1* are ochre markers.

in fig. 8. b) By coupling the pet *494-1* allele with a well-characterized amber suppressor gene (20) located on a nuclear chromosome.

The suppressed pet 494-1 strains were virtually indistinguishable from the corresponding wild-type with respect to respiratory ability, cytochrome content, and intramitochondrial levels of cytochrome *c* oxidase subunit III. We conclude, therefore, that the pet *494* locus very probably does not code for the missing mitochondrially-made subunit III but for a cytoplasmically-made protein. This as yet unidentified protein seems to control the synthesis or the integration of the mitochondrially-made subunit. This model is schematically illustrated in fig. 9. It is tempting to speculate that the polypeptide X

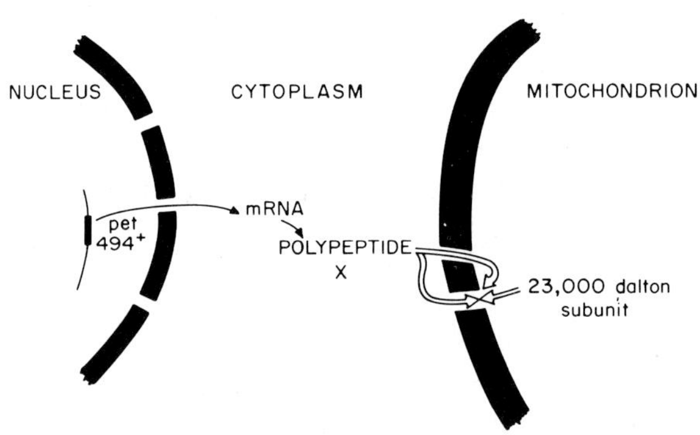

Fig. 9. Possible mechanism of the pet *494-1* mutation.

of fig. 9 is in fact one of the four cytoplasmically-made subunits of cytochrome *c* oxidase. While earlier analyses of these four polypeptides by SDS-polyacrylamide gel electrophoresis suggested that they were still present in mutant pet 494-1, a slight alteration of size or charge of one of these polypeptides would have gone undetected. We are now attempting to characterize the cytoplasmically-made subunits from mutant pet 494-1 by isoelectric focusing and by searching for alterations of these subunits in intragenic revertants.

Conclusion

The foregoing discussion shows that the seven polypeptides of cytochrome oxidase constitute a small world which mirrors in many ways the events in the larger arena of mitochondrial biogenesis. Needless to say, much remains to be done. First, we would like to know as much as possible about the chemistry of the individual polypeptides since we believe that this knowledge will be indispensable for a thorough understanding of how the enzyme functions and how it is assembled in vivo. Second, the potential of subunit-specific antisera which are now available has yet to be realized. Third, it is imperative that the heme-carrying subunits be unequivocally identified. Fourth, we would like to know how oxygen controls the assembly of the functional enzyme and which subunits are missing in the anaerobically-grown cells. Finally, we are still confronted with the formidable yet necessary task of reconstituting the active enzyme from its individual components.

Acknowledgements

These studies were supported by grants from the U.S. Public Health Service (GM 16320 and GM 18197), from the U.S. National Science Foundation (GB-40541X), from the U.S. Department of Agriculture (NY(C) 181406) and the Swiss National Science Foundation (3.2350.74).

References

1. P. Borst, Ann. Rev. Biochem. 41 (1972) 333.
2. G. Schatz and T.L. Mason, Ann. Rev. Biochem. 43 (1974) 51.
3. W. Sebald, T. Hofstötter, D. Hacker and T. Bücher, FEBS Letters 2 (1969) 177.
4. G.S.P. Groot, W. Rouslin and G. Schatz, J. Biol. Chem. 247 (1972) 1735.
5. J.L. Coote and T.S. Work, Eur. J. Biochem. 23 (1971) 564.
6. P.O. Weislogel and R.A. Butow, J. Biol. Chem. 246 (1971) 5113.
7. T.L. Mason and G. Schatz, J. Biol. Chem. 248 (1973) 1355.
8. M.S. Rubin and A. Tzagoloff, J. Biol. Chem. 248 (1973) 4269.

9. W. Sebald, H. Weiss and G. Jackl, Eur. J. Biochem. 30 (1972) 413.
10. H. Weiss, Eur. J. Biochem. 30 (1972) 469.
11. R.O. Poyton and G. Schatz, J. Biol. Chem. 250 (1975) 752.
11a. R.A. Capaldi and G. Vanderkooi, Proc. Nat. Acad. Sci. U.S. 69 (1972) 930.
12. G. Schatz, G.S.P. Groot, T.L. Mason, W. Rouslin, D.C. Wharton and J. Saltzgaber, Fed. Proc. 31 (1972) 21.
13. R.O. Poyton and G. Schatz, J. Biol. Chem. 250 (1975) 762.
14. G. Eytan and G. Schatz, J. Biol. Chem. 250 (1975) 767.
15. E. Ebner, T.L. Mason and G. Schatz, J. Biol. Chem. 248 (1973) 5369.
16. E. Ebner, L. Mennucci and G. Schatz, J. Biol. Chem. 248 (1973) 5360.
17. B. Ono, G.R. Fink and G. Schatz, J. Biol. Chem. 250 (1975) 775.
18. H.M. Goodman, J. Abelson, A. Landy, S. Brenner and D.J. Smith, Nature 217 (1968) 1019.
19. F. Sherman, S.W. Liebman, J.W. Stewart and M. Jackson, J. Mol. Biol. 78 (1973) 157.
20. F. Rasse-Messenguy and G.R. Fink, Genetics 75 (1973) 459.

Dr. MOUNOLOU:
What is the stoichiometry of iodine binding to cytochrome *c* oxidase?

Dr. SCHATZ:
At very high iodine levels, enzymic iodination may result in the incorporation of at least 5 iodine atoms per molecule of heme *a*. However, these data are not very meaningful since at high iodine concentrations free iodine is formed which may diffuse to buried subunits. Under the usual conditions of enzymatic iodination, less than 1% of the enzyme molecules are iodinated. This is a serious drawback of the enzymic iodination procedure.

Dr. MOUNOLOU:
How does iodination affect enzymic activity?

Dr. SCHATZ:
Even at high iodine levels we do not observe any decrease of enzymic activity.

Dr. APEL:
How do you define an enzyme "subunit"?

Dr. SCHATZ:
As with any membrane enzyme, this question cannot be answered in a clear-cut fashion since one has decide "where the enzyme ends and the membrane begins". Also, a given polypeptide may not be required for the oxidation of ferrocytochrome *c* in vitro, but may be essential for coupled electron transfer at site III in vivo. At present, we only argue that the seven polypeptides we find in our preparations are not merely contaminants. The

main reasons are as follows: a) They are present in constant proportions in cytochrome *c* oxidase preparations isolated from *S. cerevisiae* and *Neurospora crassa* in different laboratories. b) Antibodies against one or more of the small subunits precipitate all seven subunits from crude mitochondrial lysates. c) Antibodies against subunit II or against subunit VI, or against combinations of the cytoplasmically-made subunits, inhibit the activity of the solubilized enzyme. d) A mutant (pet 494-1) specifically lacking cytochrome *c* oxidase also lacks specifically subunit III. Of course, the final identification of an enzyme subunit requires reconstitution of the active enzyme from its components. This has not yet been done.

Dr. RAISONNIER:

Which subunits carry the heme and or the copper?

Dr. SCHATZ:

We have weak evidence suggesting that heme *a* is bound to one or more of the small subunits. No information is as yet available on the copper-binding subunits.

Dr. HOOBER:

Is polypeptide III, the one affected by mutant pet 494-1, one of the polypeptides whose synthesis requires oxygen?

Dr. SCHATZ:

Yes and no. During long-term growth in the absence of oxygen, all three large subunits (I—III) are no longer formed. However, during short-term anaerobic labeling, only subunits I and II fail to become labeled.

Dr. WEIL:

In some of your mutants, the subunits synthesized by one genetic system can accumulate in the absence of the synthesis of subunits made by the other system. How do you visualize the coordinate control of synthesis of the various subunits?

Dr. SCHATZ:

This is, of course, the central problem of our research. So far it has become clear that most, if not all cytoplasmically-made subunits of cytochrome *c* oxidase can accumulate independently of the mitochondrially-made subunits. The reverse, however, is not true. We feel, therefore, that cytoplasmically-made proteins control the accumulation of mitochondrially-made proteins.

CHAPTER IV.2

CYTOPLASM—CHLOROPLAST INTERRELATIONSHIP DURING BIOGENESIS OF CHLOROPLAST MEMBRANE IN *CHLAMYDOMONAS REINHARDI*

I. OHAD

Department of Biological Chemistry, the Hebrew University of Jerusalem, Jerusalem, Israel

The biogenesis of chloroplast membranes in mutants of *Chlamydomonas reinhardi* was used in several laboratories as a model system for the study of biological membrane formation (1—8). The greening process of the yellow heterotrophically dark grown y-1 mutant has several features which render it particularly suitable for the investigation of chloroplast membrane biogenesis.

Using a variety of techniques experiments were devised in an attempt to gain information on some of the major aspects of chloroplast membrane biogenesis such as the role of light in the control of membrane protein synthesis, the interrelation between chloroplast and cytoplasm in this process and the way of addition of new components to newly formed membranes and mechanism of membrane growth. The results of these experiments which were carried out during the last six years can be summarized as follows. The photosynthetic membranes consist of two major classes of proteins. These are proteins produced by the cytoplasmic ribosomes (L proteins) and proteins produced by the chloroplast ribosomes (5, 6, 9—11). The synthesis of L protein once induced to maximal rate can continue in the dark for a limited amount of time (up to about 90 min) and it seems that the cessation of its synthesis in the dark is due to the depletión of an unstable mRNA transcribed from nuclear DNA. Accumulation of L protein is a light requiring process. The chloroplast-made proteins can be subdivided into two classes, proteins whose synthesis is sensitive to both chloramphenicol and rifampicin and proteins sensitive only to chloramphenicol indicating that some of the chloroplast-made membrane proteins are coded by RNA of cytoplasmic (nuclear) origin (14). Both types of the chloroplast-made proteins can be synthesized in the light as well as in the dark provided that L protein synthesis proceeds concomitantly or it has been accumulated in excess to the chloroplast proteins in the growing membranes (11, 15). A similar situation is found with respect to chlorophyll synthesis. Although the enzymatic system responsible for the formation of protochlorophyll(ide) is present in the cells during the linear phase of the greening (1, 2, 16), chlorophyll will not be formed in the light unless membranes containing L protein and presumably lipids are not available either through simultaneous synthesis or preaccumulation. The enzymatic complex comprises at least one unstable enzyme coded by an unstable RNA of cytoplasmic provenience. However, this RNA can be formed in the dark and the enzymatic machinery required for protochlorophyll(ide) synthesis is ready to supply the necessary amounts of chlorophyll as soon as light and L protein are available (16). Thus chlorophyll can be added to the membranes in any order, that is, in absence of, concomitant with or after addition of chloroplast made proteins and the composition of membranes can be altered with respect to chlorophyll and chloroplast proteins relative to cytoplasmic L protein over a limited range of concentrations (10, 11).

The effect of the modulation of the membrane composition has dramatic effects on its structural and functional properties. Membranes lacking chloroplast made proteins are photosynthetically inactive (5, 12, 15, 17), have a low apparent buoyant density and high relative fluorescence (10). Addition of the missing proteins activates the impaired membranes (5, 15, 17) increases their buoyant density and drastically reduces the fluorescence (10). Addition of L protein and chloroplast-made proteins (activation proteins) to the growing membranes will not increase the photosynthetic activity unless additional chlorophyll is synthesized and incorporated into the membranes (11). Modulation of the relative proportions of the different components, activation proteins, chlorophyll and L protein will result in a modulation of the specific photosynthetic activity (ATP formed/mg chlorophyll/hr) of the membranes and of the efficiency of utilization of the incident illumination for electron transfer and photophosphorylation (15, 17).

The assembly of the newly formed components into a membrane appears to occur by their random intussusception into pre-existing membranes resulting in their growth as a continually homogenous phase (4, 10). Recent results indicate that proteins of cytoplasmic origin are associated with photosystem II activity whereas proteins of chloroplast origin are associated with the formation of the active center of photosystem I.

The ability of the receiving membrane consisting of L protein(s) and lipids to accept the other components in any sequence and to modulate its organization and function accordingly stresses the flexibility of this structure and points to the existence of a high affinity for the structural interaction of its building blocks. This might serve as a basis for the process of "target recognition", by the components formed in different subcellular compartments of the specific membrane for which they have been devised by the cell.

References

1. I. Ohad, P. Siekevitz and G.E. Palade, J. Cell Biol. 35 (1967) 521.
2. I. Ohad, P. Siekevitz and G.E. Palade, J. Cell Biol. 35 (1967) 553.
3. I. Goldberg and I. Ohad, J. Cell Biol. 44 (1970) 563.
4. I. Goldberg and I. Ohad, J. Cell Biol. 44 (1970) 572.
5. G. Eytan and I. Ohad, J. Biol. Chem. 245 (1970) 2479.
6. J.K. Hoober, J. Biol. Chem. 245 (1970) 4327.
7. S. Schor, P. Siekevitz and G.E. Palade, Proc. Nat. Acad. Sci. U.S. 66 (1970) 174.
8. S.J. Surzycki, U.W. Goodenough, R.P. Levine and J.J. Armstrong, Symp. Soc. Exp. Biol., Vol. 24, Cambridge University Press (1970) p. 13.
9. J.K. Hoober and G. Blobel, J. Mol. Biol. 41 (1969) 121.
10. G. Eytan and I. Ohad, J. Biol. Chem. 247 (1972a) 112.
11. G. Eytan and I. Ohad, J. Biol. Chem. 247 (1972b) 122.
12. J.K. Hoober, P. Siekevitz and G.E. Palade, J. Biol. Chem. 244 (1969) 2621.
13. J.K. Hoober and W.J. Stegeman, J. Cell Biol. 56 (1973) 1.
14. R. Jennings and I. Ohad, Arch. Biochem. Biophys. 153 (1972) 79.
15. D. Wallach, S. Bar-Nun and I. Ohad, Biochim. Biophys. Acta 267 (1972) 125.
16. I. Goldberg, Ph.D. Thesis, Hebrew University, Jerusalem (1971).
17. S. Bar-Nun, D. Wallach and I. Ohad, Biochim. Biophys. Acta 267 (1972) 138.

CHAPTER IV.3

PLEIOTROPIC MODIFICATIONS IN A MUTANT OF *SCHIZOSACCHAROMYCES POMBE* LACKING OLIGOMYCIN-SENSITIVE ATPase

A. GOFFEAU, F. LABAILLE and A.M. COLSON

Laboratoire d'Enzymologie, Université de Louvain, Place Croix du Sud, 1, Science 14 B, 1348 Louvain-la-Neuve, Belgium

A stable pleiotropic respiratory-deficient mutant of a "petite-negative" yeast *Schizosaccharomyces pombe* has been obtained by X-irradiation. Cytochrome *c* and c_1 and a pigment absorbing at 554 nm are present. Reduced amounts of cytochromes *b* 560 are detected in the mutant. The cytochrome aa_3 absorption peak is not observed and the cyanide-sensitive cytochrome *c* oxidase activity is markedly decreased. Mitochondrial ATPase is present but no oligomycin sensitivity could be detected. A membrane peptide of about 33,800 molecular weight, precipitated in the wild type by an antibody against the purified ATPase (F_1) is not detected in the mutant. Functional mitochondrial protein synthesis is indicated by inhibition of growth by chloramphenicol, by detection of cytochrome *b* and by significant acriflavin, ethidium bromide and chloramphenicol-sensitive leucine incorporation. The addition of antimycin A inhibits growth on glucose in the mutant but not in the wild type. The mutation which results in at least two distinct mitochondrial deficiencies, cytochrome oxidase and ATPase, is produced by a modification of one single nuclear gene or a few very closely linked genes.

We define a pleiotropic mutation as a mutation in one single gene producing several distinct enzymatic deficiencies.

The existence of such a mutant is not in contradiction with the one gene—one enzyme cornerstone of molecular biology. It simply indicates that the modified gene product controls the full expression of several enzymes. Mutants showing multiple deficiencies in the inner mitochondrial membrane are frequent in yeast. Since their first description 10 years ago by Sherman and Slonimski in *Saccharomyces cerevisiae* (1), nuclear respiratory-deficient pleiotropic mutants have not been much studied. Only recently, it became technically possible to analyze deficient mitochondrial membranes at a molecular level. Another major difficulty is that nuclear pleiotropic respiratory-deficient mutants of *S. cerevisiae* are often very unstable; an additional mutation, the cytoplasmic rho minus, is induced spontaneously with high frequency. The biochemical study of such double mutants is very difficult, especially when both mutations affect the same enzymatic activities. An easy way out of this difficulty is to use a "petite-negative" yeast species such as *Schizosaccharomyces pombe* where no cytoplasmic rho minus can be induced (2). Nuclear respiratory-deficient mutants which do not grow on glycerol and do not respire are however easily obtained in this species. After a mutagenic treatment with nitrosoguanidine, 5% of the survivors are respiratory deficient (3). Almost all of them do not exhibit a cyt aa_3 absorption spectrum and a large percentage show in addition some deficiency in the mitochondrial ATPase (4).

We have studied in detail such a mutant called M126 obtained by X-irradiation. No reversion to the wild strain could be induced with or without mutagenic treatment suggesting that the mutation is non-punctual. No cytochrome aa_3 absorption peak can be detected in this strain under any tested condition. Other cytochromes are however present. Cytochromes c and c_1 are present as well as cytochrome b_{560}, and an unusual deficient cytochrome b absorbing at 553 nm (5). As expected, cytochrome c oxidase activity of the mutant is very low, less than 2% that of the wild type. It is however important to note that this small residual activity is significant and cyanide-sensitive, suggesting a modification at the level of the regulation or expression of cytochrome oxidase rather than a drastic modification in one of its nuclear structural genes. Significant antimycin A-sensitive succinate: cytochrome c oxido reductase is also present in the mutant.

ATPase activity is detected in mutant mitochondria. However, no peak of activity is observed at pH 8.5 in contrast to the wild type. The mutant ATPase activity has been extracted by sonication and purified up to specific acitivity of 80. The five components of the wild type ATPase are recognized in the sodium dodecyl sulfate acrylamide gel electrophoresis pattern of the mutant enzyme. Like the wild type enzyme, the purified mutant enzyme is cold-sensitive, is inhibited by ADP, requires magnesium and is inhibited by several ATPase inhibitors including the antibiotic Dio-9 (6). However no oligomycin-sensitivity could be detected in the mutant homogenates, intact mitochondria or sonicated-submitochondrial particles in repressed or derepressed conditions. Other inhibitors such as venturicidin A and dicyclohexylcarbodiimide which like oligomycin acts only when the mitochondrial ATPase is associated to membrane factors are also not effective in the mutant. These data suggest that the mitochondrial ATPase or F_1 is not deficient, but that some membrane factor(s) which directly or indirectly control the expression of oligomycin sensitivity is deficient in the mutant. The following experiment supports this hypothesis. Wild type and mutant cells were labelled with tritiated leucine. Radioactive submitochondrial particles were extracted with 0.2% deoxycholate +0.2 M KCl. The extract was incubated with an antibody against purified ATPase (F_1). The precipitate was dissociated and analyzed by sodium dodecyl sulfate polyacrylamide gel electrophoresis. A radioactive peak which was very prominent in the wild type was not detected in the mutant membrane. The molecular weight of this peptide is about 33,800. Whether this peptide is really associated to the ATPase and is involved in the oligomycin sensitivity is not completely clear yet. It must be noted that in order to be able to detect the absence of any given peptide in a mutant, the mutation must be drastic such as non-sense, a deletion or a frame shift. This suggestion is supported by the absence of reversion of M126 to the wild type.

Four observations indicate that the mitochondrial protein synthesis functions in the mutant. 1) Significant cycloheximide resistant, chloramphenicol-sensitive incorporation of radioactive leucine is observed in the mutant. Because of high cycloheximide resistance of *Schizosaccharomyces pombe*, the relative extent of this inhibition is low but similar to that of the wild type. 2) Cytochrome b_{560} which in other species appears to be synthesized by the mitochondrial protein synthesis is present in the mutant (7). 3) The mitochondrial peptides of cytochrome oxidase seem present in the mutant as shown by the following experiment carried out in collaboration with Alex Tzagoloff of the Public Health Research Institute of the City of New-York. Wild type and mutant cell were labelled with tritiated leucine in the presence of 10^{-3} M cycloheximide. The radioactive submitochondrial particles were extracted with 0.5 M deoxycholate plus 0.5 M KCl. The extract was incubated with an antibody obtained against cytochrome oxidase purified from the wild type and the precipitate was dissociated and analyzed by sodium dodecylsulfate polyacrylamide gel electrophoresis. The same three radioactive bands were detected in wild type and mutant membranes and have molecular weight similar to those reported for the mitochondrial product of cytochrome oxidase in other species (8,9). 4) A chloramphenicol sensitive process is not only carried out in the mutant but is essential for its growth. Indeed, in glucose medium chloramphenicol which only slightly repress growth in the wild type prevents completely the growth of the mutant after a few residual divisions. Addition of antimycin A to the mutant also inhibits growth indicating that a functional cytochrome *b* is required for survival of the mutant but not of the wild type.

In conclusion, although genetic studies have demonstrated that we are dealing with a single gene nuclear mutation, four deficiencies are observed at the present time. Cytochrome aa_3 is undetectable, the ATPase has lost its oligomycin-sensitivity, a functional cytochrome *b* is required for growth on glucose and a peptide of 38 000 molecular weight which seems to be associated with the mitochondrial ATPase is not detected in the mutant membrane.

Nothing can be said however at the present time about the primary mutated gene product. We believe that we should resist the temptation to conclude that the 38 000 molecular weight is the modified gene product which produces the three other observed deficiencies until some direct evidence can be produced to demonstrate that point.

References

1. F. Sherman and P.P. Slonimski, Biochim. Biophys. Acta 90 (1966) 1.
2. H. Heslot, C. Louis and A. Goffeau, J. Bacteriol. 104 (1970) 482.
3. A.M. Colson, C. Colson and A. Goffeau, Methods Enzymol. 32B (1974) 838.
4. A. Goffeau, A.M. Colson, Y. Landry and F. Foury, Biochem. Biophys. Res. Commun. 48 (1972) 1448.
5. A. Goffeau, A.M. Colson, J. Delhez, F. Foury, F. Labaille, Y. Landry and O. Mohar, in A. Tzagoloff (Ed.) Membrane Biogenesis: Mitochondria, Chloroplasts and Bacteria, Plenum, New York (in press).
6. A. Goffeau, Y. Landry, F. Foury, M. Briquet and A.M. Colson, J. Biol. Chem. 248 (1973) 7097.
7. H. Weiss, Eur. J. Biochem. 30 (1972) 469.
8. M.S. Rubin and A. Tzagoloff, J. Biol. Chem. 248 (1973) 4275.
9. T.L. Mason and G. Schatz, J. Biol. Chem. 248 (1973) 1346.

Dr. SCHATZ:

Why is growth of mutant sensitive to CAP or antimycin? Is there a need for residual respiration?

Dr. GOFFEAU:

There seems to be indeed a need for intramitochondrial ATP as well as for mitochondrial oxidations in mutants which do not exhibit detectable antimycin A sensitive oxygen uptake in vivo.

Dr. WILLIAMSON:

Doctor Whitaker in the University of Sussex has evidence that suggests that mitochondrial DNA in the petite negative yeast *Kluyveromyces lactis* has a function other than those concerned with respiration.

Dr. GOFFEAU:

Nothing in our data supports this conclusion. *Schizosaccharomyces pombe* wild type 972 h^- grows well on glucose in the presence of 20 μg/ml ethidium bromide.

Dr. MORPURGO:

Have you looked for double-resistant mutants to both CAP and antimycin in your mutant strain?

Dr. GOFFEAU:

No.

CHAPTER IV.4

CELLULAR EFFECTS OF MITOCHONDRIAL INHIBITION BY ACRIFLAVINE IN *SACCHAROMYCES CEREVISIAE*

I.H. EVANS and D. WILKIE

Department of Botany and Microbiology, University College London, London, U.K.

Yeast cells which are sensitive to the membrane-reacting drug chlorimipramine (CI) show increased CI-tolerance when pre-grown in low concentrations (1—2 μg/ml) of acriflavine. This resistance to CI has limited transmissibility and daughter cells, budded off in the absence of acriflavine, soon revert to CI-sensitivity. The exception is seen in those cells where permanent lesion in mitochondrial DNA (petite mutant) has resulted from acriflavine treatment. About 5% of the petite mutants have stable CI-resistance.

It was concluded that transcription products of mitochondria are normally involved in specifying cell membrane characteristics. These products are reversibly blocked by the specific action of acriflavine and permanently lost in certain petite mutants.

Introduction

Linstead, Evans and Wilkie (1) showed that certain petite mutants from chlorimipramine (CI) sensitive strains of yeast, had acquired CI resistance. From genetic analysis of this resistance and membrane binding studies of sensitive and resistant cells, they concluded tentatively that mitochondrial transcription was involved in the control of cell membrane characteristics. To test this hypothesis, CI-sensitive strains were treated with acriflavine since it has been established that this dye specifically inhibits mitochondrial RNA synthesis (2).

Experimental procedure and results

Tests on solid medium

A number of laboratory strains were used in which the concentration of acriflavine required to inhibit growth on a non-fermentable (glycerol) medium, varied between 0.2—0.5 μg/ml. This was taken as a measure of the mitochondrial sensitivity to acriflavine since all strains were able to grow in the presence of 500 μg/ml on fermentable (glucose) medium. The strains were pre-grown in liquid medium containing acriflavine in the concentrations 0.0, 1.0, 1.25 and 1.5 μg/ml. Cells were harvested, re-suspended and drop-inoculated using a multiple inoculating device (3), onto yeast extract, glucose agar with and without acriflavine. Concentrations of acriflavine were the same as those used in the pre-culture of the cells. All untreated cultures failed to grow on plates containing 40 μg/ml CI but all acriflavine-treated strains made some growth at this concentration of CI. In the case of strains A7F, A15, A285, D6, D6-81 and D18, growth was still proceeding at 80 μg/ml CI, the highest concentration used. When these resistant cultures were sampled and constituent cells tested for the petite condition, only strains D6 and A15 showed the presence of respiratory competent cells (about 3% in each case). Since these agar plates had been incubating for upwards of 48 h, the protracted exposure to acriflavine, even at low concentration, had converted most cells to the petite condition. The respiratory competent cells which had grown at the high concentrations of CI, when subcultured in the absence of acriflavine, returned to CI sensitivity. Furthermore, it was found that when acriflavine-induced petite mutants of D6 were tested, only 5% had heritable resistance to CI.

Tests in liquid medium

Cells of strain A15 were pre-grown for varying lengths of time in the presence of acriflavine (1.2 μg/ml) and in the absence of the acridine. The medium contained yeast extract (1%) and glucose (2%) as in the plate experiments. Cells were harvested and aliquots of 5×10^6 cells inoculated into flasks containing 20 ml of medium and CI in concentrations ranging from 10—80 μg/ml. There was no detectable increase in optical density of untreated cells at a concentration of 30 μg/ml CI. Cells pretreated for 2 and 4 hr with acriflavine likewise showed no detectable growth at this concentration of CI. However, cells grown for 12 hr and longer in acriflavine showed appreciable increase in O.D. at concentrations of CI up to 50 μg/ml. When these CI resistant cultures were sampled, it was found that pretreatment with acriflavine for 12 hr had induced about 95% petite. These experiments were repeated several times with similar results.

Studies with isolated cells

Cells of strain A15 were micromanipulated onto agar blocks containing yeast extract and 0.2% glucose, with and without 2.5 μg/ml acriflavine. As daughter cells were budded off, they were transferred to other blocks containing yeast extract, 0.2% glucose and 40 μg/ml CI. All untreated daughter cells were unable to form buds at this concentration of CI, but in all cases, cells budded off in the presence of acriflavine were able to grow and form small clones of cells on the CI medium. The number of cells in these clones varied from about 5—15. It was noted that most of these cells were the progeny of the initial micromanipulated cell. These clones were broken up and individual cells, mostly with attached buds, were isolated and grown into colonies on yeast extract, glucose medium. The majority of these colonies were respiratory competent and, as expected, had reverted to CI sensitivity. Essentially similar results were obtained using strain A7F.

Discussion

Evidence that the inhibitory effect of CI on cell growth is attributable to reactivity of the drug with cell membrane is fully discussed (1). Further evidence that mitochondria control aspects of cell membrane function was provided in the present studies in which it was found that petite mutants of D6 were unable to utilize the sugar galactose. Since the parent strain was able to use galactose as sole carbon source, this indicates that permeation of the sugar was affected in the mitochondrial mutants.

The general findings show that cells formed in the presence of acriflavine have an altered response to CI. This alteration is not inherited by daughter cells subsequently formed in the absence of acriflavine, except in the case of certain petite mutants. In view of the known specificity of action of acriflavine, the conclusion is drawn that inhibition of mitochondrial transcription brings about changes in cell membrane characteristics. If this is true and applies to cells generally, the primary involvement of mitochondria in the neoplastic condition may be considered. Alteration in the cell membrane can clearly lead to change in the repression/derepression pattern of nuclear genes as well as in cell surface properties which are features of cancer cells. Since mitochondria are prime targets for many drugs and inhibitors, particularly DNA-reacting compounds (3,4), there is a distinct possibility that this could initiate the neoplastic condition. This possibility is under investigation here.

Acknowledgement

This work is supported by a grant to D.W. by the Cancer Research Campaign.

References

1. D. Linstead, I. Evans and D. Wilkie, in A.M. Kroon and C. Saccone (Eds.) The Biogenesis of Mitochondria, Academic Press Inc., New York (1974) p. 179.
2. H. Fukuhara and C. Kujawa, Biochem. Biophys. Res. Commun. 41 (1970) 1002.
3. D. Wilkie, Med. Biol. Illustr. 22 (1972) 119.
4. I. Evans, D. Linstead, P.M. Rhodes and D. Wilkie, Biochim. Biophys. Acta 312 (1973) 323.

Dr. BOGORAD:

Does the binding work only with whole living cell or do broken cell preparations bind?

Dr. WILKIE:

Drug binding has been demonstrated also in the case of cell membrane fractions.

Dr. SCHATZ:

Can you exclude toxification of imipramine via respiratory chain?

Dr. WILKIE:

Cellular inhibition by imipramine is dependent of blockage of the respiratory chain provided there is a supply of glucose to provide ATP. Remember, the yeast cell is a facultative anaerobe and can grow and divide without any respiratory chain (e.g. in petite mutants).

PART TWO

Nucleocytoplasmic interactions in differentiating cells

SECTION V

INTRACELLULAR INTERACTIONS DURING CELL DIFFERENTIATION

CHAPTER V.1

NUCLEOCYTOPLASMIC INTERACTIONS IN CELL DIFFERENTIATION

J. BRACHET

Department of molecular Biology, University of Brussels, Belgium, and
Laboratorio di Embriologia molecolare, C.N.R., Arco Felice, Naples, Italy

After a short review of the work done on nucleate and anucleate fragments of unicellular organisms (*Amoebae, Acetabularia*), the case of the sea urchin and amphibian eggs is examined in more detail: special emphasis is placed on the respective roles of the cell nucleus and the cytoplasm in hormone-induced maturation of amphibian oocytes and in sea urchin egg fertilization. The importance of the synthetic processes which already occurred, under nuclear control, during oogenesis is stressed. The shift between predominant RNA synthesis in the oocyte and predominant DNA synthesis during cleavage is discussed; the respective roles of the nucleus and the cytoplasm in this shift are examined.

Activation of the nuclear genes becomes obvious at gastrulation. Present knowledge about the molecular mechanisms of primary induction (neural induction) are presented. This induction leads to organogenesis, not to cell differentiation. The latter occurs later in development and is discussed in the special cases of the embryonic pancreas and regenerating lens. The relative importance of DNA replication and DNA transcription in such processes is examined, as well as the probable role played by signals received at the plasma membrane level.

Finally, the effects of cell fusion on the activity of chromatin and on the presence of the differentiated phenotype, as well as those of such chemicals as BrdUr and the cyclic nucleotides on cell differentiation, are presented and discussed.

Cell differentiation is clearly the result of specific gene activation; but the molecular mechanisms of this specific activation remain obscure.

Survival and biochemical activities of anucleate cytoplasm

When almost half a century ago, I attended the first lecture of the Cytology course, I was greatly surprised when our teacher, Prof. Pol Gerard, told us that anucleate fragments of Protozoa or Invertebrate eggs always survive for some time. In a naive way, I thought that taking away the nucleus should be the same as cutting the head off a man. This surprise was so great that I decided to work in Prof. A. Dalcq's laboratory, in order to study the reasons of this strange phenomenon. Today, I know why a cell does not die as soon as its nucleus has been taken away; but there remains much to be done in this field and I still like to cut eggs and unicellular organisms into two, in order to compare nucleate and anucleate halves.

What Prof. Gerard told us is indeed true and survival of anucleate cytoplasm, for a shorter or longer time, is a universal phenomenon: *Acetabularia* detains the world record of longevity in the absence of the nucleus (100 days or even more), but amoebae, *Stentor*, sea urchin and amphibian eggs can survive several days after removal of the nucleus. Mammalian cells do not escape this general rule, as shown by reticulocytes (where loss of the nucleus is a natural phenomenon) and various kinds of cells which, when cultivated in the presence cytochalasin B, eject their nucleus together with a minute amount of cytoplasm.

Biochemical analysis has shown that, in all cases, synthesis of specific proteins continues in the absence of the nucleus. But there is no cytoplasmic RNA synthesis in anucleate fragments of cells, except in DNA-containing cell organelles: mitochondria and chloroplasts. Since synthesis of specific cytoplasmic proteins requires information coming from the nuclear genome, one cannot escape the conclusion, that I proposed for *Acetabularia* in 1959 that stable mRNA molecules, of nuclear origin, must be stored in the cytoplasm. That this conclusion is correct has been demonstrated, as we shall see, in the case of anucleate fragments of sea urchin eggs.

Acetabularia is unique, among all organisms studied so far, by its exceptional potentialities for survival and regeneration in the absence of the nucleus (1,2). This is undoubtedly partly due to a certain degree of autonomy in the chloroplasts, which can replicate their DNA in the absence of the nucleus. However, these chloroplasts, despite their DNA content, are largely dependent upon the nuclear genes, as will be demonstrated by Prof. H.G. Schweiger in his paper.

In the following, I shall very briefly summarize what is known about nucleocytoplasmic interactions in what I think remains the best biological system available for the study of differentiation, embryonic development. More detailed information can be found in two recent books (3,4).

Embryonic development

Ovogenesis

This very early period of development, which takes place in the maternal organism, has the utmost importance: many constituents, in particular RNAs and proteins, which are necessary for embryonic life, are synthesized by the growing oocyte. Ovogenesis should therefore be regarded as a period of intensive synthetic and genetic activity.

The oocyte synthesized a huge number of ribosomes, enough in fact to reach the feeding stage of the larva. This is possible because the ribosomal genes are "amplified" at the very beginning of ovogenesis: there are about 1000 times more genes for the 28S and 18S rRNAs in an amphibian oocyte than in a somatic cell of the same species; this is reflected, at the morphological level, by the presence of more than 1000 nucleoli in the large nucleus (germinal vesicle: GV).

The GV contains more extrachromosomal (nucleolar) ribosomal DNA than chromosomal DNA; but the latter is exceptionally active in transcription. It is present in the lampbrush chromosomes, which form innumerable loops, where the genetic material is continuously transcribed by very active RNA polymerase. The result is of course the synthesis of a great variety of mRNAs and proteins. However, ribosomes are in such excess that only 10% of them are in the form of polysomes; this explains why foreign mRNAs (haemoglobin mRNA for instance) are translated very efficiently when they are injected into amphibian oocytes (5).

It should be pointed out that the huge size reached by the eggs of many animal species (all Vertebrates except the Mammals) is not really the result of this intensive synthetic activity in the oocyte: it is due to the accumulation of phosphoproteins in the yolk platelets; these phosphoproteins are not synthesized by the oocyte, but in the liver of the female as a result of hormonal stimulation. A large precursor of the yolk phosphoproteins (vitellogenin) is synthesized by the liver, under the influence of oestrogens and secreted in the blood stream; vitellogenin is then taken up by micropinocytosis into the oocyte, where it is transformed into the final yolk phosphoproteins (phosvitin and lipovitellin): the latter soon crystallize and are transformed into morphologically visible yolk platelets (6).

Since there is good reason to believe that the synthesis of vitellogenin in the liver is the result of a hormone-induced gene activation and since the lampbrush chromosomes are obviously very active in transcription, it can be concluded that oogenesis is essentially controlled by the activity of the nuclear genes.

Maturation (7—9)

When the oocyte has reached its full size, it must undergo maturation, that is the end of meiosis: the giant GV breaks down, the nucleoli disappear, the lampbrush chromosomes condense and, finally, the polar bodies are eliminated. The whole process is often incomplete: maturation may stop at some stage and will be completed after fertilization or parthenogenesis: for instance, in the Amphibians (which will be the main object of this section), only the first polar body has been expulsed in the unfertilized egg.

Maturation is induced by hormones (pituitary gonadotrophins, progesterone or, in starfish oocyte, the purine base, methyladenine). In the Amphibians, pituitary hormones stimulate progesterone synthesis in the follicle cells which surround the oocyte. Progesterone must act on the cell membrane, since it is inefficient in injection. Once in the cytoplasm, the hormone binds to a receptor, which is localized in the pigment granules (melanosomes). It then induces the synthesis of several antagonistic factors: a maturation inducing factor (MIF), which is present in the endoplasm of the hormone-stimulated oocyte; it induces maturation after injection into a recipient oocyte. A pseudomaturation inducing factor (PIF) also induces, after injection, GV breakdown; but the maturation quickly becomes very abnormal and ends up in cytolysis. Finally, a cytostatic factor (CSF) has been found at the end of maturation, when the first polar body has been eliminated: it stops mitosis if it is injected into one of the blastomeres of a cleaving egg. It is possible that PIF and CSF are identical since both exert inhibitory effects in contrast to MIF. All these factors are probably proteins while MIF stimulates protein synthesis in injected recipient oocytes, PIF inhibits it almost completely.

There is no DNA synthesis during maturation, except mitochondrial DNA replication; the latter is not affected quantitatively by progesterone treatment or by removal of the GV. This shows that the multiplication of the mitochondria is completely independent from the events which occur in the nucleus after progesterone treatment, and that it is even independent from the presence or absence of the GV. Although chromosomal and ribosomal DNAs are not replicated during maturation, the whole machinery for DNA synthesis is built up at that time: if DNA is injected into a hormone stimulated oocyte, it is quickly replicated (10); it is not replicated after injection into an unstimulated oocyte. Recently, it has been shown that new DNA polymerase and ribonucleotide reductase activities can be detected after amphibian oocytes have been treated with progesterone. A new histone-kinase, which might possibly play a role in chromosome condensation, also becomes detectable during maturation.

Protein synthesis is stimulated a few hours after progesterone treatment. An important aspect of this increase in protein synthesis is that it occurs even if the GV has been removed from the oocyte prior to hormone stimulation. It has also been demonstrated that the very important MIF is produced in normal amounts when enucleated oocytes are treated with progesterone. Protein synthesis, in such enucleated and stimulated oocytes, is not only quantitatively, but even qualitatively, normal.

Other important changes occur, in a normal way, when enucleated oocytes are treated with progesterone: retraction of the microvilli which were very abundant in the membrane of the oocyte, migration towards the vegetal pole of the large nucleoprotein granules which are needed for the formation, at a much later stage of development, of the germ cells. However, chromosomes do not condense when nuclei taken out of adult brain are injected into enucleated hormone stimulated oocytes; they undergo condensation if such nuclei are injected into oocytes which have been treated with progesterone, but not enucleated, or if nuclear sap taken from the GV is injected into the enucleated and stimulated oocytes.

On the whole, one can conclude that the nucleus, during maturation, no longer plays the prominant role it had during ovogenesis: it obeys, in a passive way, to the changes which have been induced in the cytoplasm by a short contact of progesterone with the cell membrane.

Fertilization (11)

In sea urchin eggs, fertilization induces a fast and considerable increase in protein synthesis. Unfertilized eggs can easily be separated into nucleate and anucleate halves by centrifugation in a density gradient: if anucleate fragments are treated with a parthenogenetic agent (hypertonic sea water, for instance), they undergo the same increase in protein synthesis as normally fertilized eggs. Thus the increase of protein synthesis at fertilization cannot be due to a synthesis of new mRNA molecules by the nucleus: it is best explained by assuming an "unmasking", probably resulting from increased proteolytic activity, of pre-existing, inactive mRNA molecules: the latter must necessarily be of maternal origin; synthesized during oogenesis under the direction of the lampbrush chromosomes, they are stored in an inactive form in the cytoplasm of the unfertilized egg. There is also good evidence that the ribosomes are different in fertilized and unfertilized eggs: the existence, in the latter of an inhibitor of protein synthesis, which is attached to the ribosomes explains in part why they contain very few polysomes.

Work in the laboratory of P. Gross et al. (12) has clearly demonstrated that anucleate fragments of sea urchin eggs contain, in a masked form, specific mRNA molecules: this has been established for the mRNAs of the histones and tubulins. The only RNA synthesis which can be detected in activated anucleate fragments of sea urchin eggs is mitochondrial; but it has been shown that this neosynthesized mitochondrial RNA is unable to bind to cytoplasmic ribosomes and to form active polysomes. It is therefore not responsible for the increase in protein synthesis. Activated anucleate fragments of sea urchin eggs are capable of adding a -CCA terminal sequence to the tRNAs and poly A sequences to the pre-existing mRNAs. On the other hand, mitochondrial DNA cannot undergo replication in such anucleate fragments, in contrast with what we saw about maturation.

The general conclusion drawn about maturation remains valid for fertilization: the nucleus plays a passive rather than an active role and the main events occur in the cytoplasm at that time.

Cleavage

This is a period of intensive and fast multiplication of the nuclei: in fact, DNA replication (which already begins at fertilization) can be faster, in *Drosophila* eggs, than in a bacterium (it takes about 8.5 min). The speed of the mitoses, in cleaving eggs, is due to the fact that there is no G_1 phase and that the G_2 phase of the cell cycle is very short. This is possible because, as we have seen, the whole machinery for DNA replication has already been synthesized during maturation; DNA polymerase migrates from the cytoplasm into the nuclei and the mRNAs for histones and tubulins already pre-exist in the cytoplasm of the unfertilized egg. However, there is, during cleavage, a considerable synthesis of ribonucleotide reductase and ornithine decarboxylase, two enzymes which are believed to play an important role in the control of DNA replication. The synthesis of these two enzymes is not affected by actinomycin D; this suggests that it is directed by pre-existing cytoplasmic templates.

In both amphibian and sea urchin eggs, cleavage is unaffected by actinomycin D treatment: the treated eggs can reach the blastula stage. Anucleate unfertilized eggs, activated by parthenogenesis, can also reach a somewhat abnormal blastula stage. On the other hand, inhibitors of protein synthesis arrest very quickly all cell division.

These experiments show that intensive DNA replication and numerous mitoses are possible in the absence of RNA synthesis; the latter is small, but detectable, during cleavage and it is probable that the mRNAs which are synthesized during this period are required for the next stage of development, gastrulation.

During cleavage, the genetic material is too busy replicating as fast as possible and it has little time for transcription it is still the cytoplasm which plays the leading role and controls the rate of cell division.

Cleavage raises an interesting point for molecular embryologists: experimental embryology has clearly established the existence, in the so-called mosaic eggs, of germinal localizations: these are special regions or plasms which will give rise, at a later stage, to a specific organ or tissue. For instance, in Ascidian eggs, a "myoplasm" accumulates first at the vegetal pole of the fertilized egg, then in the posterior-vegetal blastomeres; only the cells which contain this myoplasm will differentiate, at the tadpole stage, in muscle cells. Is such a germinal localization the result of a segregation, in the myoplasm of a specific myosin mRNA, which already existed, in a more diffuse way, in the unfertilized egg? Or, is the material which is segregated in the myoplasm a factor which will specifically de-repress the myosin genes in the nuclei? This is a modern version of the old controversy between preformation and epigenesis, which interested so much the old naturalists of the XVIIIth century and which reflects the even-lasting opposition between conversatives and progressists. The oocyte, with its powerful synthetic activities directed by the nuclear genes, is typically a "progressive" system; its treatment with progesterone produces a "conservative revolution", which lasts until the end of cleavage: preformed, stored mRNAs are much more important than newly-synthesized mRNAs. The "counter-revolution" will start as soon as gastrulation and neural induction begin.

Gastrulation and neural induction (12)

Gastrulation is a period of intensive migration of cell populations, which is made possible by the contractivity of the micro-tubules and microfilaments. This is the time when new mRNA molecules are synthesized when the synthesis of rRNA is resumed, when energy production increases at the expense of the glycogen reserves, etc. The properties of the cell membranes change also in amphibian eggs: while concanavalin A is hardly bound to the cell surface of the cleaving cells, it binds strongly to the membrane of the cells which are undergoing important morphogenetic movements (13).

As soon as the investigated chordomesoderm enters in contact with the overlying ectoderm, induction of the nervous system begins: this primary induction is very probably due to a protein, as shown by Tiedemann (14) and by Faulhaber (15). Unfortunately, we do not know whether this inducing protein pre-exists before gastrulation and, if so, whether it has a specific localization in the embryo. What is certain, however, is that induction is a derepression of the genome and that new populations of mRNAs are synthesized, but experiments with actinomycin D show that this RNA synthesis is necessary for the response of the ectoderm to the inducing stimulus and not for the inducing activity itself: an actinomycin D treated organizer (chordomesoderm) is still active, while actinomycin D treated ectoderm can no longer be transformed into nervous system.

The result of this primary induction is organogenesis, that is formation of a neural tube; but this neural tube is not yet made of neurones. Organogenesis must be followed by cell differentiation, at the morphological, physiological and biochemical levels: a neurone can be recognized under the microscope, or by its electrophysiological properties, or by the presence of specific proteins (protein S_{100}) or enzymes (cholinesterase, for instance). Such specific biochemical markers are particularly useful in the analysis of cell differentiation during embryogenesis. We shall limit ourselves here to two cases, which have been particularly well studied: the differentiation of the pancreas and that of the eye-lens. They are just examples selected in order to show how the problem of embryonic differentiation can be attacked.

Two examples of embryonic cell differentiation: pancreas (16,17)

The formation of most organs in Vertebrates is the result of secondary inductions: there must be an interaction between inducing tissue (usually of mesodermal origin) and a reacting tissue in order to get organogenesis. In the case of the pancreas, the reacting tissue is an undifferentiated epithelium of endodermal origin; under the influence of the surrounding mesenchyme, it differentiates into exocrine and endocrine cells. Both kinds of cells contain excellent biochemical markers: insulin and glucagon for the endocrine cells, trypsin, chymotrypsin, carboxypeptidase, amylase, etc. for the exocrine cells.

The differentiation of the epithelial rudiment can be obtained by a "trans-filter" induction: a Millipore filter can be inserted between the epithelium and the mesenchyme without empeding the induction. This seems to be a general phenomenon, since through Millipore filters can be obtained in the cases of the nervous system (the "primary" induction), eye-lens, kidney, salivary glands, etc. Filters which do not allow the diffusion of macromolecules, cellophane membranes for instance, in general prevent the inducing agent from reaching the target cells.

In the particular case of pancreas, it has been shown that the Millipore filter, after trans-filter induction, contains a protein which originates from the mesenchyme cells. This protein is the inducing agent; it is still active if one binds it to small Sepharose beadles, which cannot penetrate into the cells: the experiment shows that the inducing protein acts on the cell membrane. After the cell membrane has received the signal given by the inducing protein, DNA synthesis begins in the pancreatic rudiment and the number of cells increases. According to the plane in which the cells divide (parallel or perpendicular to the long axis of the rudiment), they will differentiate into endocrine or exocrine cells. Their differentiation is characterized by the sequential synthesis of the pancreatic enzymes and hormones: the different hydrolases are synthesized one after the other, in a perfectly programmed order. This clearly demonstrates that in structural genes responsible for the synthesis of each enzyme are sequentially activated; there is no evidence for the existence of an operator gene, which would activate a whole battery of structural genes controlling the simultaneous synthesis of all the pancreatic enzymes.

When the eye-lens is removed from an adult Urodele, it regenerates at the expense of the superior rim of the iris (Wolffian regeneration). Lens differentiation can be followed by the appearance of specific proteins, the α and β-cristallins; their localization can be observed, under the microscope, by immunofluorescence. Regeneration of the lens involves a succession of steps: first, dedifferentiation of the iris cells, which loose their pigment. It is followed by DNA synthesis and a wave of mitoses, which increases the size of the lens rudiment. Synthesis of rRNA then increases the number of the ribosomes; it is quickly followed by mRNA synthesis and finally by the appearance of the cristallins (in sequential order) in the eye-lens cells. Synthesis of the cristallins and morphological differentiation of the lens fibers go hand in hand. A similar sequence of events is found when the differentiation of the eye lens is studied in the embryo, where it is induced by the eye cup (which is part of the brain), instead of the adult.

In these two examples, the first response to the inducing stimulus is a wave of mitoses, thus an increase in the number of available cells. This is certainly a fairly general phenomenon; immunologists, in particular, attach great importance to the formation of clones of cells, deriving by successive mitoses from a single cell which has received, at the level of the plasma membrane, an antigenic stimulus. However, for all rules, there are exceptions: the most striking one is *Acetabularia*, where extensive differentiation (cap formation) normally occurs when the cell is still uninucleate; as we have seen, it can even take place in the absence of the nucleus.

Holtzer et al. (18) who have done very important work on cell differentiation (in particular, of cartilage cells) during embryogenesis, has introduced two important and interesting concepts in the field. The first one is that of "luxury" proteins as opposed to "housekeeping" proteins. Luxury proteins are those which are specific of a given kind of differentiated cell: cristallins in the eye-lens, chondrin in cartilage cells, myosin and actin in muscle cells, haemoglobin in erythrocytes, melanin in pigment cells, etc. Their production must necessarily result from the activation, in certain cells, of the appropriate structural gene; the same gene is repressed and is thus not expressed phenotypically in all the other cells of the organism. Housekeeping proteins, in contrast, are those which are common to all cells and which are necessary for their survival (for instance, histones, ribosomal proteins, respiratory enzymes, etc.).

The second concept proposed by Holtzer is that of "quantal" mitoses: in such a mitosis, DNA would be equally distributed in the two daughter cells; but genetic activity would be different in these two cells. In one of them, the structural gene for a specific luxury protein would be derepressed and the cell would differentiate; in the other cell, this gene would remain inactive and mitotic proliferation would go on. This concept is useful not only for embryologists, but also for immunologists. There is some factual evidence for the reality of the quantal mitosis concept: for instance, when spermatogonia divide, one of the daughter cells only becomes a spermatocyte and gives rise to spermatozoa; the other will remain a spermatogonium and will go on dividing. Quantal mitoses also occur in the differentiation of the exocrine and endocrine cells of the pancreas, as mentioned above. It is generally assumed that quantal mitosis is the result of a signal received by the cell membrane, which contains a large number of receptors of all kinds: this is easy to conceive in the cases of spermatogonia and pancreatic cells where the fate of the cell depends upon its micro-environment: the latter will change if a cell moves out of a row of undifferentiated dividing cells and acquires a new position, thus fresh "positional information".

Analysis of differentiation in cells cultivated in vitro

It is outside the scope of the present review article to summarize all the work that has been done by the many workers who have studied the in vitro differentiation of embryonic cells.

Only 3 aspects of this intensive work will be shortly discussed here: experiments on cell fusion (somatic hybridization), on the effects of bromodeoxyuridine (BrdUr) on cell differentiation and on the effects of the cyclic nucleotides (cAMP and cGMP) on cell growth and differentiation.

Cell fusion can be obtained by treating in vitro cultivated cells with inactivated Sendai virus or chemicals which affect the integrity of the cell membrane. As shown by Harris (19) fusion between a chick erythrocyte (a metabolically inactive cell) and a metabolically very active cell, such as a HeLa cancer cell re-activates the chicken erythrocyte nucleus: it swells, as the result of an intake of HeLa cell cytoplasmic proteins, and then soon begins to replicate its DNA and to synthesize all kinds of RNAs. Nuclear activity in this case, is thus clearly dependent upon the cytoplasm of the HeLa cell. This situation is very similar to that found by Gurdon in anucleate unfertilized eggs of *Xenopus*: if a nucleus taken from an adult cell (brain, for instance) which never divides but actively synthesizes rRNA in situ, is transplanted in an enucleated unfertilized egg, it first swells; this is the result of an intake of egg cytoplasmic proteins; RNA synthesis stops and the previously prominent nucleoli can no longer be seen. A few hours later, DNA synthesis starts in this nucleus which would never have replicated its DNA if it had remained in the adult brain. It is thus clear that, in tissue culture cells as well as in eggs, it is the cytoplasm which decides whether nuclear DNA will be replicated or transcribed. The decision, in this all-important choice is probably made by proteins of cytoplasmic origin which migrate into the nucleus and bind to its chromatin.

Ephrussi and Sorieul (20) were the first to obtain somatic hybrids between cells which had been cultivated in vitro; together with their colleagues (M. Weiss, Davidson and others), they used somatic hybridization as a tool for the analysis of cell differentiation. The principle of their work is to fuse together a differentiated cell with an undifferentiated, embryonic one and to grow cell lines out of such hybrids. While the hybrid cells used by Harris were, in general, heterokaryons (two different nuclei are present in a common cytoplasm), those used by Ephrussi were synkaryons (the two nuclei of different origins fuse together and then undergo mitotic divisions). In such synkaryons, the general rule is that the differentiated character is lost: it looks as if a repressor, which prevents the phenotypic expression of the differentiated character had been produced by the undifferentiated cell nucleus.

On the biochemical level, luxury proteins are no longer synthesized by the somatic hybrid: for instance, if a pigmented melanoblast is fused together with an embryonic fibroblast, the pigment granules soon disappear. The enzymes needed for melanin synthesis (tyrosinase, DOPA oxidase), which were very active in the melanoblast, can no longer be detected in the hybrid. Similar results have been obtained with many kinds of differentiated cells: after somatic hybridization with undifferentiated cells, liver cells no longer produce albumin and lose some of their specific enzymes; neurones no longer contain the specific S_{100} protein which characterizes them, etc.

The situation becomes more complex when the analysis takes into account what happens to the chromosomes during the culture of the hybrid cells. During the replication of the synkaryons, chromosomes are progressively lost in the cytoplasm, where they degenerate. A frequent consequence of this chromosome elimination is the reappearance of the differentiated phenotype: pigment granules and enzymes responsible for melanin synthesis, for instance, again become detectable in the hybrid cells. It has also been shown that, in such experiments, gene dosage plays an important role: if the melanocyte cell line has a higher degree of ploidy then the undifferentiated fibroblast with which it has been fused, pigmentation is retained in the hybrid.

One can hope that continuation of this important work will lead to a better understanding of the role played by individual chromosomes and genes in cell differentiation. On the other hand, studies on the heterokaryons obtained soon after cell fusion, should give us more information about the mechanisms which allow the cytoplasm to orientate nuclear DNA towards replication or transcription. *Xenopus* enucleated unfertilized eggs are of course another very valuable material for such an analysis.

The thymidine analogue bromo-deoxyuridine (BrdUr) has the curious and interesting property of inhibiting selectively the synthesis of the luxury proteins (18). In order to be effective, the inhibitor should be added to cells which are on the verge of differentiation, but are still undifferentiated: for instance, unpigmented melanoblasts cultivated in vitro will form neither melanin nor the enzymes needed for its synthesis in the presence of BrdUr. Addition of small amounts of BrdUr, which are insufficient to inhibit DNA replication and cell proliferation inhibits the synthesis of chondrin in chondroblasts, of myosin in myoblasts, of haemoglobin in erythroblasts, etc. The result is that the cells can multiply, but remain in the undifferentiated state. The effects of BrdUr are easily reversible by washing: if the drug is removed, the cells soon differentiate normally. The addition of thymidine to BrdUr prevents its effects and cell differentiation proceed unimpaired. For reasons which are not clear, BrdUr does not markedly inhibit embryonic differentiation when it is added to developing eggs; it has, however, been reported that when very early chick embryos are treated with the drug, the blood islets do not form erythrocytes. It may be that certain conditions, attachment of a monolayer of cells to a solid substratum for instance, are required for the BrdUr effect; these conditions are met in tissue cultures, but not in whole embryos.

The mode of action of BrdUr on cell differentiation is still a matter for discussion: what has just been said about the differences between tissue culture cells and embryos speaks in favor of the idea that BrdUr might affect the cell surface. But the fact that the BrdUr effects are completely inhibited by thymidine is a very strong argument for the view that it acts on DNA itself: it is well established that BrdUr is incorporated into DNA if it is added during differentiation, at least one round of cell multiplication is needed: in other words, BrdUr is ineffective unless conditions which allow its incorporation into DNA are met. It has been suggested, a few years ago, by Holtzer that BrdUr might inhibit a selective amplification of the structural gene responsible for the synthesis of a given luxury protein: if so, the erythroblast should contain a larger number of haemoglobin genes than the other cells of the organism. Elegant experiments of molecular hybridization have clearly shown that this is not the case: the number of haemoglobin genes is small (4 to 5 per haploid genome) and is the same in haemoglobin producing cells as in all the other cells of the organism. A recent analysis of BrdUr incorporation into DNA, under the conditions where cell differentiation is suppressed, has led to the conclusion that BrdUr affects probably a small number of genes only; but why the structural genes for luxury proteins would be a preferential target for BrdUr remains a mystery. It has also been suggested that the binding of regulatory proteins to BrdUr containing DNA might be modified: this could inhibit the transcription of certain genes. Another recent report adds to the present state of confusion: a BrdUr-dependent strain of mammalian cells has been isolated; in its cells practically all the thymines are substituted by bromo-uracil; nevertheless, the cells are capable of division and proliferation. They are thus viable despite the fact that their DNA is completely abnormal.

A third promising approach for the study of cell differentiation lies in the effects of the cyclic nucleotides cAMP and cGMP on cell proliferation and differentiation. Much work is now being done on normal and cancer cells (usually cells transformed by an oncogenic virus): without going here into details, one can say that the addition of cAMP or of substances which, like theophyllin, increase the cAMP content of the cells inhibits mitotic activity and favours differentiation. Cells which have a high mitotic rate are, in general, richer in cAMP than cells which divide rarely or not at all (21,22). Cyclic GMP has the opposite effects: it stimulates cell proliferation and inhibits cell differentiation. These experiments lead to the conclusion that the choice between mitosis and differentiation might be controlled by the cAMP/cGMP ratio in the cells. Factors which, like the addition of serum to tissue culture cells, stimulated growth shift this ratio; they decrease the cAMP content of the cells. It should be added that, as in the case of BrdUr, embryos do not seem to obey the same rules as tissue culture cells: in still unpublished experiments, we found that the cAMP and cGMP content of amphibian embryos is high and remains remarkably constant from the ovarian oocyte to the neurula stage. Addition of cAMP dibutyryl cAMP, cGMP, theophyllin, alone or in combination to fertilized amphibian eggs exerts no effect whatsoever upon development.

The mode of action of cAMP begins to be well understood: it is synthesized, in the cell membrane, by an adenylcyclase (there is also a guanylcyclase) at the expense of ATP. The membrane also contains a phosphodiesterase, which breaks down the cyclic nucleotides. The cell thus possesses adequate enzymatic mechanisms for the control of the cAMP/cGMP ratio. Cyclic AMP acts by stimulating a number of protein-phosphokinases: they phophorylate proteins and modify their activity (if they are enzymes) or structure (for instance for ribosomal proteins or histones). It is probable that the kinases which phosphorylate the chromosomal proteins (histones and non-histone nuclear proteins) play a particularly important part: according to Bradbury et al. (23), phosphorylation of histones F_1 would be responsible for the condensation of the chromosomes prophase and the initiation of mitosis. Non histone chromosomal proteins are believed to play an essential role in the control of gene activities in eukaryotes: their phosphorylation probably modifies their affinity for specific sites of the DNA molecules (24). It is also known that cAMP affects the polymerization of tubulin molecules and the edification of the microtubules: the latter are of course the main constituents of the mitotic apparatus; they probably also play a very important role in the maintenance of the architecture of the cell and in the transport of molecules from the outside medium (Ca^{2+} ions, for instance) to the various parts of the cell.

Conclusions

Great progresses have been made, during the past 10 or 20 years, towards the solution of the problem of cell differentiation; whether the biological material used was a unicellular alga like *Acetabularia* or a developing embryo, or cells cultivated in vitro, we know much more than we did a few years ago.

However, the problem is still far from solved and we are still waiting for an answer to major questions: how does a signal received at the level of the cell membrane produce the activation or repression of a set of specific genes? What are the molecular bases of positional information for a cell? And, most important of all, how is the activation of a specific gene controlled in eukaryotic cells?

There is good reason to believe that this control is exerted by non-histone chromosomal proteins (24). More extensive biochemical work about these proteins will certainly be rewarding and it will be of great interest to test the effects of purified chromosomal non-histone proteins on cell differentiation. Another promising approach is that followed by Gilmour and Paul (25): they were able to demonstrate, with elegant methods of molecular hybridization, that when chromatin from erythropoietic cells is transcribed with bacterial RNA polymerase, the specific mRNA for haemoglobin can be detected among the transcripts; it is not detectable when liver chromatin is used as template for the polymerase instead of chromatin isolated from erythropoietic cells.

There is no doubt that the problem of cell differentiation (which is of the utmost importance for the understanding of the causes of cancer and ageing) can only be solved by a convergent attack from the biochemical, genetical and immunological angles. The recent work of F. Jacob and his colleagues on mouse teratoma (26) can be taken as a good example of what one should try to do on a variety of biological systems.

References

1. J. Hämmerling, Intern. Rev. Cytol. 2 (1953) 475.
2. J. Hämmerling, Ann. Rev. Plant Physiol. 65 (1963) 92.
3. J. Brachet, Introduction to Molecular Embryology, Springer Verlag (1974).
4. J. Brachet, Introduction à l'Embryologie moléculaire, Masson, (in press).
5. C.D. Lane, G. Marbaix and J.B. Gurdon, J. Mol. Biol. 51 (1971) 73.
6. R.A. Wallace and D.W. Jared, Science 160 (1968) 91.
7. L.D. Smith and R.E. Ecker, Curr. Topics Dev. Biol. 5 (1970) 1.
8. J. Brachet, E. Baltus, A. De Schutter, F. Hanocq, J. Hanocq-Quetier, E. Hubert, S. Iacobelli and G. Steinert, Mol. Cell. Biochem. 3 (1974) 189.
9. J. Brachet, Ann. Biol. 13 (1974) 5.
10. J.B. Gurdon, M.L. Birnstiel and V.A. Speight, Biochim. Biophys. Acta 174 (1969) 614.
11. A. Monroy, Chemistry and Physiology of Fertilization, N.Y. Holt (1965).
12. K.W. Gross, M. Jacobs-Lorena, C. Baglioni and P. Gross, Proc. Nat. Acad. Sci. U.S. 70 (1973) 26.
13. D. O'Dell, R. Tencer, A. Monroy and J. Brachet, Cell Diff., (in press).
14. H. Tiedemann, Curr. Topics Dev. Biol. 1 (1966) 85.
15. I.W. Faulhaber, Roux' Arch. 171 (1972) 87.
16. W.J. Rutter, N.K. Wessels and C. Grobstein., Nat. Cancer Inst. Monog. 13 (1964) 51.
17. T. Yamada, Curr. Topics Dev. Biol. 2 (1967) 147.
18. H. Holtzer, H. Weintraub, R. Mayne and B. Mochan, Curr. Topics. Dev. Biol. 7 (1972) 229.
19. H. Harris, Cell Fusion, The Dunham Lectures Oxford, Clarendon Press (1970).
20. B. Ephrussi and S. Sorieul, Nature 190 (1962) 653.
21. R. Karam, P. Mamont and G.M. Tomkins, Proc. Nat. Acad. Sci. U.S. 70 (1973) 1432.
22. J. Otten, G.C. Johnson and I. Pastan, Biochem. Biophys. Res. Commun. 44 (1971) 1192.
23. E. Bradbury, R. Inglis, H.R. Matthews and T.A. Langman, Nature 249 (1974) 553.
24. G.S. Stein, T.C. Spelsberg and L.J. Kleinsmith, Science 183 (1974) 817.
25. R. Gilmour and J. Paul, Proc. Nat. Acad. Sci. U.S. 70 (1973) 3440.
26. K. Artzt, D. Bennet and F. Jacob, Proc. Natl. Acad. Sci. U.S. 71 (1974) 811.

Dr. BANNWARTH:

You told us that differentiation takes place in the anucleate cells and that regulation should take place on translational level. On the other hand BrdUr inhibits differentiation. Since this drug acts on the level of DNA synthesis does it mean, that organellar DNA synthesis is involved for the regulation or differentiation? What is the role of organellar DNA on differentiation?

Prof. BRACHET:

The work we did, a few years ago, on the effects of various inhibitors of DNA synthesis on nucleate and anucleate fragments of *Acetabularia* speaks against an important role of chloroplastic and mitochondrial DNA synthesis in morphogenesis. In this alga, chloroplastic DNA synthesis, which is needed for the multiplication of chloroplasts, might play a role in the growth of the caps; but it is not required for the initiation of cap formation. In sea urchin eggs, mitochondrial DNA is not replicated in activated anucleate fragments: the abortive cleavage undergone by such fragments is therefore not due to mitochondrial DNA synthesis.

Dr. FRANKEL:

You mentioned in your talk that cytoplasmic localization in ascidian eggs may involve preformed mRNA, or agents that induce the production of such RNA. I would like to mention an old experiment of Conklin, who showed that if the ascidian egg is centrifuged so that the internal cytoplasm is shifted, the cytoplasm will return to its original place if the centrifugation is performed at a sufficiently long time before cleavage. This implies either that the specific mRNA or inducing agents are bound to the cell surface region, or that there is some other basis of localization. I wonder whether you might comment on this observation.

Dr. BRACHET:

The centrifugation experiments of Conklin and others on eggs of many different species (mollusks, sea urchins, amphibians, etc.) have been taken by the experimental embryologists as evidence for the localization, in the cell membrane or the underlying cortex, of morphogenetic factors. This interpretation might be correct, but is not really proven: if these morphogenetic factors are mRNAs bound to proteins in the form of a ribonucleoprotein complex lighter than the ribosomes, their position would not be shifted by the very weak centrifugation forces used in Conklin's experiments. In fact, if eggs are ultracentrifuged, sedimentation of heavy cell organelles (yolk, in particular) becomes irreversible and development is impossible.

CHAPTER V.2

ACETABULARIA, A CELLULAR MODEL FOR THE STUDY OF NUCLEOCYTOPLASMIC INTERACTIONS

H.G. SCHWEIGER, H. BANNWARTH, S. BERGER and K. KLOPPSTECH

Max-Planck-Institut für Zellbiologie, D. 2940 Wilhelmshaven, W. Germany

The eukaryotic cell is characterized by a clear separation of cell nucleus and cytoplasm. The species-specific nuclear genome is stored in the cell nucleus and it is there where the first stage of gene expression, transcription, is localized. The cytoplasm plays a major role in the second stage of gene expression, namely in translation. The structural separation of the two compartments by the nuclear envelope raises the question as to what kinds of interactions relevant to gene expression exist between the nucleus and cytoplasm. In principle there are four possibilities: 1) The cell nucleus affects the cytoplasm; 2) The cytoplasm influences the cell nucleus; 3) The cytoplasm acts autonomously; 4) The cell nucleus acts autonomously.

With respect to all four problems the unicellular and uninuclear green alga *Acetabularia* is a favorable experimental organism. Simple growth conditions and the possibilities of enucleating a cell, isolating a cell nucleus and implanting a nucleus into a selected cytoplasm offer suitable prerequisites for cell biological studies.

Introduction

The term nucleocytoplasmic interaction implies the existence of two compartments which mutually influence each other. The nucleus, an essential feature of the eukaryotic cell, is separated from the cytoplasm by a complex membranous structure, the nuclear envelope.

The interrelationship between the nucleus and the cytoplasm is of special interest with respect to the control of gene expression. The nucleus contains the nuclear genome and all of the components necessary for the replication and transcription of nuclear DNA whereas the product of the transcription, the message, is translated within the cytoplasm on 80S cytosol ribosomes. In addition, mitochondria and chloroplasts contain protein-synthesizing systems which are separate from each other and from the one in the cytosol. These involve 70S ribosomes. The organelles also have organelle-specific DNA and their own replicating and transcribing systems.

Investigations on nucleocytoplasmic interactions are mainly concerned with the influence of the nucleus on the cytoplasm and vice versa. However, regulatory mechanisms may also operate exclusively within either the nu-

TABLE 1

Nucleocytoplasmic interactions.

1. Nucleus affects cytoplasm as revealed by
   a) enucleation
   b) implantation of a homologous nucleus
   c) implantation of a heterologous nucleus
   d) tracing nuclear products in the cytoplasm

2. Cytoplasm affects nucleus as revealed by
   a) experimental removal of cytoplasm
   b) introduction of material into cytoplasm
   c) implantation of a nucleus into a different cytoplasmic environment

3. Cytoplasm affects cytoplasm as revealed by
   a) regulatory functions after enucleation
   b) experimental modification of the cytoplasm in an anucleate cell

4. Nucleus affects nucleus as revealed by
   regulatory functions in an isolated nucleus

cleus or the cytoplasm. All four kinds of regulatory processes will be discussed below, the major object of this contribution being to present an outline of different experimental approaches to each category (table 1) and to demonstrate the unique value of *Acetabularia* for these studies.

Acetabularia

The unicellular and uninuclear green alga *Acetabularia* has been a favorite organism in cell biology for about 50 years (1,2,3). It develops a stalk which may be as long as 200 mm in some species. In the final stages of its life cycle, it produces a cap which is characteristic for the species. Throughout the vegetative phase, the single nucleus is located in the rhizoid (4) and can be easily removed by amputation. This operation gives rise to viable anucleate fragments which are capable of growth and differentiation. The nucleus can also be isolated and implanted into the cytoplasm of a cell of the same or a closely related species (1,2). The simplicity of all of these operations makes *Acetabularia* well suited for studies of nucleo-cytoplasmic interactions.

Nucleus affects nucleus

For investigations of autonomous nuclear regulation, it would be highly advantagous to be able to work with isolated nuclei, especially since in situ experiments have permitted only indirect conclusions. This has not been possible to date and thus very little is known about this class of regulatory phenomena.

Considerable effort has been expended in attempts to keep isolated *Acetabularia* nuclei functionally intact (Werz, unpublished). In tests of various isolation media, it could be shown that the addition of substantial amounts of sucrose improved nuclear viability. Nuclei which had been kept as long as 30 min in isolation were shown to have retained their morphogenetic capacity when they were implanted into anucleate stalk fragments (Werz, unpublished; 1,5,6). A survival time of 10 min was reported by Brändle and Zetsche (7) who used a similar medium containing sucrose. A substantial improvement in these viability times was recently achieved with the demonstration that nuclei remained morphogenetically competent after being stored for 24 h in an artificial medium (fig. 1). The medium used contained 0.067 M phosphate buffer, pH 7.8, 0.005 M magnesium acetate, 0.3 M sucrose and 0.1% bovine serum albumin (8).

The viability of a nucleus, which had been kept in isolation for 24 hr, was certified by its ability to induce growth and cap development in non-growing anucleate basal fragments. Moreover, when the implanted nucleus and cell fragment were from different species, the cap formed was that of the species from which the nucleus came. Cyst formation was also observed in some cases.

This recent technical development should soon permit direct investigations of intranuclear processes involved in gene expression and intranuclear metabolism.

Cytoplasm affects cytoplasm

Intracytoplasmic regulation can be conveniently studied in *Acetabularia* because the cell is so easily enucleated. Hämmerling's (4) early work showed that young enucleated *Acetabularia* remain capable of growth and morpho-

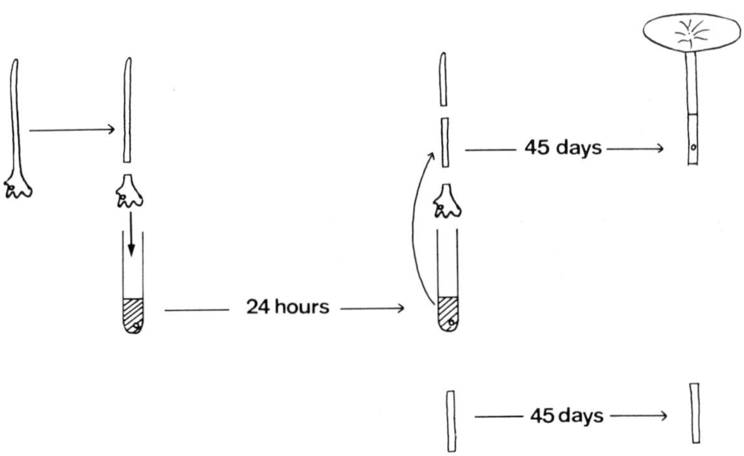

Fig. 1. The morphogenetic capability of a nucleus which was kept in an artificial medium for 24 hr.

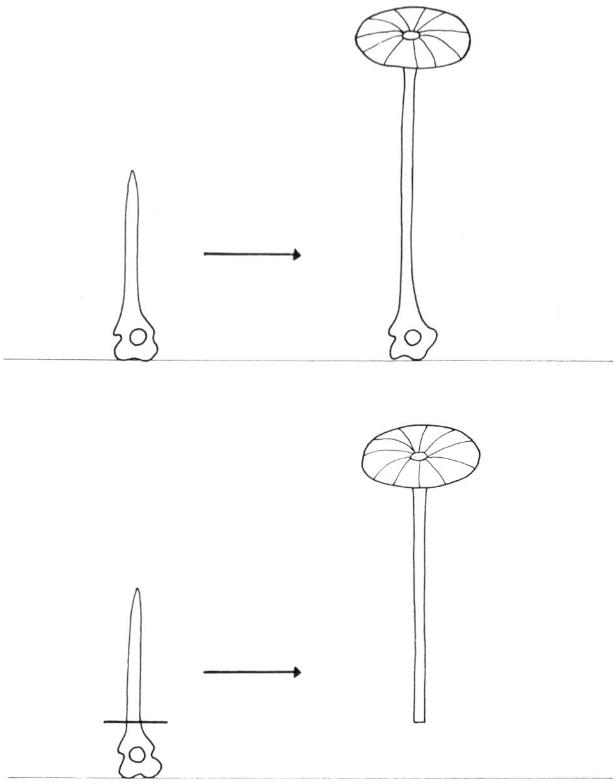

Fig. 2. Schematic presentation of the morphogenetic capability of a nucleate and an enucleated *Acetabularia* cell.

genesis and this has been confirmed repeatedly (fig. 2). A number of conclusions can be drawn from these results: 1) All of the components necessary for growth and morphogenesis are present in the cytoplasm of a young cell. 2) The information for all of the many steps of differentiation is transferred from the nucleus to the cytoplasm at about the same time and well in advance of specific morphogenetic events. Certainly the sequence of events during differentiation is not determined by an ordered release of mRNA from the nucleus. 3) The morphogenetic information can be stored in a stable form in the cytoplasm. 4) The cytoplasm displays temporal organization. In other words, it is capable of synthesizing specific proteins at required times in needed amounts. 5) In *Acetabularia*, the regulatory mechanisms underlying differentiation do not include selective gene activation.

Biochemical aspects of differentiation in *Acetabularia* have attracted considerable attention. The activities of a number of enzymes were shown to increase at discrete developmental stages in the absence of the nucleus. In the case of an alkaline phosphatase, this increase was initiated about 10 days after enucleation (9). A similar increase was also observed for a neutral phosphatase (10). Unfortunately, in none of the cases studied so far has it been definitely shown that the increase in enzyme activity is due to de novo synthesis or that the enzyme is coded by the nuclear genome.

For a number of enzymes which are involved in the synthesis of carbohydrate components of the cell wall, a close relationship between the increase in enzyme activity and morphogenesis was also demonstrated (11, 12). The activity of UDPG pyrophosphorylase starts to increase shortly prior to the initiation of cap formation. A similar increase was observed in enucleated cells. From the observation that the increase can be inhibited by puromycin it has been concluded that the increase is due to de novo synthesis of the enzyme protein.

Among the enzymes whose activities would be expected to increase at a distinct stage of development are those which are involved in the synthesis of DNA precursors. At the beginning of the generative phase, the number of nuclei in a cell increases from one to about 10^7 within a period of about 200 hr (13). This occurs after the cap has reached its maximum diameter. Since there is good reason to assume that the degree of ploidy is rather low before nuclear division, it is clear that a very rapid synthesis of DNA must take place. Studies on the enzyme thymidine kinase have shown that there is a substantial increase in the activity of this enzyme immediately prior to the period of nuclear multiplication (13). This activity increase occurs in the absence of the nucleus and is inhibited by chloramphenicol and puromycin, but not by cycloheximide (14). Evidently, thymidine kinase, an enzyme associated with preparation for nuclear DNA synthesis, is regulated within the cytoplasmic compartment and this regulation requires protein synthesis on 70S ribosomes. Two questions are raised: where does the signal for this regulation originate and what is the source of the mRNA for the protein involved? If the nuclear genome provides the genetic information, we must postulate that mRNA leaves the nucleus and migrates via the cytosol into the organelles where it is translated. How the mRNA finds its way into the organelle is yet another matter. In any event, regulation cannot take place at the level of nuclear transcription. Alternatively, if the information originates in the organelle genome and is translated there, we would have a situation in which the organelle genome codes for a protein which is directly or indirectly concerned with a nuclear function at a particular developmental stage.

Cytoplasm affects nucleus

In principle, there are two ways of investigating cytoplasmic effects on the nucleus. First, one may look for such an effect and then try to identify the active cytoplasmic factor. Second, one might try to find a substance such as a protein which migrates from the cytoplasm into the nucleus and then attempt to discover its function in the nucleus (15).

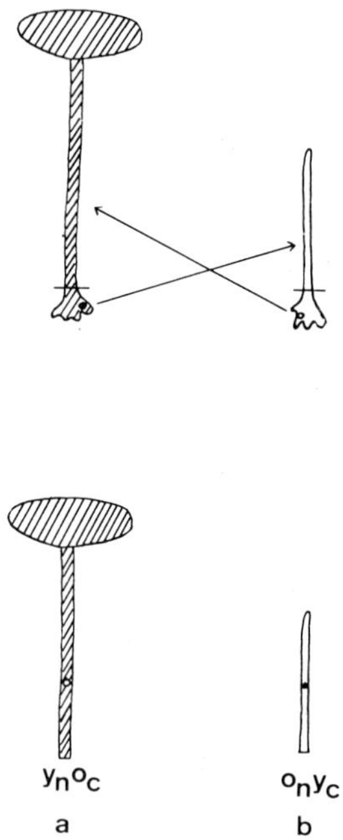

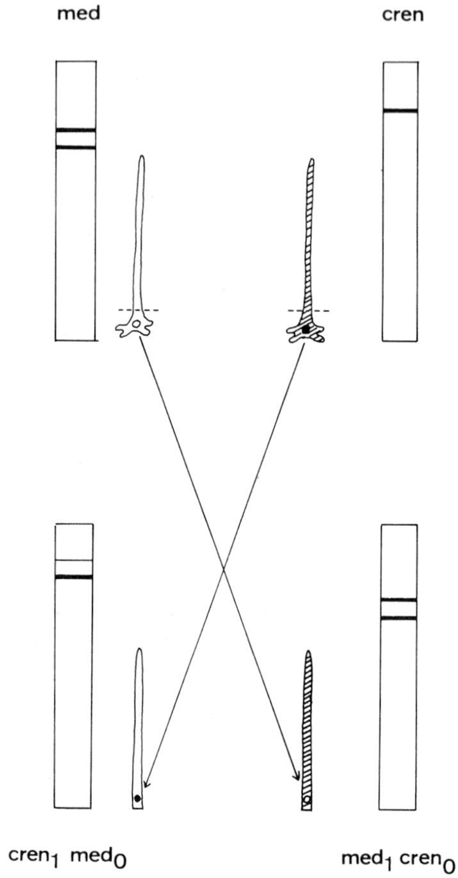

Fig. 3. Nuclear exchange between a young and an old cell. $y_n o_c$ = young nucleus in an old cytoplasm. $o_n y_c$ = old nucleus in a young cytoplasm.

Fig. 4. Lactic dehydrogenase isozyme pattern before and after nuclear exchange between a cell of *Acetabularia mediterranea (med)* and *A. crenulata (cren). $cren_1$ med_o* = nucleus of *cren* in cytoplasm of *med. med_1 $cren_o$* = nucleus of *med* in cytoplasm of *cren.*

With respect to the first of these possible approaches, a test system is now available. The system is based on an experiment of Hämmerling (16, 17) in which he showed that the grafting of a young rhizoid onto an old stalk shortened the time needed for the onset of cyst formation; the reciprocal transplants (old rhizoid onto young stalk) delayed the time of cyst formation. Recently, it has been learned that the nucleus and perinuclear region have a different morphology in old and young cells (18,19). If a young nucleus is isolated and implanted into an old cytoplasm, it soon acquires the morphology of an old nucleus (fig. 3). An old nucleus, on the other hand, develops the characteristic appearance of a young one after it is implanted into a young stalk (18,19). These experiments clearly demonstrate that the cytoplasm is capable of affecting the nucleus. The task now is to find the cytoplasmic agent which is responsible for this effect.

Nucleus affects cytoplasm

One simple way to investigate the influence of the nucleus on the cytoplasm is to study the effect of enucleation. However, in *Acetabularia* no dramatic changes in morphogenesis can be observed after the removal of the nucleus. Similarly, most of the biochemically defined changes associated with the cell's development are not affected by the loss of the nucleus.

Two other techniques have proved useful for studies of this problem; the grafting of a nucleated rhizoid onto an anucleate stalk (20) and the implantation of an isolated nucleus into an enucleated cell (1,6,7; Werz, unpublished). These methods are most elegant if the donor and recipient cells belong to different species and if the cytoplasm is a non-growing anucleate basal fragment. Then, growth, morphogenesis characteristic of the species from which the nucleus came and even cyst formation are induced in the basal fragments (21,2,7).

This type of experiment has been successfully extended to the molecular level. Gel electrophoresis of *Acetabularia* homogenates yielded enzyme banding patterns in the specifically stained gels which were characteristic for the species. These patterns differed in number and position of the bands. If nuclei were exchanged between cells differing in enzyme pattern, the enzyme pattern of the heterologous combination changed after a few weeks from the original cytoplasmic pattern to the one characteristic of the species from which the nucleus originated. With this type of experiment, it was shown that the enzymes lactic dehydrogenase and malic dehydrogenase (22,23,5) are coded by the nuclear genome (fig. 4). These methods can also be used to localize the genetic information of organelle proteins. If it could be shown that a given organelle protein is specified by the nuclear genome, it would mean that either the mRNA or the protein itself has to migrate into the organelle. If it could be shown further that the protein coded in the nucleus is synthesized on the organelle's 70S ribosomes (24), it would signify that both nuclear and organellar systems are involved in the expression of a single gene.

Besides the lactic dehydrogenase and the malic dehydrogenase a number of chloroplast membrane proteins (25) and at least some of the chloroplast ribosomal proteins (26) were shown to be specified by the nuclear genome, this despite the fact that the chloroplast has a genome of its own as well as the enzyme systems needed for transcription and translation. The situation is particularly complex in the case of the chloroplast ribosomal proteins. Nuclear coding of some of these proteins means that the ribosomes are "hybrids" with respect to the source of genetic information for their structure. Moreover, the organelle rRNA is known to be transcribed on the organelle genome (27, 28). This "hybridization" is interesting from an evolutionary point of view since in general the protein components of the chloroplast ribosomes seem to have undergone more structural changes with time than the rRNA has.

A more direct approach to the problem of nuclear influences on the cytoplasm requires a search for nuclear products in the cytoplasm. Hämmerling's (4,1) experiments demonstrated that there had to be materials in the cytoplasm of nuclear origin which induced species-specific morphogenesis. He called these materials "morphogenetische Substanzen". Although direct proof is still lacking, it is generally assumed that Hämmerling's hypothetical substances are a type of mRNA. The existence of "morphogenetische Substanzen" can be proven in the cytoplasm for weeks after enucleation (29). If these substances are indeed mRNA, their stability is surprising, especially since there are rather high levels of RNase in the cytoplasm of an *Acetabularia* cell (30). It is possible that the mRNA is protected against RNase attack because it is bound to proteins and especially if it exists in the cytoplasm as RNP particles (31). Several classes of such particles exist in *Acetabularia*. They sediment at about 20S, 40S and 110—120S (32—35). According to Zubarev and his coworkers, all of these classes of particles are functionally similar. We feel that further investigation is needed to definitely exclude the possiblity that, e.g. the 120S particles may be virus particles. Since both 40S and 120S particles appear to be synthesized in the absence of the nucleus (35) and virus-like particles have been seen in *Acetabularia* (36), some caution seems advisable.

The 40S particles have been isolated and injected into non-growing anucleate basal fragments by Zubarev and his colleagues. They report that the 40S particles induced limited growth and morphogenesis under the conditions of their experiments (33, 34). Discussions of the significance of this report should await confirmation from other laboratories, especially since it seems unlikely that all of the mRNA for many proteins with different molecular weights could belong to a single class of RNP particles having a common molecular weight.

Another approach to identifying the postulated "morphogenetische Substanzen" starts with the assumption that they are mRNA. It is known from other systems that most of the mRNA species contain terminal stretches of poly A (37). This property has been successfully exploited to isolate material with mRNA properties (38). Preliminary studies of the polyadenylated RNA in *Acetabularia* suggest it is also identical with mRNA (39). The amount of polyadenylated RNA increased in nucleated, but not in enucleated cells and is correlated with the length of the cell. It is almost evenly distributed within the cell. This polyadenylated RNA migrates electrophoretically in two broad bands which are clearly separated from rRNA. Interestingly, there is a substantial increase in the labeling of the poly A-RNA in regenerating cells, but this increase is significantly lower than the one observed for incorporation of label into RNA (40). The difference might reflect the fact that nucleated cell is normally synthesizing mRNA at almost its full capacity and rRNA to a lesser extent. Then a stimulus to produce RNA at a maximal rate, as is given when the apex is removed and the cell starts to regenerate, would have a greater effect on the rate of rRNA synthesis than mRNA synthesis.

A chapter on nuclear products in the cytoplasm must include discussion of the rRNA of the 80S ribosomes. Recent experiments have established the existence of 80S cytosol ribosomes in *Acetabularia* (41). The corresponding rRNAs sediment with S-values of 18S and 26.5S. Labeling techniques have been used to estimate the rate of rRNA synthesis in normally growing and regenerating cells (40), as well as the transport and stability of the 80S ribosomes (42). The extreme stability of the 80S ribosomes deserves particular mention. It has been estimated that the half life of the rRNA in the ribosomes is roughly 80 days. This stability makes it possible to both estimate the rate of rRNA synthesis directly from the labeling results and calculate the number of genes which code for rRNA. These genes appear to be highly redundant (table 2).

Very recently, direct evidence for rRNA gene redundancy was obtained from electron micrographs of nucleolar material prepared by the Miller (43) technique. The nucleolar material from *A. major* contains long stretches of DNA which show many active regions of RNA synthesis. These active regions are equal in length and are evenly distributed along the DNA molecule alternating with non-active spacer regions (fig. 5) (44). The length of an active region corresponds approximately to the length expected for an rRNA gene. This means these regions are not long enough to permit loss of infor-

TABLE 2

Number of rRNA cistrons in one cell of *Acetabularia major.* (Theoretical considerations).

Experimental result:
Rate of synthesis in one cell
100 ng rRNA in 60 days
or
3.4×10^7 nucleotides in 1 sec

Assumptions:
1) 100 polymerase molecules on 1 cistron
2) 1 polymerase molecule polymerises 10 nucleotides per second
from 1 and 2 follows:
rate of polymerisation on one cistron
1×10^3 nucleotides per second

Conclusion:
About 3.4×10^4 cistrons are needed per cell
in order to explain the rate of synthesis observed

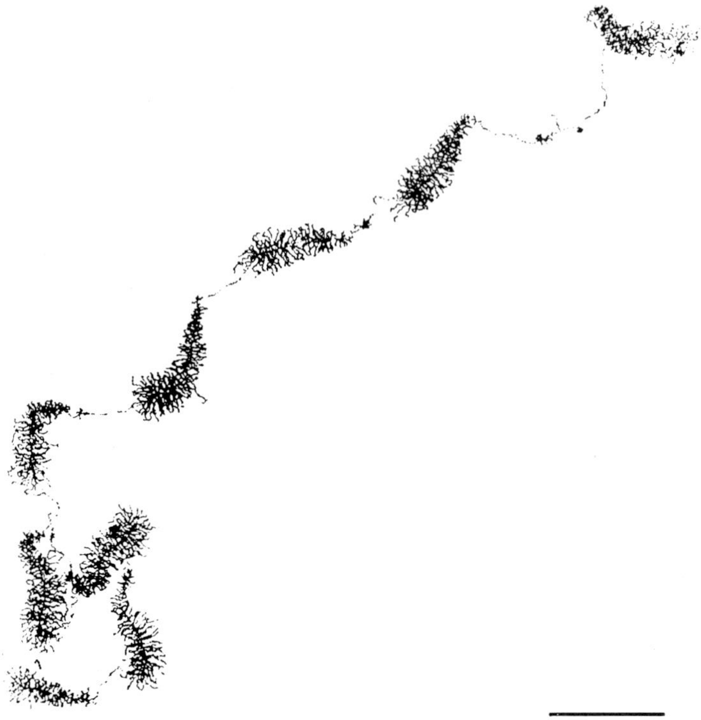

Fig. 5. Cistrons of rRNA from a nucleus of *Acetabularia major*. The "active regions" are separated by "spacers". Note the extralong spacer in the upper right part. Bar represents 1.56 μm.

mation during rRNA maturation. Such a loss of material has been reported to occur during rRNA maturation in other types of cells (45). Approximate estimates of rRNA redundancy based on these electron microscopic studies agree reasonably well with the figures derived from the rate of rRNA synthesis. The number of rRNA cistrons in a growing *A. major* cell is approximately 3×10^4. Similar results have been obtained with *A. mediterranea* using a slightly modified technique (46, 47). The amount of duplex DNA corresponding to this number of genes and the inactive spacer regions is approximately 3×10^{-13} g. This represents a substantial proportion of the total DNA contained in the *A. major* nucleus.

A full understanding of the regulation of gene expression has to consider a network of mutual interrelationships whose investigation needs, at least in a first step, a clean dissectioning of the individual mechanisms involved. The unicellular and uninuclear green alga *Acetabularia* offers the possibility to investigate separately all four interrelationships between the cell nucleus and the cytoplasm as well as autonomous functions confined to only one of the two compartments. The large size of this organism allows the application of cell biological as well as biochemical methods to a limited number of cells.

Acknowledgement

The authors are grateful to Dr. Marlene Karakashian for valuable help in preparing the manuscript.

References

1. J. Hämmerling, Ann. Rev. Plant Physiol. 14 (1963) 65.
2. H.G. Schweiger, Curr. Top. Microbiol. Immunol. 50 (1969) 1.
3. S. Puiseux-Dao, Acetabularia and Cell Biology, Logos Press Ltd., London (1970).
4. J. Hämmerling, Biol. Zentralbl. 51 (1931) 633.
5. L. Sandakhchiev, R. Niemann and H.G. Schweiger, Protoplasma 76 (1973) 403.
6. J. Hämmerling, Biol. Zentralbl. 74 (1955) 545.
7. E. Brändle and K. Zetsche, Planta 111 (1973) 209.
8. S. Berger, R. Niemann and H.G. Schweiger, Protoplasma (in press).
9. T. Spencer and H. Harris, Biochem. J. 91 (1964) 282.
10. E.L. Triplett, A. Steens-Lievens and E. Baltus, Exp. Cell Res. 38 (1965) 366.
11. K. Zetsche, Biochim. Biophys. Acta 124 (1966) 332.
12. K. Zetsche, Z. Naturforsch. 23 b (1968) 369.
13. H. Bannwarth and H.G. Schweiger, Proc. Roy. Soc. London, Ser. B 188 (1975) 203.
14. H. Bannwarth and H.G. Schweiger (in preparation).
15. P. Clerin, T. Vanden Driessche and J. Brachet, This volume.
16. J. Hämmerling, Int. Rev. Cytol. 2 (1953) 475.
17. J. Hämmerling, Biol. Zentralbl. 59 (1939) 158.
18. S. Berger and H.G. Schweiger, J. Cell Sci. 17 (1975) (in press).
19. S. Berger and H.G. Schweiger, Proc. 2nd Europ. Symp. Cell Cycle, Innsbruck (1973) 49.
20. J. Hämmerling, Z. induktive Abstammungs- und Vererbungslehre 81 (1943) 114.
21. W. Reuter, G. Werz and H.G. Schweiger, Protoplasma 68 (1969) 351.
22. W. Reuter and H.G. Schweiger, Protoplasma 68 (1969) 357.
23. H.G. Schweiger, R.W.P. Master and G. Werz, Nature 216 (1967) 554.
24. K. Apel and H.G. Schweiger, Eur. J. Biochem. 38 (1973) 373.
25. K. Apel and H.G. Schweiger, Eur. J. Biochem. 25 (1972) 229.
26. K. Kloppstech and H.G. Schweiger, Exp. Cell Res. 80 (1973) 69.
27. H.G. Schweiger and S. Berger, Biochim. Biophys. Acta 87 (1964) 533.
28. S. Berger, Protoplasma 64 (1967) 13.
29. J. Hämmerling, Roux' Arch. Entwicklungsmech. 131 (1934) 1.
30. H.G. Schweiger, Planta 68 (1966) 247.
31. A.S. Spirin, Eur. J. Biochem. 10 (1969) 20.
32. K. Kloppstech, Protoplasma 75 (1972) 479.
33. A.B. Alexeev, T.N. Zubarev, S.L. Stvolinskij, A.A. Jazykov and M.I. Betina, Dokl. Akad. Nauk SSSR 215 (1974) 712.
34. A.B. Alexeev, S.L. Stvolinsky, L.A. Tikhomirova, A.A. Yazykov and T.N. Zubarev, Protoplasma 83 (1975) 168.
35. K. Kloppstech and H.G. Schweiger (in preparation).
36. S. Berger, W. Herth, W.W. Franke, H. Falk, H. Spring and H.G. Schweiger, Protoplasma (in press).
37. G.A. Schultz, Developm. Biol. 31 (1973) 2.

38. H. Aviv, I. Boime, B. Loyd and P. Leder, Proc. Nat. Acad. Sci. U.S. 69 (1972) 1408.
39. K. Kloppstech, Protoplasma 83 (1975) 177.
40. K. Kloppstech and H.G. Schweiger, Differentiation 1 (1973) 331.
41. K. Kloppstech and H.G. Schweiger, Biochim. Biophys. Acta 324 (1973) 365.
42. K. Kloppstech and H.G. Schweiger, Protoplasma 83 (1975) 27.
43. O.L. Miller and B.A. Hamkalo, Int. Rev. Cytol. 33 (1972) 1.
44. S. Berger and H.G. Schweiger, Protoplasma 83 (1975) 41.
45. J.E. Darnell, Bacteriol. Rev. 32 (1968) 262.
46. M. Trendelenburg, H.Spring, U. Scheer and W.W. Franke, Proc. Nat. Acad. Sci. U.S. 71 (1974) 41.
47. C.L.F. Woodcock, J.E. Stanchfield and R.R. Gould, Plant Sci. Lett. 4 (1975) 17.

Dr. FRANKEL:

Is there any biochemical evidence of nuclear activity under conditions in which most metabolic activities are shut down but morphogenetic substances are known to be produced, namely in the dark?

Dr. SCHWEIGER:

The RNA components of the 80S ribosomes, e.g., are synthesized in the dark by about 80% of the light production.

Dr. FRANKEL;

What is the evidence that the giant primary nucleus of mature *Acetabularia* cells is diploid?

Dr. SCHWEIGER:

The low degree of ploidy of the primary nucleus of the mature *Acetabularia* cell is indicated by: a) the difficulty to find Feulgen positive regions in the nucleus; b) the lack of labelling of nuclear DNA in the maturing cell.

Dr. OHAD:

Does the nucleus isolated for transplantation contain remnants of the cytoplasm or chloroplasts and if so, is there any correlation between this extra nuclear material and the survival of the nucleus in vitro?

Dr. SCHWEIGER:

There is a certain amount of cytoplasm including chloroplasts sticking to the nucleus during the isolating procedure. In a good preparation this remaining cytoplasm should not exceed 0.1% of the nuclear volume. In the final stage of development of the vegetative nucleus this percentage may be increased by a perinuclear cytoplasmic layer, which in our opinion is functionally related to the nucleus rather than to the rest of the cytoplasm.

Dr. DELSENY:

It has been reported that the release of mRNA in mammalian cells from nucleus is dependent of ATP and protein factors from cytoplasm. Did you try to incubate isolated nuclei with ATP or with cytoplasmic extracts before transplantation and to test for their morphogenetic potentialities. Do you think it would be an interesting approach?

Dr. SCHWEIGER:

We did not try so far. However, similar experiments are in our program.

Dr. WILLIAMSON:

Do you have evidence that the thymidine kinase made in the anucleate fragments is the same enzyme as is made in the nucleated cell? Also is it possible to block nuclear reproduction by chloramphenicol?

Dr. SCHWEIGER:

The only indication that the enzymes in both cases are identical comes from a study on the distribution of the enzyme activity between supernatant and sediment. The percentage is the same in both cases. Chloramphenicol has not been used in blocking nuclear reproduction as far as I know.

SECTION VI

FROM THE NUCLEUS TO THE CYTOPLASM

CHAPTER VI.1

INFORMATION TRANSPORT FROM THE NUCLEUS TO CYTOPLASM

O.P. SAMARINA

Institute of Molecular Biology, Academy of Sciences of the USSR, Moscow, USSR

The different steps in information transfer in eucaryotic cells were studied: 1) Some aspects of the organization of transcription unit. 2) Characteristics of pre-mRNA as the primary product of transcription: the localization and some properties of its messenger sequences; together with some details of the structure of its non-informative part. 3) Characteristics of the specific RNP particles which act as carriers for the pre-mRNAs in the cell nucleus while they undergo processing and are transferred to the nuclear membrane. 4) The fate of mRNA in the cytoplasm judged on the comparison of nuclear and cytoplasmic RNP containing mRNA.

Information flow in eucaryotic cells includes several steps and levels, the detailed mechanism of which is not yet sufficiently clear.

To understand the information transfer the following questions need to be answered: 1) How is the genome organized? 2) How is RNA transcribed from the genome? 3) What happens to the RNA after transcription? In what form and how is it transferred across the nuclear membrane? and finally: 4) What is the fate of mRNA in the cytoplasm?

We do not pretend to be able to answer these questions, but we will try to elucidate the present state of our knowledge concerning some of them.

Genome organization

A brief description of three points related to the organization of the genome:

1) It has been shown for yeast chromosome (1), and less conclusively for *Drosophila* chromosome (2) that each eucaryotic chromosome contains one linear DNA molecule.

2) Using polytene chromosomes, it has been established that the number of complementation groups is roughly equal to the chromomere number. This means that each chromomere contains probably one, or only a very few, structural genes.

3) The DNA content of one chromomere is much larger than is required for one gene (3,4). The question therefore arises as to whether the excess of DNA in a chromomere is transcribed and what its function is.

Daneholt in 1972 (5) showed that RNA synthesized in one puff of a Balbiani ring, which comprises one chromomere, contains information for only one, or at least a very few proteins, and has a high molecular weight.

Thus it seems that the whole chomomere is transcribed. If this is the case, one alternative arises for discussion: a) the DNA of a chromomere contains multiple repetitions of one structural gene; or b) each chromomere contains information for one, or a very limited number of structural genes together with additional information.

To discriminate between these possibilities it was necessary to find out how many copies of each gene were in the genome and in the chromomere.

This question can be answered by using the renaturation-hybridization technique. For this purpose, individual genes or their transcript-mRNA were needed. At the time a number of individual mRNAs have been isolated and their anti-mRNA synthesized with the aid of reverse transcriptase.

Investigation of the hybridization kinetics of mRNA with DNA showed that the majority of genes in the genome of eukaryotes were not reiterated. Only a few exceptions were found; for example the genes for histone mRNA are reiterated (6).

Thus is seems likely that the chromomere in most cases contains only one structural gene. The excess of DNA in chromomeres does not therefore represent structural genes.

Transcription

As mentioned earlier some data has shown that the whole chromomere is transcribed. Detailed investigation of this primary transcript may help to understand the structural organization of the transcription unit.

Several years ago, the giant nuclear pre-mRNAs with a molecular weight of several millions were discovered (7—10). It was assumed that they represented the primary products of transcription.

It is known that newly synthesized RNA contains a triphosphate group at the 5′-end. It was shown in our laboratory that only giant molecules of pre-mRNA contained this marker at 5′-ends. This means that only giant molecules are primary products of transcription (11).

This transcript seems to contain mRNA sequences although it represents only a small part of the whole molecule. To investigate the presence of sequences corresponding to mRNA, different approaches have been used:

a) In hybridization experiments, the cytoplasmic RNA (which contained mRNA) showed 20—40% of competition with nuclear pre-mRNA. On the other hand, the hybridization of cytoplasmic mRNA was completely inhibited by nuclear pre-mRNA. This nuclear pre-mRNA contained all the sequences presented in mRNA (12,13).

b) Identical results have been obtained by Lindberg and Darnell (14) using cells transformed by SV-40 viruses, and by ourselves using Polyoma viruses.

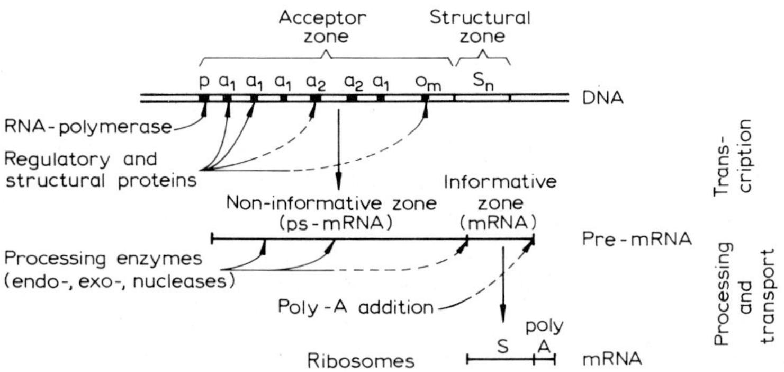

Fig. 1. A slightly modified model of transcription in eukaryotes (16) (p) Promotor; (a_1—a_m) acceptor sites; $m \geq 1$; (S_n) structural gene(s); usually $n = 1$.

The giant pre-mRNA synthesized after infection contained virus specific sequences, denoted virus mRNA.

c) Imaizumi et al. (15) investigated hybridization of globin anti-mRNA, synthesized on mRNA, with giant pre-mRNA and showed that the latter contained sequences similar to the globin mRNA.

d) Finally, Williamson et al. (16) showed induction of hemoglobin synthesis in frog oocytes upon injection of pre-mRNA.

Thus, many experiments confirmed the presence of mRNA sequences as an inherent part of pre-mRNA.

The next question is when these sequences are localized in the chromomere or in its transcript. In other words, what is the structural organization of transcription units in eucaryotes?

Several hypotheses concerning this problem have been suggested. According to Georgiev's scheme (17) the transcription unit in eucaryotes (fig. 1) consists of two types of nucleotide sequences: non-informative ones located mainly close to the promotor zone; and informative sequences or structural cistrons situated distal to the promotor. The non-informative part of the genome is usually much longer than the informative part and seems to have a regulatory function. There are many regulatory sequences common in different transcription units. This makes it possible to perform simultaneous regulation (for example switching on or off) in many different transcription units. These sequences can represent one of the sources of repetition in hybridization experiments. RNA-polymerase transcribes all the sequences. As a result the giant pre-mRNA is synthesized which contains non-informative replicas near the 5′-end and replicas from structural cistrons close to the 3′-end. During processing the non-informative part of pre-mRNA, which is called pseudo mRNA (ps-mRNA), is degraded. The informative part, that is the sequences corresponding to mRNA, is tranferred to the cytoplasm after transformation.

Localization of mRNA sequences in pre-mRNA

According to Georgiev's scheme, it was predicted that the mRNA is located near the 3′-end. This was proved in two different ways: a) recently it has been discovered that mRNA, as well as pre-mRNA, contains poly A sequences of 150—200 nucleotides in length. Later it was shown that these poly A sequences were localized at the 3′-end of giant pre-mRNA and mRNA (18—20). Since poly A sequences are connected with the mRNA, it is reasonable to assume that mRNA sequences are also localized near the 3′-end of giant precursor molecules; b) in our laboratory the following experiments with competitive hybridization of double-labelled pre-mRNA have been performed (21). Uniformly labelled in vivo pre-mRNA was labelled in vitro in 3′-nucleosides using the periodation-^{3}H borahydride reduction technique (22). This made it possible to investigate the hybridization properties of the whole molecule and its 3′-end sequences independently. It was shown that the hybridization of 3′-end sequences was inhibited by mRNA much more strongly than that of the whole molecule (table 1). This also means that mRNA sequences transported to the cytoplasm are located near the 3′-end of pre-mRNA.

The interesting result of this experiment is that hybridization of 3′-end sequences of pre-mRNA is 2—3 times higher than that of the whole molecule. Thus, besides mRNA sequences, some repetitive sequences can be connected with mRNA near 3′-end. The nature and localization of these sequences is not clear yet. It is possible that they represent terminator sequences for transcription or translation and are localized after mRNA sequences. In this case poly A blocks are added to these sequences during processing.

TABLE 1

Hybridization properties of 5′- and 3′-end fragments of pre-mRNA from Ehrlich Ascites carcinoma cells uniformly labeled in vivo by ^{14}C-orotic acid and chemically in the 3′-end position with ^{3}H (22).

	Hybridization, %		Ratio	Hybridization, %	
	Np* total	pppNp** 5′-end	^{3}H/^{14}C	^{14}C total	^{3}H 3′-end
Non-hybridized material	—	—	0.56	—	—
Hybrid	7.4	35	2.3	5.2	18.3
Hybrid in the presence of mRNA competitor	6.4	40	0.8	4.7	5.7
Competition, %	14	0	—	10	69

* Inside nucleotides.

** Starting nucleotides (recovered as pppXp after the complete alkaline hydrolysis.)

Non-informative part of pre-mRNA

According to Georgiev's hypothesis this RNA (ps-mRNA) is positioned close to the 5′-end of giant pre-mRNA. The size of this part is usually much larger than that of mRNA sequences. Owing to its large size and high complexity the structure of this part of pre-mRNA has been poorly investigated. Until now only some marker sequences have been studied. These are as follows:

1) The starting sequences which comprise the sequences containing the 5′-end of natural pre-mRNA. They are characterized by the presence of a triphosphate group in the first nucleotide. As was already mentioned, only giant pre-mRNA contained these true starting sequences. Smaller pre-mRNA molecules comprise the processing product and do not contain this marker (23). Therefore the 5′-end of pre-mRNA must be degraded during processing.

Recently success was achieved in the separation of pre-mRNA fragments containing 5′-end using hydroxyapatite chromatography (24). Isolation of the fragments of different size (from 10—15,000 to 30—60,000 daltons) seems to be possible. This technique is based on the retention of triphosphate group-containing fragments to the absorbent. Further rechromatography produced purified fragments. These were identified by their similarity in size, calculated on the basis of 5′-end triphosphate content, to the size of the input fragments. Almost 80% of terminal fragments were recovered.

2) The next marker sequences in the non-informative part of pre-mRNA are the fragments containing blocks of homonucleotides. Two kinds of such sequences have been discovered. a) Poly U sequences, about 30 nucleotides long, have been found in giant pre-mRNA (25,26). These are degraded during processing and are not components of mRNA. b) Some evidence about the existence of short (20—40 nucleotides long) poly A sequences has also been obtained (27). These sequences also belong to ps-mRNA as they have not been found in mRNA.

3) The third marker sequences in ps-mRNA are the repetitive sequences which can be identified by the hybridization test. Their predominant localization in the non-informative part of pre-mRNA was checked in experiments with pre-mRNA which had been synthesized after UV-irradiation. Presumably under these conditions incomplete molecules of pre-mRNA, enriched with ps-mRNA, are synthesized. It has been established that RNA from irradiated cells hybridizes with DNA 2—3 times more efficiently than pre-mRNA from normal cells and that mRNA does not compete in this hybridization. Therefore the non-informative part of pre-mRNA seems to contain many more reiterated sequences (28) and these are not transferred to the cytoplasm. Part if this repetition can be produced by double-stranded, hairpin-like structures (29,30). These structures are characterized by their resistance to RNase. It is possible to isolate two types of hairpin-like structures

(further referred to as 'hairpins'): a) large hairpins containing 80—100 base pairs and which are AU- base rich; b) short hairpins containing 15—20 base pairs. These are more heterogeneous in size and very GC- base rich.

The long hairpins were isolated and investigated in more detail: they are characterized by high speed of renaturation. Analysis of their renaturation kinetics reveals two main fractions: rapidly renaturing hairpins and a more heterogeneous and slowly renaturating fraction. The later comprises 20—25% of the first fraction. These two fractions seem to be transcribed from different reverted sequences scattered along the genome.

The fraction of rapidly renaturing hairpins is of great interest. Some data indicate that this fraction contains one or a few kinds of inverted repetition sequences (31). Cross renaturation of long hairpins isolated from pre-mRNA of different mouse tissues showed that this fraction seems to be identical in different tissues and therefore that there is an absence of tissue specificity.

Similar hairpin-like structures with complementary sequences were found in DNA (31). They comprise about 2% of mouse nuclear DNA. Half of them belong to the sattelite fraction but the rest (~1% of total DNA) are transcribed. The size of these structures is comparable with that of the rapidly renaturing long hairpins of pre-mRNA. The DNA hairpins renatured at a very fast rate (with the $C_0t = 10^{-6}$—10^{-7}) and it was therefore very difficult to isolate them in a denatured state. It was also found that this DNA hairpins hybridized very efficiently with rapidly renaturing hairpins isolated from pre-mRNA. Thus, the active hybridization of hairpins isolated from RNA and DNA together with their similarity in size, indicated that the long hairpins of the pre-mRNA represent the transcripts of the inverted repetitions of DNA.

The next interesting finding is the presence of complementary sequence in the hairpins and mRNA. About 1/3 of the hairpin sequences are hybridized with mRNA at the saturation concentration of the latter (table 2).

It is possible that the ability of mRNA for hybridization with pre-mRNA and with DNA can be explained by the presence of these complementary sequences (31).

TABLE 2

Hybridization of pre-mRNA hairpins and mRNA.

C_0t of hairpins	Amount of mRNA added (ratio to hairpins)	RNase stable material (%)	
		total	minus background*
1.5×10^{-4}	9×10^2	15	10
1.5×10^{-4}	94×10^2	32	27
5×10^{-3}	24×10^2	30	23
5×10^{-3}	96×10^2	32	25

* Background, the amount of self-reassociated hairpin sequences.

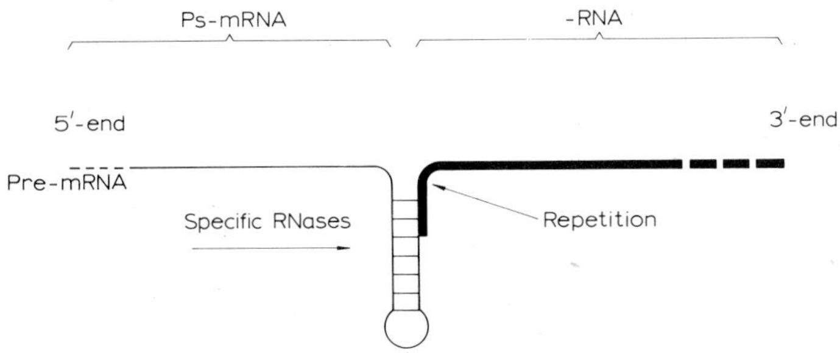

Fig. 2. Hypothetical scheme of the double-stranded region functioning by separating the messenger sequences and ps-mRNA in pre-mRNA.

The location of these sequences in pre-mRNA has not yet been elucidated. It is reasonable to assume, as shown on fig. 2, that they can separate the mRNA and ps-mRNA parts of the precursor molecules and may be recognized by the specific RNases involved in processing. During this processing the ps-mRNA and some parts of the hairpins are degraded. Part of the latter may be preserved. If this is the case, these sequences must be positioned at the 5′-end of mRNA and must be common to many mRNAs because of their ability to hybridize with pre-mRNA (fig. 2).

Processing

Let us consider shortly the forms in which pre-mRNA exists in the nucleus, undergoes processing and is transferred to the nuclear membrane. All the processing steps take place in the cell nucleus. During this, ps-mRNA is destroyed and mRNA sequences are selected ready to be transported to the cytoplasm.

Little is known at the moment about the precise mechanism and the enzymes involved in this process, but there is a great deal of evidence suggesting that processing takes place while the pre-mRNA is combined with proteins.

In 1965 nuclear RNP containing pre-mRNA was discovered (32). Almost all pre-mRNA exists in this form. The following general scheme of the structural organization of these particles was postulated: the particles have a polysome-like structure (fig. 3) where the pre-mRNA chain is combined with a number of specific macroglobular protein particles and is distributed mainly on the surface of these particles. These protein particles are referred to as informofers. Each informofer is combined with a chain of pre-mRNA of molecular weights about 200,000 daltons, forming monomeric 30S particles which usually are extracted from nuclei without RNase inhibitor (fig. 3). Informofer itself consists of a number (25—50) of very similar protein subunits which have a molecular weight of about 40,000 and are called infor-

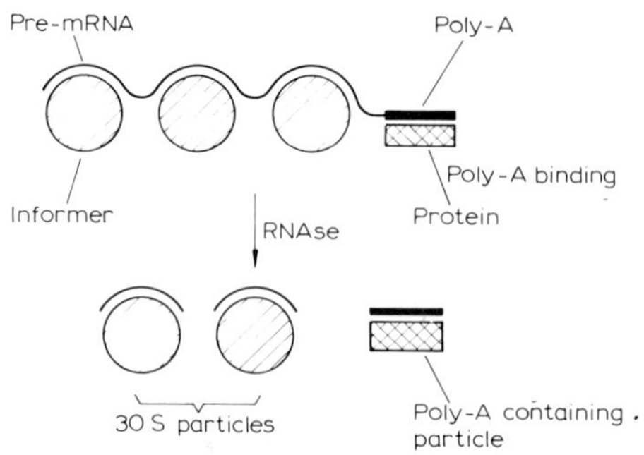

Fig. 3. The probable structural organization of the heavy nuclear complexes containing pre-mRNA with poly A and two kinds of the protein particles.

matin. This protein has recently undergone extensive discussion as, by using SDS polyacrylamide gel electrophoresis some additional minor components have been revealed. Especially high heterogenity can be observed if the particles are not sufficiently purified. However, in purified samples the quantity of the minor components is not more than 10%. About 90% of material consists of homogeneous protein-information which is sometimes seen on polyacrylamide gel as two bands localized very close together. Thus, it seems that the main protein of informofer is informatin, and that 30S particles may contain some additional proteins either as an impurity or as the enzymes which are involved in pre-mRNA processing. It is possible that in different experimental conditions the amount of these functional proteins may vary significantly.

The second statement about the existence of informofer as a macroglobular entity gained evidence from the survival of informofer after the complete removal of RNA from the particles by dissociation with salt. During this process the protein part of some particles is conserved and retains the same S values as the native particles. The stability of informofers sharply decreases after the removal of RNA; some of them are destroyed but preliminary iodination seems to help stabilize them. After dissociation of iodinated particles about half of the informofers can be recovered.

All sequences of pre-mRNA (ps-mRNA as well as mRNA) were bound with informofers. The binding of ps-mRNA with informofer is clear as this RNA comprises the bulk of pre-mRNA. The binding of mRNA with informofer was shown by detecting that in cell nuclei virus specific mRNA was also present in informofer complexes (34). Secondly, it was shown that RNA containing poly A blocks, presumably mRNA sequences, also complexed with informofers (35). It is interesting that poly A itself is not bound to informofers but to some other protein and forms particles of a different kind. The sedimentation coefficient of poly A containing particles is 14S instead of 30S for monoparticles containing informofers. The molecular weight of the main protein which is complexed with poly A is much higher than that of information (36). Thus, polyparticles comprise the mixed structures in which long pre-mRNA sequences are bound to informofers with the poly A sequences at the 3′-end bound to a distinct protein (35) (fig. 3).

Thus, pre-mRNA during all processing stages is complexed with informofers. These structures seem to provide everything that is needed for processing: pre-mRNA is specifically packed on the surface of informofers, which not only greatly diminishes their linear size, but also gives optimal exposure of pre-mRNA for the action of specific processing enzymes. Transformation of pre-mRNA, via addition of poly A blocks to its 3′-end, also takes place in these structures.

Informofers accompany mRNA to the nuclear membrane. Then the processed mRNA is transferred to the cytoplasm. The mechanism of this process is unknown. To understand it, it is necessary to find out whether nuclear particles as such are transported with mRNA to the cytoplasm. Practically the question concerns the identity or difference between nuclear and cytoplasmic RNP and their protein.

There are two kinds of mRNA-containing particles in the cytoplasm: 1) free informosomes, initially described by Spirin (37); and 2) polysome-bound mRNP.

Unfortunately free informosomes have not yet been isolated in a pure enough state. In some cases their presence in the cytoplasm can be explained by leakage from the nuclei. However in other cases their presence has been sufficiently established, for example K. Scherrer and his coworkers examined free cytoplasmic 20S RNP particles from duck erythroblasts containing 9S hemoglobin mRNA. These particles did not contain informatin (38).

Polysomal mRNP can be obtained in a more or less pure state. This makes it possible to investigate its proteins. For example homogenous 14S mRNP released from polysomes of mouse reticulocytes has been isolated. It contained two main proteins and both of them were different from informatin when analysed by polyacrylamide gel electrophoresis (39). The same results were obtained for mRNP isolated from polysomes of duck (40) and rabbit (41) erythroblasts. Differences in the proteins of nuclear particles and mRNP from polysomes have been also confirmed by immunological technique using antiserum to 30S particles or to informatin (42). Thus, differences between the protein of polysomal mRNP and nuclear particles have been shown by a variety of methods.

If this is the case, one can assume that mRNA, which is retained in nuclei after processing, is transported with informofer only to the nuclear membrane. Then it is released from the informofers and injected into the cytoplasm. Electron-microscopic observation also showed the disappearance of particles when mRNA passed the nuclear membrane (43).

The fate of mRNA in the cytoplasm is not yet clear. One can assume that mRNA combines with ribosomes directly on the outer side of the nuclear membrane before or after binding with some proteins. If so, the role of the informosome-like particles is unclear. However investigation of the kinetics of mRNA labelling in different cytoplasmic structures shows indirectly that these free particles may represent an intermediate step in mRNA transport from the nuclei to polysomes in the cytoplasm (44).

References

1. T.D. Petes, C.S. Newlon, B.E. Byers and W.L. Fangman, Cold Spring Harbor Symp. Quant. Biol. 38 (1974) 9.
2. R. Kavenoff and B.U. Zimm, Cold Spring Harbor Symp. Quant. Biol. 38 (1974) 1.
3. B.H. Judd, M.W. Shen and T.C. Kaufmann, Genetics 71 (1972) 139.
4. W. Beerman, in I.B. Zbarsky and G.P. Georgiev (Eds.), Cell Nucleus: Morphology, Physiology and Biochemistry. Nauka, Moscow.
5. B. Daneholt, Nature New Biol. 240 (1972) 229.
6. L.H. Kedes and M.L. Birnstiel, Nature New Biol. 230 (1971) 165.
7. G.P. Georgiev and V.L. Mantieva, Biokhimiya 26 (1971) 1095.
8. K. Scherrer, L. Marcaud, F. Zajdela, I. London and F. Gross, Proc. Natl. Acad. Sci. U.S. 56 (1966) 1571.
9. J.R. Warner, R. Soeiro, C. Birnboim, M. Girard and J. Darnell, J. Mol. Biol. 19 (1966) 349.
10. G. Attardi, H. Parnas, M. Hwang and B. Attardi, J. Mol. Biol. 20 (1966) 145.
11. A.P. Ryskov and G.P. Georgiev, FEBS Letters 8 (1970) 186.
12. G.P. Georgiev, in Cell Nucleus Metabolism and Radiosensitivity, Taylor and Francis, London (1966) p. 79.
13. K. Scherrer and L. Marcaud, J. Cell. Physiol. 72, Suppl. I (1968) 181.
14. U. Lindberg and J.E. Darnell, Proc. Nat. Acad. Sci. U.S. 65 (1970) 1089.
15. T. Imaizumi, H. Diggelman and K. Scherrer, Proc. Nat. Acad. Sci. U.S. 70 (1973) 1122.
16. R.H. Stevens and A.R. Williamson, J. Mol. Biol. 78 (1973) 517.
17. G.P. Georgiev, J. Theoret. Biol. 25 (1969) 473.
18. M. Edmonds, M.N. Vaughan and N. Nakazato, Proc. Nat. Acad. Sci. U.S. 68 (1971) 1336.
19. J.E. Darnell, R. Wall and R.L. Tushinsky, Proc. Nat. Acad. Sci. U.S. 68 (1971) 1321.
20. G.R. Molloy, M.R. Sporn, D. Kelly and R.P. Perry, Biochemistry 11 (1971) 3256.
21. Ch. Coutelle, A.P. Ryskov and G.P. Georgiev, FEBS Letters 12 (1971) 21.
22. S. Leppla, B. Bjoraker and R.M. Rock, in L. Grossman and K. Moldave (Eds.) Methods in Enzymology, Vol. 12B. Academic Press New York (1968) p. 236.
23. G.P. Georgiev, A.P. Ryskov, Ch. Coutelle, V.L. Mantieva and E.R. Avakyan, Biochim. Biophys. Acta 259 (1972) 259.
24. G. Bajszar, O.P. Samarina and G.P. Georgiev, Mol. Biol. Reports 1 (1974) 305.
25. R.H. Burdon and A. Shenkin, FEBS Letters 24 (1972) 11.
26. G.R. Molloy, W.L. Thomas and J. Darnell, Proc. Nat. Acad. Sci. U.S. 69 (1972) 3684.
27. H. Nakazato, M. Edmonds and D.W. Kopp, Proc. Nat. Acad. Sci. U.S. 71 (1974) 200.
28. V.L. Mantieva and V. Arion, Molek. Biol. USSR 3 (1969) 294.
29. A.P. Ryskov, V.R. Farashyan and G.P. Georgiev, Biochim. Biophys. Acta 262 (1972) 568.

30. W. Jelinek and J.E. Darnell, Proc. Nat. Acad. Sci. U.S. 69 (1972) 2537.
31. G.P. Georgiev, A. Varshavsky, A. Ryskov and R. Church, Cold Spring Harb. Symp. Quant. Biol. 38 (1974) 869.
32. O.P. Samarina, A.S. Asrijan and G.P. Georgiev, Proc. Nat. Acad. Sci. USSR 163 (1965) 1510.
33. O.P. Samarina, E.M. Lukanidin, J. Molnar and G.P. Georgiev, J. Mol. Biol. 33 (1968) 251.
34. E.M. Lukanidin, E.S. Zalmanzon, L. Komaromi, O.P. Samarina and G.P. Georgiev, Nature New Biol. 238 (1972) 193.
35. O.P. Samarina, N.A. Aitkhozhina and J. Besson, Mol. Biol. Reports 1 (1973) 193.
36. J. Molnar and O.P. Samarina, Mol. Biol. Reports (in press).
37. A.S. Spirin, Eur. J. Biochem. 10 (1969) 20.
38. E. Gander, A. Stewart, C. Morel and K. Scherrer, Eur. J. Biochem. 38 (1973) 443.
39. E.M. Lukanidin, G.P. Georgiev and R. Williamson, FEBS Letters 19 (1971) 152.
40. C. Morel, B. Kayibanda and K. Scherrer, FEBS Letters 18 (1971) 84.
41. B. Lebleu, G. Morboix, H. Huez and J. Temmerman, Eur. J. Biochem. 19 (1971) 264.
42. E.M. Lukanidin, S. Olsnes and A. Pihl, Nature New Biol. 240 (1972) 90.
43. A. Monneron and W. Bernhard, J. Ultrastruct. Res. 27 (1969) 266.
44. N. Leytin, P. Podobed and M. Lerman, Biokhimija USSR 37 (1972) 65.

Dr. XXX

What is the new evidence available on the hypothesis that one band (chromomer) contains only the structural gene?

Dr. SAMARINA:

Initially Judd showed that the number of complementation groups in the particular part of the X-chromosome is roughly equal to the number of bands there. Similar data were obtained by Gvozdev working on another part of the X-chromosome and by Hochman, who also investigated one of the chromosomes of *Drosophila*, namely autosome 4. Certainly this is only the working hypothesis and it must be investigated more precisely.

Dr. XXX

Have you some idea of the in vivo stability of the particles? How long do they live in the nucleus?

Dr. SAMARINA:

It was shown that the particles contain all nuclear pre-mRNA, both short- and long-lived fractions. Therefore the life time of particles coincides with the life-time of pre-mRNA.

The protein moiety of the particles is much more conservative. Some data indicated the possibility of the reutilization of the informofers as such, or after their dissociation into informatin subunits.

CHAPTER VI. 2

NUCLEAR-CYTOPLASMIC INTERRELATIONS IN EUCARYOTES

M.Ya. TIMOVEEVA and A.A. NEYFAKH

Institute of Molecular Biology, Institute of Developmental Biology, Moscow, USSR

Cell compartmentalisation into the nucleus and cytoplasm which appeared in the course of evolution has resulted in uncoupling of two main steps of transfer of genetic information, viz. transcription and translation. These two processes are uncoupled not only in space but also in time: a latent period would occur between the time of mRNA synthesis in nuclei and template protein synthesis in cytoplasm. This temporal shift is of special importance in differentiating cells.

In evolution, upon transition from procaryotes to eucaryotes three major alterations occurred in the system of transfer of genetic information: 1) complication of the gene structure due to elongation of its regulatory part which is many times longer than the sequences in the structural region and due to appearance of the protein components of the chromosome; 2) complication of the processing mechanism of transcripts (pre-mRNA) synthesised on the gene to make mature mRNAs appearing in the cytoplasm; 3) complication of translation due to appearance of a new mechanism providing storage (masking) of mRNAs in the cytoplasm and specifically controlling the time of translation of different mRNAs.

In the nucleus, giant RNAs are synthesised on the transcription units; they form complexes with informofers consisting of specific protein. In the course of processing these complexes are attacked by enzymes of methylase and endonuclease type. The point of attack seems to be determined by the primary and secondary RNA structure and the protein component of the complex. This problem remains to be solved (cf. chapter of Dr. Samarina, this volume).

Nuclear RNAs (pre-mRNAs) evidently possess specific affinity for the chromatin which maintains them within the nucleus up to the completion of processing. This is clear from the data on RNA behaviour in the course of mitosis (1). In *Amoeba* (2), fish embryos (3), and mammalian fibroblast culture (4), as testified by radioautography technique, RNA synthesised before the mitosis is evenly distributed in cytoplasm during cell division and comes back to the nuclei of the daughter cells after division is completed. RNAs appearing in the cytoplasm during mitosis seem not to be subject to processing since they contain a fraction of giant molecules (more than 40S) which is not revealed in the cytoplasm of interphase cells (5).

The rate of realisation of genetic information is determined by the rates of transcription, processing and RNA release into cytoplasm, mRNA transportation to the translation site and translation itself. In different situations any of these processes may be limiting but the leading role is thought to be played by processing and mRNA transportation to the translation apparatus. This suggestion is proven by the data obtained in developing systems.

In loach embryos (teleost fishes), RNA synthesis is revealed very early, before midblastula but mainly pre-tRNAs are accumulated at this period. After the midblastula stage the activation of processing of these pre-tRNAs into mature tRNAs takes place, and tRNA synthesis is activated in total (6). It is evident that the processing is a limiting step here and its activation is a controllable link. In salivary glands of *Drosophila*, according to the data of Greenberg (7), the presence of methionine is a prerequisite for the activation of the pre-RNA processing; without this amino acid not only methylation but endonuclease action also are inhibited. In our experiments on the trout eggs we have shown that after the embryo is separated from the yolk, rRNA processing is considerably decreased and this change is not abolished by methionine addition (8). In loach embryos, separation from the yolk before the midblastula stage also inhibits the processing (about 75% of the label is found in pre-tRNAs for a long time and only 25% of the label is transferred to the mature tRNAs (6). Methionine does not abolish this inhibition. However, if the embryo is separated from the yolk after the midblastula stage then both the pre-tRNA processing and total tRNA synthesis are normally activated (6). These facts agree with the findings that the embryos separated from the yolk before the midblastula show ability for cell division but neither for protein synthesis activation nor differentiation. If the separation is carried out after the midblastula the embryos acquire this ability. It is possible that some substances transferred from the yolk at the midblastula stage (e.g., ribosomes (9)) are necessary for the pre-tRNA processing and it cannot be excluded that just the absence of new tRNAs inhibits the activation of protein synthesis and differentiation in embryos which were prematurely separated from the yolk.

It is known that mRNAs released into cytoplasm are stored for a long time; as an example we can consider the mRNA synthesised during oogenesis whose translation in amphibians is continued up to the stage of tadpole (10). It is possible that gradual utilisation of masked mRNAs (mmRNAs) is controlled via ribonucleoprotein complexes, informosomes (11). The protein of the latter ones is distinct from that of nuclear informofers discovered by Georgiev.

The importance of proteins synthesised on maternal mRNAs during oogenesis and continuing to be synthesised during embryogenesis seems to be underestimated till now. In hybrids of amphibians and fishes, using electrophoretical mobility (12) and heat stability (13) as a genetic marker of paternal proteins, it has been shown that the expression of genes controlling numerous enzymes proceeds rather late, in the middle or even in the end of embryo development. However, in the embryo a change of the isozyme set of some proteins (e.g., lactate dehydrogenase) occurs earlier, hence it possibly concerns mRNA templates accumulated during oogenesis (14). Moreover, the early biochemical differentiation manifested in uneven distribution of enzymatic activities in different parts of the early embryo can proceed on mRNAs transcribed during oogenesis. For example, aldolase whose expression on the genes of the embryo itself starts in the middle of embryonic development (13) is distributed unevenly before this point: its activity in the somites is 10 times as high as in the cells of the head part of embryos (15). Hence, during early stages of development biochemical differentiation can be mediated not by differential gene activity, that is the case as later stages, but apparently by differential activation of translation on the preformed templates.

Thus the problem of nuclear-cytoplasmatic interrelations in eucaryotes includes not only the "classical" problem of specific recognition and switching-on of the genes but also some other problems of specific recognition which play an important and sometimes leading role in realisation of the genetic information. The control of processing is evidently determined by specific nucleotide sequences, poly U and poly A regions, reversible repetitions or "palindromes" forming "hairpins" in the RNA secondary structure, repeated sequences etc. They all can be used as specific recognition sites for protein binding, critical points for action of the processing enzymes, recognition sites for factor responsible for translational control and determining time of template functioning within the polysomal complexes etc.

References

1. A.A. Neyfakh, Priroda 7 (1970) 44.
2. A.L. Yudin and A.A. Neyfakh, Exp. Cell Res. 82 (1973) 210.
3. A.A. Neyfakh and A.A. Kostomarova, Exp. Cell Res. 65 (1971) 340.
4. A.A. Neyfakh, N.B. Abramova and A.M. Bagrova, Exp. Cell Res. 65 (1971) 345.
5. N.B. Abramova and A.A. Neyfakh, Exp. Cell Res. 77 (1973) 136.
6. M.Y. Timofeeva and I.A. Solovjeva, FEBS Letters 33 (1973) 327.
7. J.R. Greenberg, J. Mol. Biol. 46 (1969) 8598.
8. N.L. Melnikova, M.Y. Timofeeva, N.N. Rott and G.M. Igmatjeva, Ontogenes 3 (1972) 85.
9. M.A. Aitkozhin, N.V. Belitsina and A.S. Spirin, Biochimiya 29 (1964) 169.
10. B.R. Hough, P.H. Jancey and E.H. Davidson, J. Exp. Zool. 185 (1973) 357.
11. A.S. Spirin, Curr. Topics Dev. Biol. 1 (1966) 1.
12. D.A. Wright and F.H. Moyer, J. Exp. Zool. 167 (1968) 197.
13. A.A. Neyfakh, M.A. Glushankova, N.S. Korobtzova and A.A. Kusakina, Develop. Biol. 34 (1973) 309.
14. N.M. Botvinik and A.A. Neyfakh (in press).
15. A.A. Neyfakh, A.A. Kusakina and A.M. Burakova (in press).

CHAPTER VI.3

DIRECT VISUALIZATION OF INFORMATION TRANSFER EVENTS*

C.L.F. WOODCOCK and J.E. STANCHFIELD

Department of Zoology, University of Massachusetts, Amherst, Mass. 01002, U.S.A.

and

R.R. GOULD**

Biological Laboratories, Harvard University, Cambridge, Mass 02138, U.S.A.

Data concerning the size and morphology of ribosomal RNA cistrons in two unicellular plants (*Acetabularia mediterranea* and *Chlamydomonas reinhardi*) are presented in the context of a general discussion of methods of direct visualization of events in the DNA — RNA — protein sequence, and the prospects for further progress.

Introduction

Of the several methods for preparing biological material for the electron microscope, three have been most widely used for observing structures involved in the transfer of information from DNA to protein. Thin sectioning, especially combined with autoradiography is useful for establishing the site of events within the cell, and following the migration of components. Negative staining of isolated components is valuable for smaller structures such as ribosomes, but the larger nucleic-acid—protein complexes tend to be too condensed to yield much information by this method. However, the method introduced by Miller and Beatty (1) for observing transcription provides a much more informative picture of the DNA—RNA polymerase—RNA complex than any other technique. Only with this method is the transcription complex sufficiently dispersed to see the individual RNP strands still attached to the DNA, and to measure cistron and spacer lengths, and polymerase density. With the few exceptions mentioned below, the technique has been used only for studies of transcription. It therefore seems worthwhile to

* Supported in part by Grants from the Research Council, University of Massachusetts, and GM-06637 from the USPHS.

** Present address: Laboratory of Molecular Biology, University of Wisconsin, Madison, Wisconsin, U.S.A.

explore the possibility of using similar methods to examine some of the other complexes involved in the decoding of genetic information.

Limitations of the technique

Basically, the method consists simply of centrifuging the material to be examined through a formaldehyde fixing solution onto an electronmicroscope grid, followed by drying and contrasting (1,2). However, success is also dependent on certain other conditions being fulfilled. Firstly, the structures in question must be present in sufficient quantity. In the case of rRNA synthesis, the probability of seeing "active" cistrons is greatly increased where these genes are amplified, as in certain oocytes (1). Although ribosomal cistrons have been recognized in HeLa cells (2) and rat liver cells (3) their frequency is low. For plant material, where cell generation times, and metabolic rates are generally slower than in animals, this becomes a severe problem. Secondly, rapid extraction and preparation are necessary to prevent the degradation of delicate structures. In the case of transcription, delay could result in the termination of nascent RNA strands, and/or their digestion by RNases. Such complexes are also susceptible to shear forces. The material to be used must therefore be available without the use of harsh cell-breaking methods or prolonged centrifugation. A third consideration is that the material is centrifuged onto the grid at a relatively low speed (about 3000 *g*). This force would be insufficient to sediment structures which do not occur in large aggregates. However, this limitation should not be regarded as fundamental to the technique. Lastly, one of the key features of the method is the dispersal in a low ionic strength environment of structures which are normally condensed in vivo. Complexes which depend for their integrity on the presence of salt or divalent cations will therefore be fundamentally altered. With these limitations in mind, the various structures and steps involved in the transfer of genetic information will be considered.

Chromatin structure and replication

Eukaryotic chromatin typically appears in electron micrographs as relatively structureless fibers of varying diameter (4,5). When prepared by the Miller—Beatty technique, however, a distinct substructure in the fibers becomes visible (fig. 1). The fibers are seen to consist of spherical particles, about 70 Å in diameter, interconnected by thin strands about 20 Å wide (6—8). Evidence is accumulating that this substructure reflects the in vivo condition, and is a case of improved preservation rather than artefact (9,10). Further progress on the chemistry of the particles, and structural organization at the chromatin fiber level can be anticipated.

To date, direct observation of replication in eukaryotes has been restricted to deproteinized DNA (11) but it should be interesting to see if replication forks can be visualized in whole chromatin by the Miller—Beatty technique.

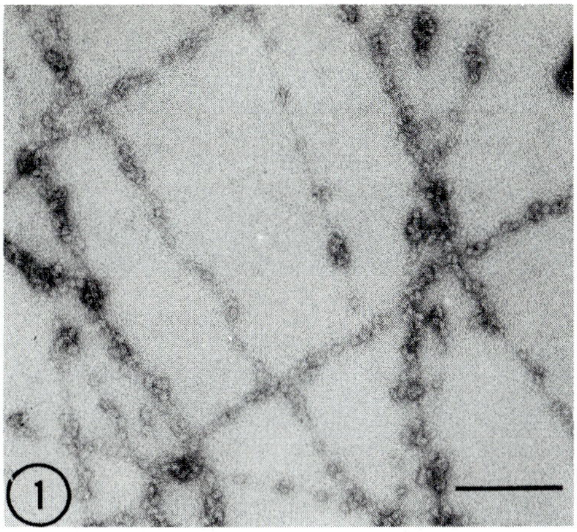

Fig. 1. Chromatin fibers from chicken erythrocyte nuclei. The particulate substructure is clearly evident. Nuclei were isolated following haemolysis in 0.05% Triton X-100, 0.15 M NaCl pH 8.0, allowed to swell in distilled water, and centrifuged through a 0.1 M sucrose cushion containing 10% formalin pH 8.5 onto a carbon coated grid (2). Uranyl acetate negative staining. Scale is 0.1 microns.

Perhaps the most interesting aspect of such micrographs will be the extent to which the DNA remains complexed with protein at the fork regions. The most important technical consideration here will be that of quantity, and a system in which rapid and continuous replication is proceeding will be necessary (11).

Transcription

The synthesis of pre-rRNA has been extensively studied in a number of animal systems (1—3) but has not previously been observed in plants. We now report the size and morphology of ribosomal cistrons from two unicellular plants, *Chlamydomonas reinhardi* (figs. 4, 5), and *Acetabularia mediterranea* (figs. 2, 3). Similar results from *A. mediterranea* are described elsewhere in this volume by Schweiger (12) and Trendelenburg et al. (13). In the case of *Acetabularia*, the amplification of ribosomal genes (12) in the large nucleus increases the probability of observing active cistrons. *Chlamydomonas* was selected for study because of its relatively rapid generation time (about 8 hr), and the availability of a readily-lysed cell wall-less mutant cw15 (18). Details of the preparation methods are included in the figure legends.

From figs. 2—5 it is clear that the morphology and arrangement of ribosomal genes and spacers in the two lower plants is similar to that described for animals (3). The length of the ribosomal genes (table 1) in both plants is close to the minimum necessary to produce a pre-rRNA strand which can be cleaved into one "25S" and one "18S" rRNA molecule (combined molecular

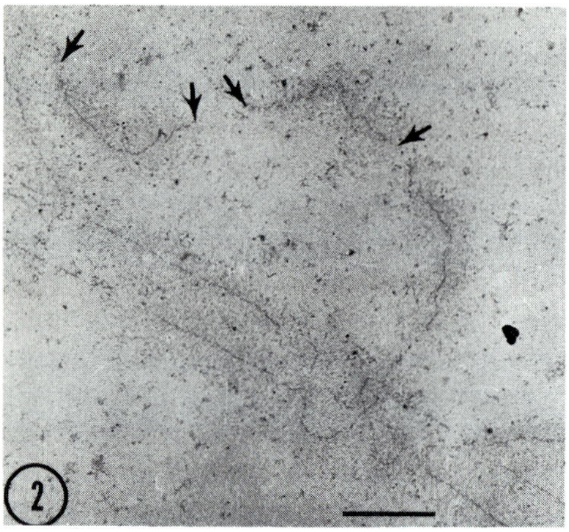
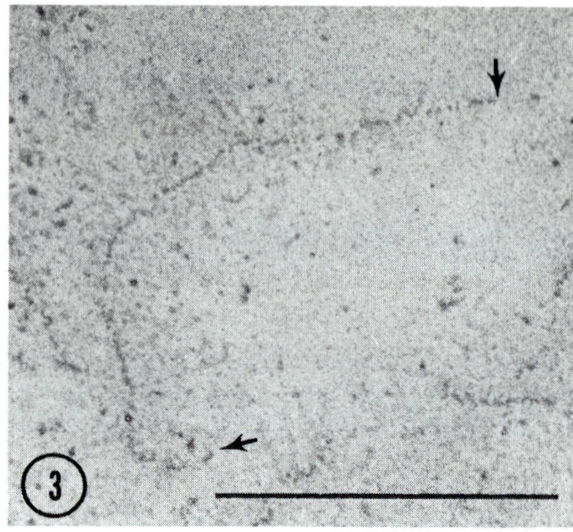

Figs. 2, 3. Portions of nucleolar core material from *Acetabularia mediterranea*. Nuclei were dissected out, washed, lysed in 0.2% "Joy" detergent, 0.05 M phosphate buffer pH 8.0 (18), and prepared for the electron microscope as in fig. 1, except that positive staining in 1% phosphotungstic acid in 70% ethanol (2) was used. Arrows indicate initiation and termination points of the rRNA genes. Scales are 1 micron.

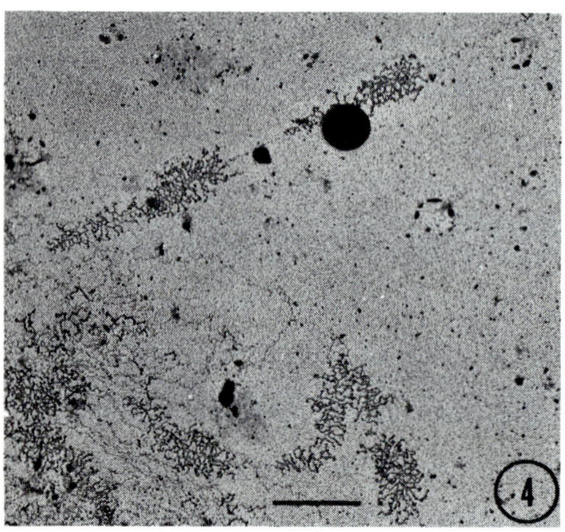
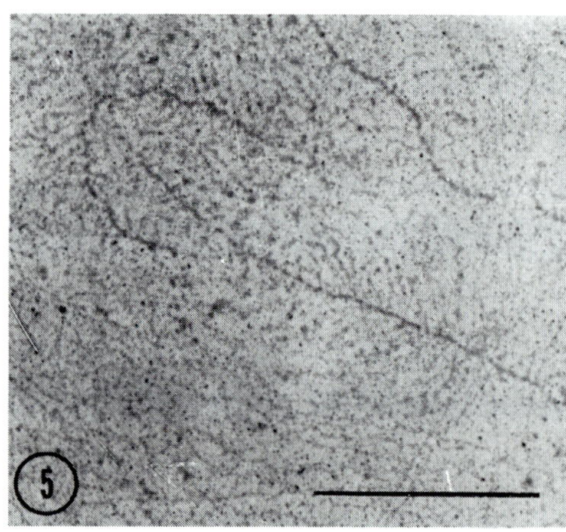

Figs. 4, 5. rRNA synthesis in *Chlamydomonas reinhardi*. Cells of the cw15 strain were lysed in 1% Triton X-100 pH 8.5, and a crude nuclear pellet obtained (18). Nuclei were then swollen in distilled water, and prepared for the microscope as described above. In fig. 4, some of the active rRNA genes have been freed from the main mass of chromatin. The black, spherical object is a starch grain. Fig. 5 shows a more typical preparation in which the nucleolar region is enmeshed in chromatin, but the active rRNA cistrons recognizable by the dense line of polymerases and associated RNP strands. Scales are 1 micron.

TABLE 1

Organism	Condition of nucleolar genes	Matrix length (microns)	Spacer length	Spacer/ matrix
*Chironomus thummi**	Chromosomal	2.2	0.45	0.20
*Acheta domesticus**	Extrachromosomal	5.6	5.5	0.98
Triturus sp.*	Extrachromosomal	2.2—2.9	0.80—2.2	0.33—0.92
Xenopus sp.*	Extrachromosomal	2.0—3.1	0.67—2.1	0.33—0.81
Rat liver*	Chromosomal	5.2	0.5	0.1
HeLa Cells*	Chromosomal	3.5	3.5	1.0
Acetabularia mediterranea	Chromosomal ?	1.85 S.E. 0.06	~ 0.5	~ 0.25
Chlamydomonas reinhardi	Chromosomal	1.9 S.E. 0.06	~ 1.0	~ 0.5

S.E. is standard error.
* From table 1 of Trendelenburg et al. (3).

weight, 2×10^6 daltons). This is in contrast to the system in many animals where 20—40% of the original pre-rRNA transcript must be discarded during processing (14). The significance of this difference between animal and plant systems is discussed elsewhere (18).

The transcription of pre-mRNA has not been convincingly visualized in eukaryotes except in the special case of lampbrush chromosomes (2). For normal cells, the problem here is one of quantity; at any given time, only a very small portion of the genome will be transcribing pre-mRNA and the chances of seeing this in a spread preparation are low. Moreover, in cases of putative pre-mRNA transcription in eukaryotes (2) and the undisputable cases in prokaryotes (15), it is clear that the polymerases are much more widely spaced than on ribosomal RNA genes. This also would make the unequivocal identification of pre-mRNA synthesis difficult.

The processing of RNA

Before it becomes functional as either rRNA or mRNA, the initial transcript (in the form of an RNP complex) is processed by specific cleavages, and in some cases, methylations. For rRNA, this sequence of events has been well understood at the biochemical level for some time (16), but little is known of the position in the nucleus/nucleolus where each step is per-

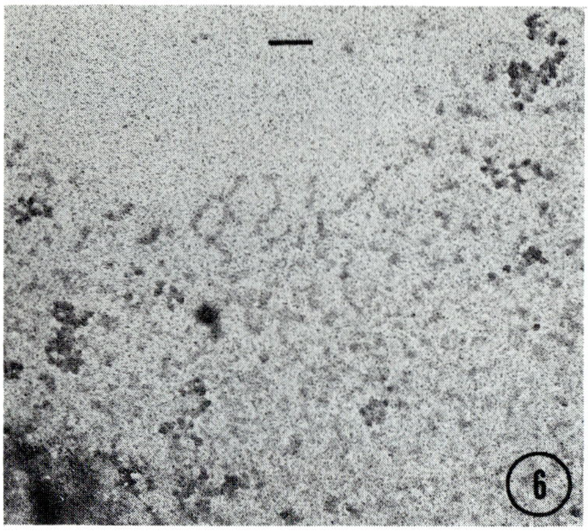

Fig. 6. Portion of nucleolar cortex material from *Acetabularia mediterranea* prepared as in fig. 1. Fibers and groups of ribosome-sized particles are seen. Scale is 0.1 micron.

formed, or of the structural organization of the RNP during the process (17). When nucleolar cortex material from *Acetabularia* is examined by the spreading technique, a variety of fibrous and particulate structures can be seen (fig. 6). It should now be possible by combining electron microscopy with biochemical analysis to correlate each structure with specific steps in the rRNA processing sequence, and so build up a picture of the organization at the RNP level. Similar progress in the case of pre-mRNA is unlikely, primarily because of the low frequency of these particles.

Translation

Direct observation of eukaryotic translation events is made difficult at present because polysomes do not sediment at the low speed centrifugation currently in use. Even if it becomes technically possible to overcome this, it is doubtful whether new structural information will be gained. The elegant demonstration of DNA—RNA—ribosome complexes in bacteria (15) did not yield much useful information about the translation process itself.

Conclusion

There is every prospect that the Miller—Beatty technique which has made such an impact on our knowledge of transcription and gene organization can be successfully applied to other nucleic acid—protein complexes. It is now possible to approach questions concerning the structural organization of chromatin, its behavior during replication, and the processing of pre-rRNP by such methods.

References

1. O.L. Miller, Jr. and B.R. Beatty, Science 164 (1969) 955.
2. O.L. Miller, Jr. and A.H. Bakken, Acta Endocrinol. 168 (1972) 155.
3. M.F. Trendelenburg, U. Scheer and W.W. Franke, Nature New Biol. 245 (1973) 167.
4. H. Ris and D.F. Kubai, Ann. Rev. Genet. 4 (1970) 263.
5. J. Huberman, Ann. Rev. Biochem. 42 (1973) 355.
6. A.L. Olins and D.E. Olins, J. Cell Biol. 59 (1973) 252a.
7. A.L. Olins and D.E. Olins, Science 183 (1974) 330.
8. C.L.F. Woodcock, J. Cell Biol. 59 (1973) 368a.
9. R.D. Carlson and A.L. Olins, Fed. Proc. 33 (1974) 1538.
10. C.L.F. Woodcock, D.A. Maguire and J.E. Stanchfield (in preparation).
11. H.J. Kriegstein and D.S. Hogness, Proc. Nat. Acad. Sci. U.S. 71 (1974) 135.
12. H.G. Schweiger, this volume (1975).
13. M.F. Trendelenburg, H. Spring, U. Scheer and W.W. Franke, Proc. Natl. Acad. Sci. U.S. 71 (1974) 3626.
14. R.P. Perry, T.-Y. Cheng, J.J. Freed, J.R. Greenberg, D.E. Kelley and K.D. Tartof, Proc. Nat. Acad. Sci. U.S. 65 (1970) 609.
15. O.L. Miller, Jr., B.R. Beatty, B.A. Hamkalo and C.A. Thomas, Cold Harbor Symp. Quant. Biol. 35 (1970) 505.
16 R.P. Perry, in A. Lima-da-Faria (Ed.) Handbook of Molecular Cytology, North-Holland, Amsterdam (1969) p. 626.
17. O.L. Miller, Jr. and B.R. Beatty, in A. Lima-da-Faria (Ed.) Handbook of Molecular Cytology, North-Holland, Amsterdam (1969) p. 609.
18. C.L.F. Woodcock, J.E. Stanchfield and R.R. Gould, Plant Sci. Lett. 4 (1975) 17.

Dr. BRACHET:

Have you any information about the number of ribosomal cistrons in *Acetabularia*?

Dr. WOODCOCK:

No, but I believe Dr. Schweiger has, and will be telling us about it later.

Dr. SCHERRER:

Is it known if the primary transcript in *Acetabularia* is a high molecular weight compound?

Dr. WOODCOCK:

No. It is not known.

CHAPTER VI.4

THE ULTRASTRUCTURE OF THE NUCLEOCYTOPLASMIC INTERFACE IN *ACETABULARIA*

S. BERGER and H.G. SCHWEIGER

Max-Planck-Institut für Zellbiologie, D 2940 Wilhelmshaven, W. Germany

The nucleus and the perinuclear cytoplasm of the *Acetabularia* cell have some unusual features which make this cell interesting for studies of the fine structure of nucleocytoplasmic interactions. Among these special features are extranuclear DNA-containing structures, an extremely high number of nuclear pore complexes and a cytoplasmic layer which surrounds the nucleus and is connected by thin strands to the rest of the cytoplasm.

Gene expression requires the translocation of material from the nucleus to the cytoplasm and vice versa. Among the substances moved are mRNA and rRNA as well as precursors of RNA and DNA. The regulation of this intercompartmental traffic may be as significant for differentiation as the synthesis and translation of mRNA. Therefore it is important to have available detailed ultrastructural information about the nucleocytoplasmic interface, namely the region inside and outside the nucleus adjacent to the nuclear envelope.

In the giant cell of the uninuclear green alga *Acetabularia*, the primary nucleus is located in one of the rhizoid branches (1) where it can grow to a maximum size of 200 μm in diameter (fig. 1). It contains a skein-like nucleolus which looks like many sausage-like nucleoli in semithin sections (1—5).

Tangential sections of a fully developed nucleus reveal that the nuclear envelope contains pore complexes which occur with a maximum frequency of about 70 pores per μm^2 in *A. mediterranea*. About 85 pores per μm^2 have been observed in the *A. major* nuclear envelope (fig. 2) (6). These values for pore frequency are among the highest so far reported.

The frequency and structure of the pore complexes are also apparent in cross sections (fig. 3). The pore complexes show the typical arrangement of eight annular granules which are symmetrically distributed, usually around a central granule (inset in fig. 2). The percentage of pores showing a dense central granule is about 95% which is rather high (6). The translocation of material into and out of the nucleus is believed to take place through the pore complexes.

Dense granules are visible in the nucleus adjacent to the nuclear envelope

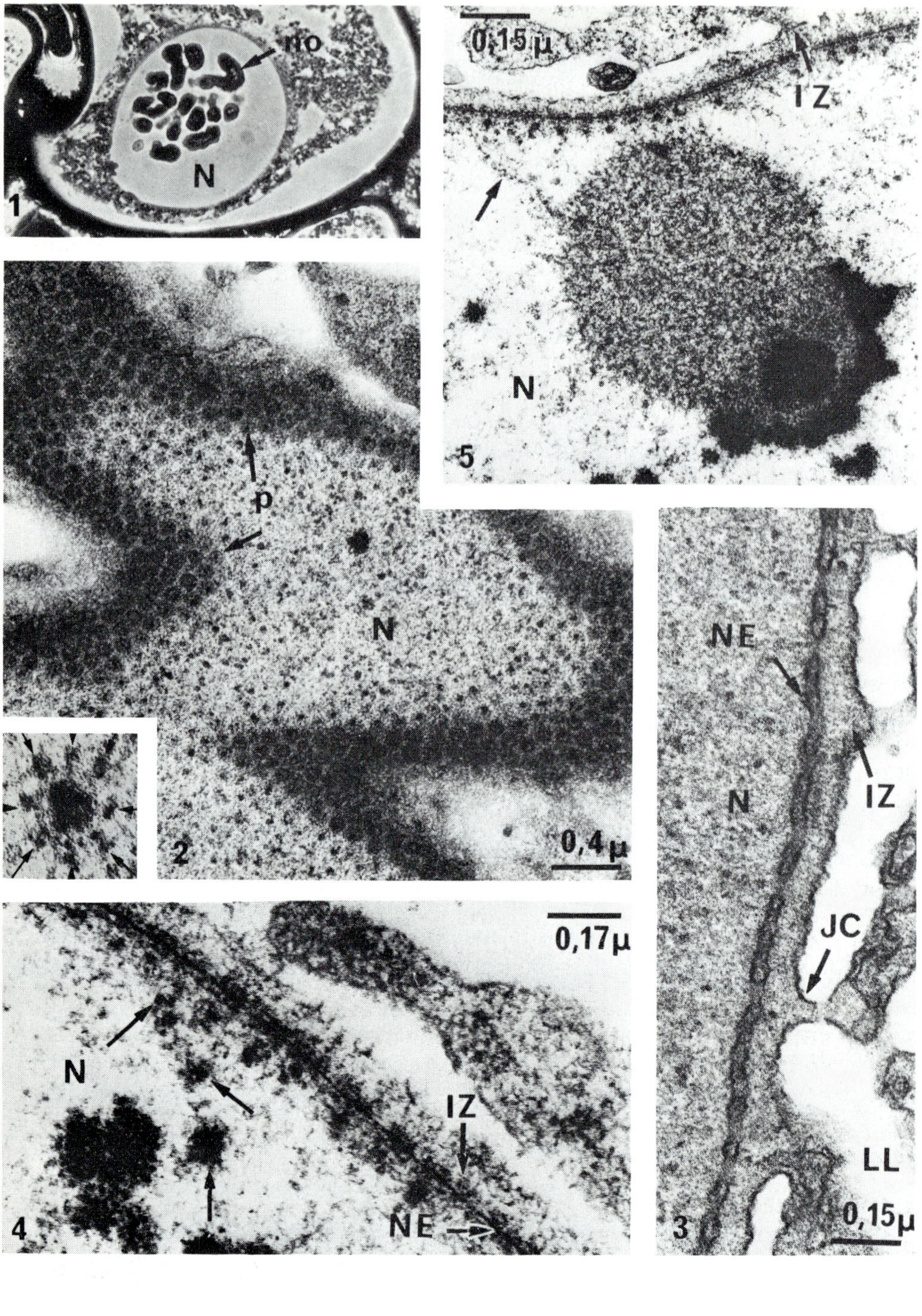
1
no
N
2
p
N
0,4 μ
3
NE
N
IZ
JC
LL
0,15 μ
4
0,17 μ
N
IZ
NE
5
0,15 μ
IZ
N

and seem to be migrating towards the pore complexes (figs. 4 and 6). Similar structures can also be detected in the perinuclear cytoplasmic layer (fig. 7). Their low number in this region may indicate that they pass through the perinuclear cytoplasmic layer rather quickly. Other structures often encountered in the nuclear periphery are characteristic aggregates of spheroidal elements and densely stained globules (fig. 5). Often fibrillar strands extend from the spheroidal elements towards the pore complexes.

The perinuclear region of the *Acetabularia* cell has a number of structural peculiarities (fig. 8) (3,5–8). The nuclear envelope is covered by a continuous layer of cytoplasm which is 120 nm wide. No organelles, ER-components, dictyosomes or microtubules have been detected in this region. Adjacent to this thin perinuclear cytoplasmic layer is a complex vacuolar system, the so-called lacunar labyrinthum. The cytoplasm which is embedded in this labyrinthum contains numerous intensely stained structures, the perinuclear dense bodies. The perinuclear cytoplasmic layer and the cytoplasm in the region of the lacunar labyrinthum are connected by junction channels of cytoplasm (fig. 9). It has been suggested that these channels mediate the exchange of material between the nucleus or the perinuclear cytoplasmic layer and the rest of the cytoplasm. Possibly the junction channels also play a role in the regulation of nucleocytoplasmic communication. Such an as-

---

Fig. 1. Semithin section through part of the rhizoid. The large single primary nucleus is always located in one of the rhizoid branches. The nucleus contains many sausage-like nucleoli. N, nucleus; no, nucleolus; cw, cell wall. Magnification 250 ×.

Fig. 2. Section tangential to the nuclear envelope. The extremely high number of nuclear pores is shown where the section grazes the nuclear envelope. The inset demonstrates the typical structure of a pore complex with eight symmetrically distributed annular granules (arrows) and one large central granule. N, nucleus; p, pore. Magnification of the inset 143,000 ×.

Fig. 3. Cross section of the nuclear envelope. The double membrane of the nucleus is interrupted by many pores. A thin cytoplasmic layer surrounds the nucleus. Junction channels connect the cytoplasmic layer around the nucleus, the so-called intermediate zone, to the rest of the cytoplasm. Part of the cisternae of the lacunar labyrinthum can be seen in the micrograph. N, nucleus; NE, nuclear envelope; LL, lacunar labyrinthum; IZ, intermediate zone; JC, junction channel.

Fig. 4. Part of the nuclear periphery. Granules seen at the periphery of the nucleolus seem to be migrating out of the nucleus (arrows). N, nucleus; NE, nuclear envelope; IZ, intermediate zone.

Fig. 5. Part of the nuclear periphery. Aggregates of densely stained granules and spheroidal elements can often be detected near the nuclear envelope. Often fibrillar strands extend from these spheroidal elements to the pore complexes (arrow). N, nucleus; IZ, intermediate zone.

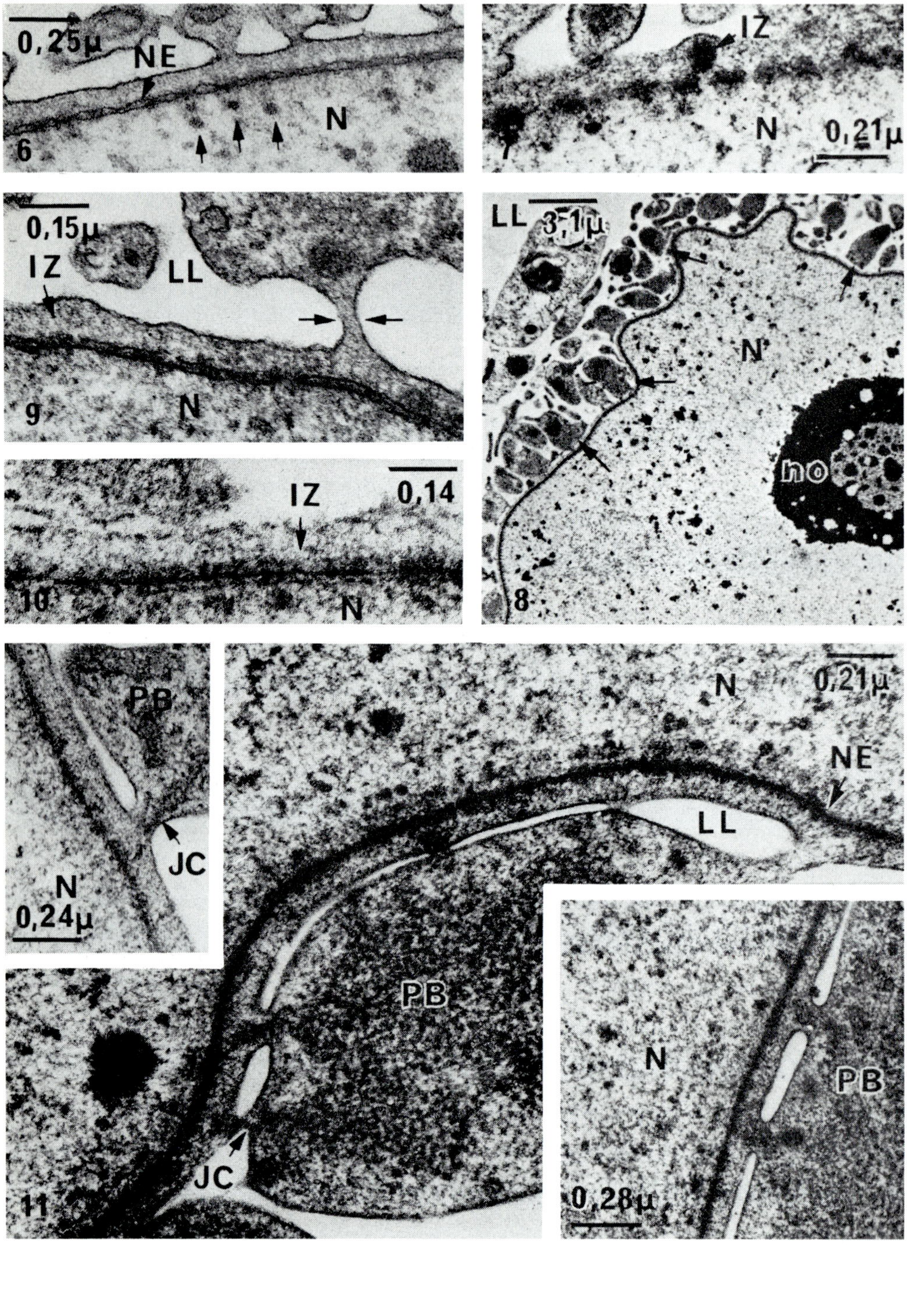
0,25μ
NE
N
6
IZ
N
0,21μ
0,15μ
IZ
LL
9
N
LL
3,1μ
N
no
8
IZ
0,14
10
N
PB
JC
N
0,24μ
N
0,21μ
NE
LL
PB
JC
11
N
PB
0,28μ

sumption is supported by the fact that their number and total cross section is significantly smaller than that of the nuclear pores.

The interior of the junction channels is filled with a filamentous meshwork (6). A similar meshwork is often recognized in the perinuclear layer (fig. 10). This fibrillar meshwork may be responsible for the fact that the thin perinuclear cytoplasmic layer adheres to the nucleus during the isolation procedure, a property which might contribute to the good survival rates of isolated nuclei. Frequently one sees granules and fibrillar strands in the junction channels, suggesting material is enroute from the nucleus to the perinuclear cytoplasm (fig. 11 and its inserts).

The cytoplasm of the lacunar labyrinthum region particularly those parts next to the junction channels, is filled with a variety of structures, among them small dense aggregates, rod- or sheet-like aggregates and vesicles (6). Most conspicuous in this region is a high number of large dense bodies (fig. 12) (6). At higher magnification, these bodies are seen to be subdivided into a granular and a fibrillar region (figs. 13 and 14). Their similarity to the nucleolar granular and fibrillar zone is striking; even the granules are about the same size as those of the nucleolus. Characteristic aggregates of large dense granules can be detected at the distal side of the perinuclear dense bodies.

These perinuclear dense bodies react like chromatin in cytochemical tests:

Fig. 6. Part of the nuclear periphery. Granules near the nuclear envelope (arrowheads) seem to be passing through the pore complexes. N, nucleus; NE, nuclear envelope.

Fig. 7. Part of the nuclear periphery. Granules characteristically seen inside the nucleus can sometimes be detected in the intermediate zone (arrow). N, nucleus; IZ, intermediate zone.

Fig. 8. Part of the nucleus. The nucleus contains many nucleoli from which granules seem to migrate via the pore complexes of the nuclear envelope. Conspicuous are the thin cytoplasmic layer around the nucleus, the numerous perinuclear dense bodies (arrows) and the highly vacuolated cytoplasm. N, nucleus; no, nucleolus; LL, lacunar labyrinthum.

Fig. 9. Part of the nuclear periphery. Thin cytoplasmic strands from the junction channels (arrows) between the intermediate zone and the rest of the cytoplasm. N, nucleus; NE, nuclear envelope; IZ, intermediate zone; LL, lacunar labyrinthum.

Fig. 10. Nuclear envelope with intermediate zone. The intermediate zone is filled with a fibrillar meshwork. N, nucleus; IZ, intermediate zone.

Fig. 11. Junction channels. The junction channels often are filled with fibrillar (upper inset) or granular material (lower inset and fig. 11). The material seems to migrate into the perinuclear dense bodies. N, nucleus; NE, nuclear envelope; PB, perinuclear dense bodies; JC, junction channel; LL, lacunar labyrinthum.

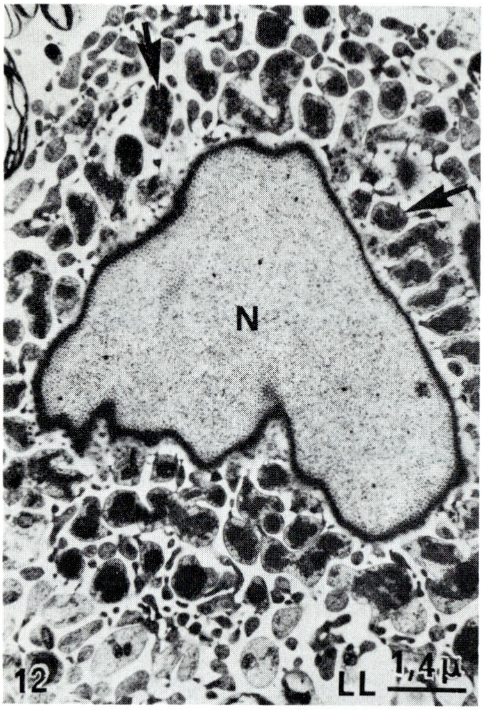
N
12
LL
1,4 µ

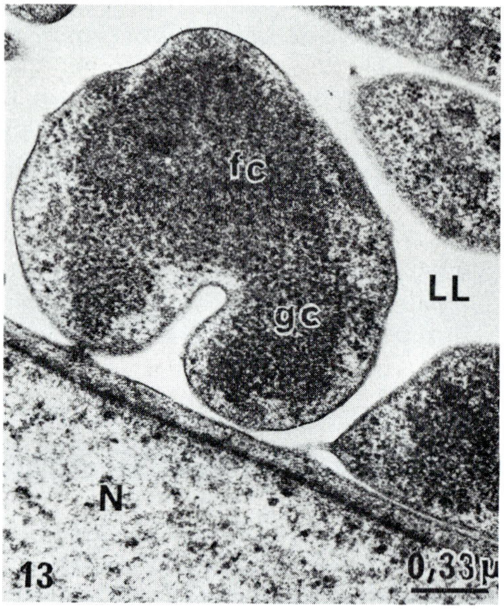
fc
gc
LL
N
13
0,33 µ

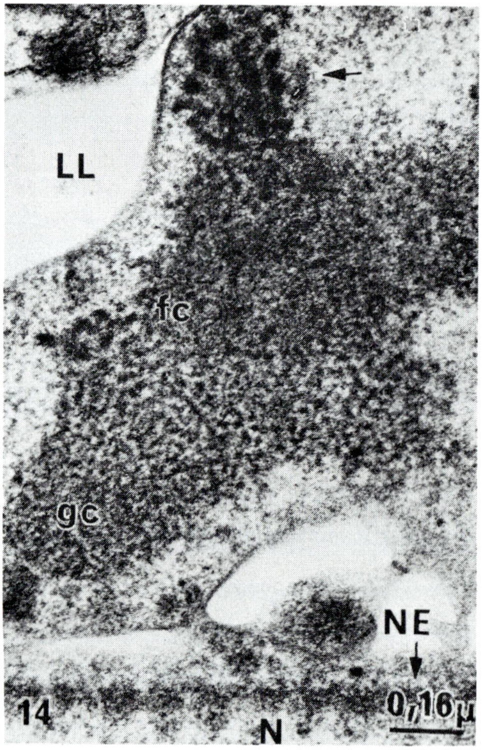
LL
fc
gc
NE
14
N
0,16 µ

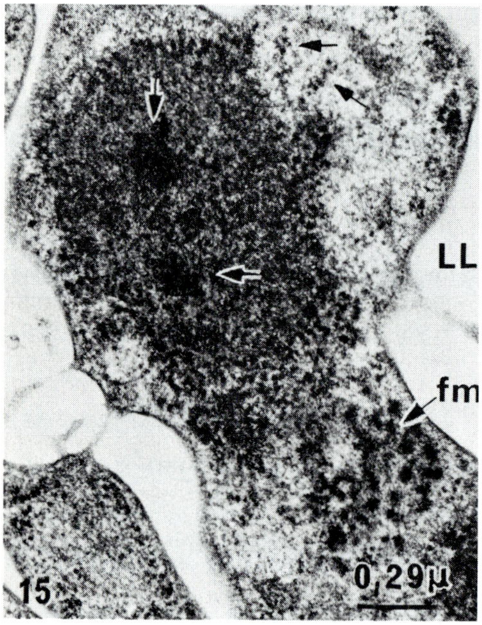
LL
fm
15
0,29 µ

they are bleached following EDTA-treatment of sections stained with uranyl acetate (6) and show a green fluorescence after treatment with acridine orange (9). Presumably they contain DNA.

Farther away from the nucleus in the region where there is an intricate branching of the lacunar labyrinthum, the perinuclear dense bodies show a somewhat altered structure (fig. 15). They contain densely stained islets in a fibrillar zone. Polysomes and ribosomes appear at the periphery of these perinuclear dense bodies.

The ultrastructure of the nuclear envelope and the adjacent nuclear and cytoplasmic regions in *Acetabularia* indicates that the border between the nucleus and the cytoplasm is somewhat ill defined, at least from a functional point of view. The existence of the perinuclear cytoplasmic layer, the cytoplasmic junction channels and the extranuclear DNA-containing structures suggest to us that the perinuclear region and nuclear envelope together represent a functional unit in nucleocytoplasmic interaction.

Acknowledgement

The authors are grateful to Dr. Marlene Karakashian for valuable help in preparing the manuscript.

Fig. 12. Section through the tip of the lense-shaped nucleus. Many perinuclear dense bodies surround the nucleus (arrows). The lacunar labyrinthum interlaces the cytoplasm. N, nucleus; LL, lacunar labyrinthum.

Fig. 13. Part of the perinuclear cytoplasm. Perinuclear dense bodies are found in the cytoplasm which is connected to the nuclear periphery by junction channels. The perinuclear dense bodies are composed of a fibrillar and a granular component. N, nucleus; gc, granular component; fc, fibrillar component; LL, lacunar labyrinthum.

Fig. 14. Perinuclear dense body. The perinuclear dense body is composed of a granular and a fibrillar component and associated with a densely stained flocculent material (arrow). N, nucleus; NE, nuclear envelope; gc, granular component; fc, fibrillar component; LL, lacunar labyrinthum.

Fig. 15. Perinuclear dense body more distant from the nucleus. The perinuclear dense body contains densely stained islets (arrowheads) in a fibrillar zone. Granular zones can be detected. Polyribosomes appear at the periphery (arrow). The perinuclear dense body is associated with a densely stained flocculent material. LL, lacunar labyrinthum; fm, flocculent material.

References

1. J. Hämmerling, Biol. Zentralbl. 51 (1931) 633.
2. K.L. Schulze, Arch. Protistenk. 92 (1939) 179.
3. C.L.F. Woodcock and G.J. Miller, Protoplasma 77 (1973) 331.
4. S. Berger and H.G. Schweiger, Protoplasma 83 (1975) 41.
5. H.G. Schweiger, S. Berger, K. Kloppstech, K. Apel and M. Schweiger, Phycologia 13 (1974) 1.
6. W.W. Franke, S. Berger, H. Falk, H. Spring, U. Scheer, W. Herth, M.F. Trendelenburg and H.G. Schweiger, Protoplasma 82 (1975) 249.
7. G. Werz, Planta 62 (1964) 255.
8. M. Bouloukhère, Thesis, Université Libre de Bruxelles (1970).
9. H. Spring, W.W. Franke, H. Falk and S. Berger, Meeting abstract, Protoplasma 83 (1975) 180.

CHAPTER VI.5

RNA TRANSPORT FROM THE NUCLEUS TO THE CYTOPLASM IN *ACETABULARIA MEDITERRANEA*

A.C. DAZY, H. BORGHI, J. GAUCHERY and S. PUISEUX-DAO

Laboratoire de Biologie Cellulaire Végétale, Université Paris VII, 75221 Paris Cédex 05, France.

The RNA transport in *Acetabularia* cells has been studied by means of pulse-chase experiments and counting of non-chloroplastic TCA insoluble radioactivity or autoradiography. In normal light conditions, transport of labelled macromolecules was very rapid and the radioactivity reached the apex within a few hours. After about 48 hr a steady state was reached. The increasing of the radioactivity in the stalk was interpreted as the release and the transport of macromolecules, presumably RNAs, from the nucleus to the cytoplasm. The release was discontinuous. The transport did not occur in the dark when the algae were kept 2 weeks in the dark before the incubation time.

Introduction

Acetabularia is a giant cell whose long siphon is very polarised. The nucleus is in the rhizoids at one end, growth and morphogenesis occur in the opposite apical end. So at least part of the RNAs released from the nucleus must migrate along the whole length of the cell to reach the apex where they are required for protein synthesis (1). Thus *Acetabularia* provides a model system to determine the rates and modes of migration of macromolecules within the cytoplasm, especially the migration of RNAs, and allows one to study the effects of some factors such as light or temperature on this process.

Materials and methods

In order to follow the migration of the RNAs from the nucleus to the apex of *Acetabularia* pulse-chase experiments have been carried out. Algae growing in the normal light—dark (L—D: 12—12) conditions were labelled with H^3-uridine for 1 hr*. Then the cells were washed and put in a culture medium containing cold uridine. In most experiments we used regenerating algae, cut, at the end of the incubation time, 1 cm above the rhizoids in

---

* Uridine H^3-5 was kindly provided by the C.E.A. Saclay, France.

order to produce a more physiologically homogeneous population of cells. For each sample, cells were frozen on dry ice and cut into 2 fragments about 5 mm long. After homogenisation of the fragments, the chloroplasts were sedimented by centrifugation and discarded. The TCA insoluble radioactivity of the super-natant was measured by liquid scintillation counting. To prevent the bacterial and chloroplastic contaminations, all experiments were done in the presence of rifampicine and chloramphenicol.

Electron micrographs and radiographs were obtained using the usual techniques (2).

Results and discussion

The TCA insoluble radioactivity increased rapidly in the stalk of regenerating algae during the first 48 hr; then it levelled off for a couple of weeks and finally decreased slowly (fig. 3).

Parallel experiments using autoradiographic techniques showed that, at the end of the incubation time, the radioactivity was localised in the nucleus, the nucleolus being more labelled than the nucleoplasm (fig. 1.1). Very few silver grains could be observed either on the basal (fig. 1.2) or on the apical cytoplasm (fig. 1.3,4). After 3 hr of chase, the nucleus (for the most part the nucleolus) was still strongly labelled (fig. 2.1) and silver grains were present on the basal cytoplasm (fig. 2.2,3) but not on the apical (fig. 2,4).

Although we have not yet completed the analysis of the molecular nature of the TCA insoluble radioactivity, these above observations allow us to ascribe the time course of the accumulation of TCA insoluble radioactivity in *Acetabularia* stalk to the release of relatively stable and labelled macromolecules (presumably RNP) from the nucleus to the basal cytoplasm, and afterwards, to the apex.

As shown on figs. 3 and 4, this release was discontinuous. It is likely that there were 2 populations of radioactive macromolecules migrating into the cytoplasm at different rates, or released by the nucleus at different times or, which is more probable, both simultaneously. This is in agreement with previous results obtained on animal cells (3) and with the known differences in processing and turnover time of the nuclear ribosomal and messenger RNAs (4,5). However, detailed biochemical analysis of the macromolecules released by the nucleus is needed to confirm this hypothesis.

Results obtained with whole algae (fig. 4b) showed the same pattern of accumulation of the TCA insoluble radioactivity. However the second phase of the release was somewhat delayed compared to regenerating algae, suggesting that the section of the stalk stimulated the release of the second population of macromolecules synthesized in the nucleus.

In order to determine whether the light, which is an important growth factor, affects the transport of the RNAs in *Acetabularia*, we performed

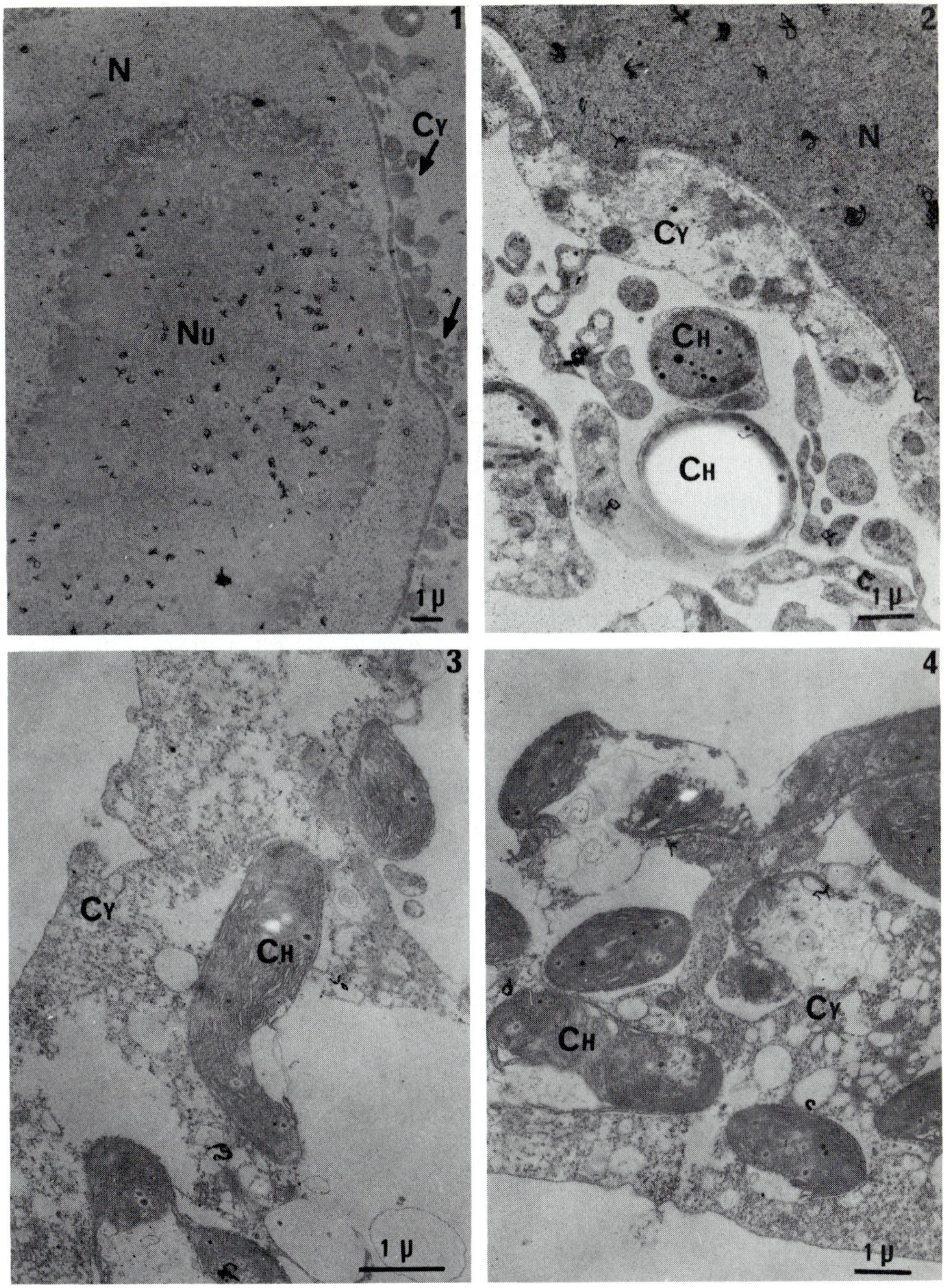

Fig. 1. Autoradiography of *Acetabularia mediterranea* after a labelling of 30 min with $5H^3$ uridine. 1, 2, Parts of the nucleus and of the nuclear periphery. 3, 4, Apical cytoplasm. Only the nucleus and especially the nucleolus are labelled. N, nucleoplasm; Nu, nucleolus; Cy, cytoplasm; CH, chloroplast.

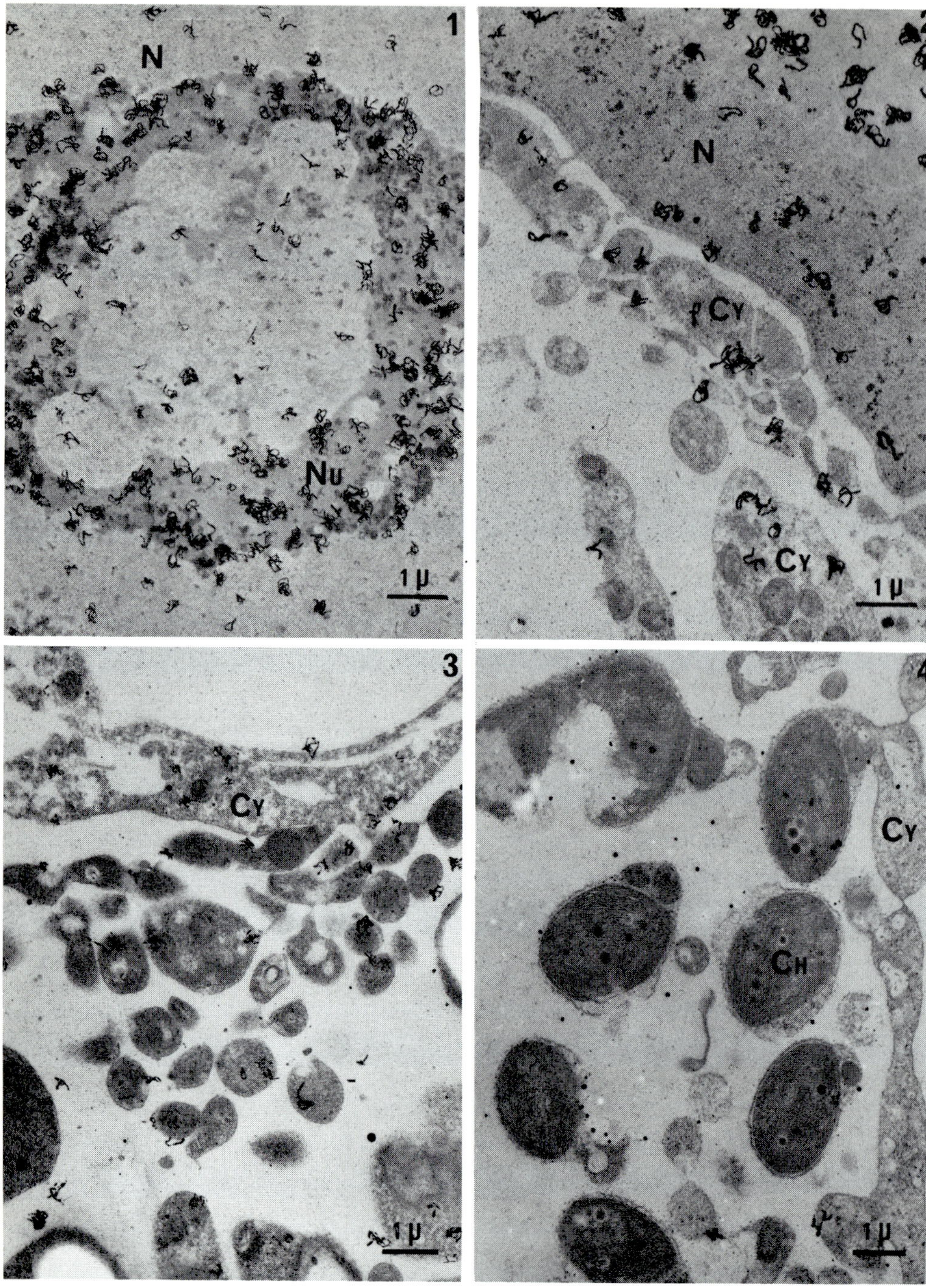

Fig. 2. Autoradiography of *Acetabularia mediterranea* after a labelling of 30 min with $5H^3$ uridine and a chase of 3 hr. 1, Part of the nucleolus. 2, Parts of the nucleus and of the nuclear periphery. 3, Perinuclear cytoplasm. 4, Apical cytoplasm. The nucleus and the nucleolus are still strongly labelled. Silver grains are present on the perinuclear cytoplasm but not on the apical cytoplasm. N, nucleoplasm, Nu, nucleolus, Cy, cytoplasm, CH, chloroplast.

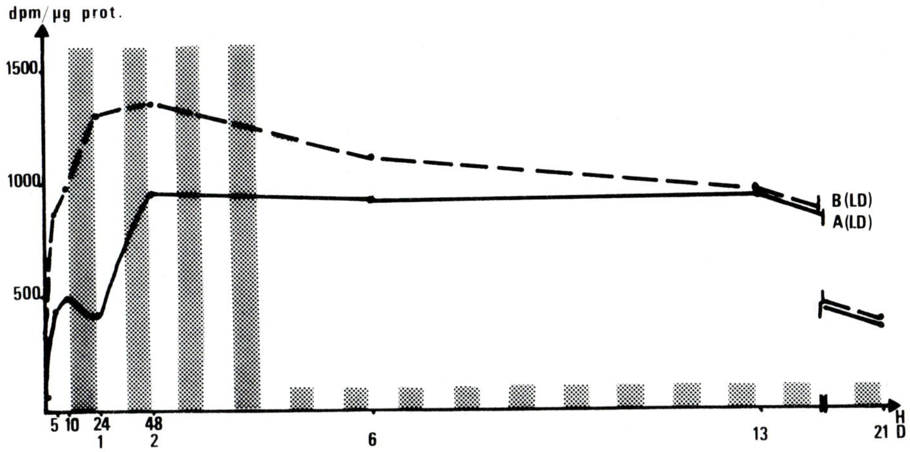

Fig. 3. Kinetic of accumulation, in the *Acetabularia* siphon of the non-chloroplastic TCA insoluble material labelled with 5H^3 uridine. Cells of *A. mediterranea* were incubated for an hour in the presence of 5H^3-uridine (40 μCi/ml — 20 Ci/mM), rifampicine (40 μg/ml) and chloramphenicol (10 μg/ml), then cut 1 cm above the rhizoids and transferred to a medium containing cold uridine. Abcissa, chase time. (H, hr; D, days). Nights are indicated by the shaded bars. Light dark cycle L—D: 12—12. (- - - - - -), radioactivity measured in the basal part (B) of the siphon (without nucleus). (———), radioactivity measured in the apical part (A) of the siphon.

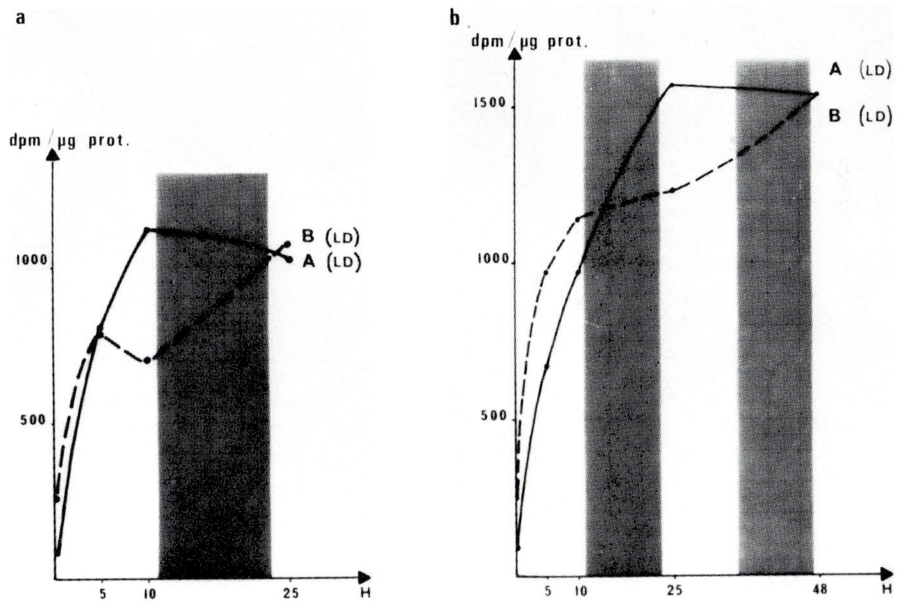

Fig. 4. Kinetic of accumulation, in the *Acetabularia* siphon of the non-chloroplastic TCA insoluble material labelled with 5H^3-uridine. For conditions, see fig. 1. a) Cells cut at the end of the incubation time (regenerating algae). b) Cells non-cut at the end of the incubation time (whole algae).

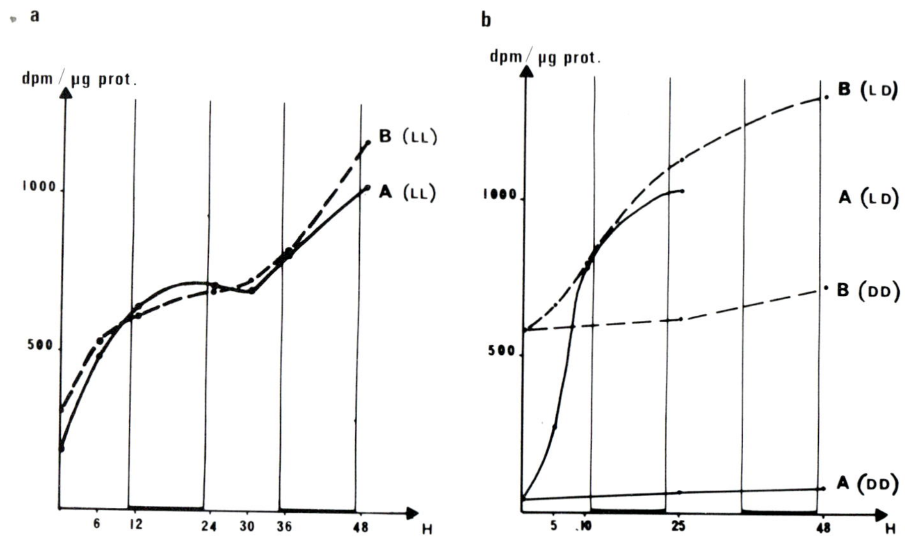

Fig. 5. Kinetic of accumulation, in the *Acetabularia* siphon, of the non-chloroplastic TCA insoluble material, labelled with 5H^3-uridine. For conditions, see fig. 1. a) regenerating algae maintained in continuous light (LL) during the chase period. b) whole algae kept in the dark for 2 weeks before the experiment. DD, cells maintained in the dark during incubation and chase times; LD, cells transfered to the light (normal LD conditions) at the end of the labelling.

experiments in continuous light and continuous darkness.

The general time course of the appearance of labelled macromolecules in the stalk was not modified if the algae were kept in continuous light after the incubation time (fig. 5a).

In contrast no accumulation of radioactivity was observed in the siphon if the algae were kept in the darkness for 2 weeks and maintained in the darkness during all the experimentation time (fig. 5b). However, the radioactive macromolecules did appear in the basal and apical cytoplasm when the algae were transfered from dark to light after the incubation. (fig. 5b).

Thus, even if the radioactive macromolecules are released from the nucleus and migrate along the siphon during the normal dark period of the L—D cycle (figs. 3,4), one may assume that light is necessarily a factor in the migration process. From these results however, it cannot be determined whether or not light affects the release of the radioactive molecules from the nucleus.

Future experiments are projected towards investigating the role of light in these processes.

The micrographs were taken on the electron microscope Philips of the E.N.S. Sèvres. 1, Rue Maurice Arnoux, Montrouge, Paris.

References

1. D. de Vitry, Bull. Soc. Chim. Biol. 47 (1965) 1325.
2. F. Larra and D. Droz, J. Microscop. 9 (1970) 845.
3. D.E. Schumm and T.E. Webb, Biochem. Biophys. Res. Commun. 48 (1972) 1259.
4. K. Scherrer, L. Marcaud, F. Zajdela, B. Breckenridge and F. Gros, Bull. Soc. Chim. Biol. 48 (1966) 1037.
5. G. Attardi, H. Parnas, M. Hwang and B. Attardi, J. Mol. Biol. 20 (1966) 145.

CHAPTER VI.6

ON THE NATURE OF THE SIGNAL INDUCING THE SYNTHESIS OF 'MORPHOGENETIC SUBSTANCES' IN THE *ACETABULARIA* NUCLEUS IN CAP REGENERATION

T.N. ZUBAREV and N.P. ROGATYKH

Institute of Genetics and Selection of Industrial Microorganisms, Moscow, USSR

Estimations are made of the velocity at which the "morphogenetic substances" travel along the *Acetabularia* stalk. For one of the populations of *A. mediterranea* cultivated under the standard conditions the above velocity was found to be 0.025 cm/hr. This value is in good agreement with the velocity of the motion of the main *Acetabularia* cytoplasm components. Thus it may be suggested that the "morphogenetic substances" are transferred from the nucleus to the apical end by the cytoplasm stream. The velocity of the signal of cap amputation, passing along the stalk and inducing the synthesis of the "morphogenetic substances" is also estimated. The signal velocity was found to be substantially greater than that of the "morphogenetic substances" not less than by a factor of ten. This result makes one suggest a special mechanism for transferring the above signal. The question of the possibility of an electrical nature of the signal is discussed.

The Hämmerling hypothesis of "morphogenetic substances" of *Acetabularia* (1) gives rise, in particular, to the question of the nature of a signal coming to the nucleus after cap amputation and inducing the synthesis of the morphogenetic substances which seem to be the RNP-complexes of nuclear origin (2,3). In this connection it seems interesting to estimate the value of the speed (v) at which the above mentioned signal travels over the *Acetabularia* stalk from the site of amputation of the cell apical part to the nucleus. To this end we can write the following equation which bounds the time interval (T) between the time of cap amputation and the beginning of regeneration of the new cap with the above speed v and the speed V of diffusion of morphogenetic substances from the nucleus to the amputation site.

$$l(v^{-1} + V^{-1}) + \Delta t_1 + \Delta t_2 = T \,, \tag{1}$$

where l is the length of the cell fragment remaining after cap amputation; Δt_1 is the time interval between the time of the signal arrival at the nucleus and the beginning of the motion of the morphogenetic substances from the nucleus to the cytoplasm; Δt_2 is the time interval between the time of the arrival of the morphogenetic substances from the nucleus to the cap amputation site and the beginning of cap regeneration.

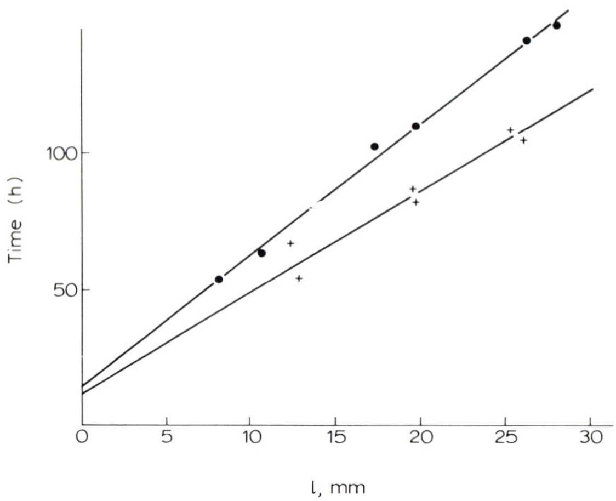

Fig. 1. Dependence of the time interval between the time of the amputation of the apical end of the *A. mediterranea* cell and the time of the appearance of the apical tip on the basal fragment length (two populations). At each point the data are averaged over 60 cells.

The time intervals T, Δt_1, Δt_2 naturally represent some consistent extremal values of these quantities.

Fig. 1 shows the results of the measurements of the $T(l)$ dependence for two populations of *A. mediterranea.* The *A. mediterranea* cells were cultivated using the conventional technique (4). The data of fig. 1 show first of all that the $T(l)$ dependence has the form of a straight line, which allows to consider the values v, V and Δt_1, Δt_2 to be independent of l.

After treating the results of fig. 1 we have

$$v^{-1} + V^{-1} \equiv a \simeq 40 \text{ hr/cm} , \tag{2}$$

$$\Delta t_1 + \Delta t_2 \equiv b \simeq 13 \text{ hr}. \tag{3}$$

In a separate experiment (5) estimations were made of the time interval (τ) between the cap amputation time and the time when the maximum amount of the morphogenetic substances come to the stalk or, more exactly, to the stalk fragment being at a distance of Δl away from the basal end of the cell. For this purpose the basal fragments of *Acetabularia*, formed after amputation of the upper half of the cell and having the length l were subjected to reamputation in different intervals of time. The fragments formed after reamputation were cultivated under the normal conditions and the number of fragments with the apical tips was registrated. Using the above notation we may write the following relationship:

$$l_c(v^{-1} + pV^{-1}) + \Delta t_1 = \tau , \tag{4}$$

where $p = \Delta l / l_c$.

Comparing (1) and (4) we may also write (instead of (4))

$$T_c - (1-p)l_c V^{-1} - \Delta t_2 = \tau \, . \tag{5}$$

Solving Eqs. (1) and (5) we have

$$V = (1-p)[a - (\tau - \Delta t_1)l_c^{-1}]^{-1} \, , \tag{6}$$

$$v = (1-p)[(\tau - \Delta t_1)l_c^{-1} - ap]^{-1} \, . \tag{7}$$

From the comparison of (7), (6) and (2), (3) we have the following estimations

$$\Delta t_1 < \tau - apl_c \, , \tag{8}$$

$$b - (\tau - apl_c) < \Delta t_2 < b \, , \tag{9}$$

$$a^{-1} < V < (1-p)(a - \tau l_c^{-1}) \, , \tag{10}$$

$$(1-p)(\tau l_c^{-1} - ap)^{-1} < v \, . \tag{11}$$

Using the experimental value of the parameters ($\tau \sim 12$ hr, $l_c \sim 1.7$ cm, $p \sim 0.1$) we have

$$\Delta t_1 < 5 \text{ hr} \, , \tag{8a}$$

$$8 \text{ hr} < \Delta t_2 < 13 \text{ hr} \, , \tag{9a}$$

$$0.025 \text{ cm/hr} < V < 0.027 \text{ cm/hr}, \tag{10a}$$

$$0.3 \text{ cm/hr} < v \, . \tag{11a}$$

Some uncertainty in the experimental values of individual parameters affects weakly the value of the speed V which may be taken as $V = 0.025$ cm/hr. Of course, this value of V may be somewhat different for different cell populations. This value of the speed is consistent with the value of the minimum speed of motion of the main components of the cytoplasm along the *Acetabularia* stalk. The latter speed we determined using the time lapse microfilming (6). Thus it may be suggested that the morphogenetic substances are transferred from the nucleus to the apical part of the cell by the cytoplasm stream. It is different from the value of the speed v because, as we have seen, $v/V > 10$, and may even be $v/V \gg 10$. Great speed values (up to 0.5 cm/hr) were recorded by the time lapse microfilming for local motions of single organelles observed by the light microscope. It seems unlikely, however, that such local motions of single organelles might play an essential role in transferring a signal about the cap amputation to the nucleus. The very great value of the speed v makes one suggest that there would be some other mechanism for transferring the above signal. For example, it may be suggested that the signal would be transferred either by the cytoplasm contraction wave generally observed in the amputation of a part of the cell, or by the electrical excitation wave propagation along the membrane structures of the *Acetabularia* stalk. The latter suggestion seems to be quite likely taking into account the semipermeable membrane of the *Acetabularia* nucleus. (The membrane potential of the *Acetabularia* nucleus measured by us in vitro (7) was found to be about −52 mV). For the verification of the possibility of the electrical nature of the signal in question we investigated the electrical activity of the *Acetabularia* cell after cap amputation. The biopotential was measured by the standard technique (8) using a chlorine-silver electrode inserted into the vacuole. It was found in these experiments that after cap amputation a characteristic series of spontaneous action potentials was registrated. An attempt was made to delay the passage of these series action potentials to the nucleus by suppressing them by inserting an additional electrode into the vacuole, which was connected with the first electrode by a feedback circuit. This gave an obvious increase in time T by about a factor 2 (9). Such an increase in the time T may be obviously obtained by replacing the speed v by the speed V in (1). The possible explanation of such a result may be that, normally, a "fast" electrical communication channel is used for transferring the signal about cap amputation. However, provided that such a signal is not produced, the nucleus will "learn" about the cap amputation by a message due to the transfer of some specific substances by the cytoplasm stream to the nucleus or due to the lack of some specific substances in the cytoplasm stream, which are transferred perhaps to the nucleus from the growing apical tip.

In any case the results obtained are, in our opinion, significant enough to continue investigations to clarify the role the electrical activity of *Acetabularia* might play in inducing the synthesis of the morphogenetic substances by the nucleus.

References

1. J. Hämmerling, Ann. Rev. Plant Physiol. 14 (1963) 65.
2. A.B. Alexeev, T.N. Zubarev, S.L. Stvolinsky, A.A. Yazykov, M.I. Betina. DAN SSSR 215 (1974) 712.
3. A.B. Alexeev, M.I. Betina, S.L. Stvolinsky, A.A. Yazykov, T.N. Zubarev. Plant Sci. Letters 3 (1974) 297.
4. M.I. Betina, A.A. Yazykov, I.V. Yanushevich, T.N. Zubarev. Fiziologia rasteniy 20 (1973) 1075.
5. A.B. Alexeev, M.I. Betina, E.V. Decterev, T.N. Zubarev, S.L. Stvolinsky, A.A. Yazykov, Fiziologia rasteniy 21 (1974).
6. L.A. Martynov, A.K. Sokolov, A.A. Yazykov, T.N. Zubarev. Protoplasma 80 (1974) 175.
7. A.A. Yazykov, N.P. Rogatykh, V.G. Yasinovsky, T.N. Zubarev. Fiziologia rasteniy 21 (1974) 726.
8. T.N. Zubarev, N.P. Rogatykh, V.G. Yasinovsky, Fiziologia rasteniy 21 (1974) 529.
9. V.G. Melkumyan, N.P. Rogatykh, V.G. Yasinovsky, T.N. Zubarev, DAN SSSR 218 (1974) 951.

SECTION VII

FROM THE PLASMA MEMBRANE TO THE NUCLEUS

CHAPTER VII.1

A DIRECT CELL-TO-CELL PATHWAY AND ITS ROLE IN MULTICELLULAR ORGANIZATION

E.L. BENEDETTI and I. DUNIA*

Institut de Biologie Moléculaire du C.N.R.S. et de l'Université Paris VII, 2 Place Jussieu, 75221-Paris Cedex 05, France.

Low-resistant pathways between animal cells represent an important tool for cell-to-cell communication, electrical and metabolic coupling. The passageway is associated with a structural and chemical differentiation of the junctional plasma membranes (nexus or gap junction).

Low-resistant junctions and metabolic cooperation may be established between cells that are genetically different. This fact probably correlates with the existence of a common protein component in the junctional membrane, independently from the origin of the cell type. Data on the assembly of low-resistant junctions, in developing tissues, are discussed.

The constitution of close aqueous compartments during the evolutionary sequences has probably involved the adsorption of amphipatic compounds at air/liquid interface and the casual orientation of the molecules forming a primitive membrane (1). But the superiority of a biological compartment or cell over the primitive "liposome" probably arose when the limiting membrane acquired stereospecificity and selective permeability. A further evolutionary advantage was set when living compartments could act as receiving-transmitting units and operate a more efficient control and adjustment of both external and internal environment.

Presumably this essential step of the evolving living matter could intervene only when the acquired molecular complexity of the surface membranes permitted specific interactions and coupling between functional units.

We are only beginning to understand how the regulation of multicellular organization is brought about; nevertheless a consistent experimental evidence suggests that, in a cell community, order, rank and survival are primarily dependent upon cell recognition, adhesiveness and cell-to-cell communication. Cell contact and interaction between plasma membranes, triggered by cAMP, are required during morphogenetic stages and differentiation even of free-living vegetative myxamoebae (2,3).

---

* The authors dedicate the present work to the memory of Denise Luzzati with whom they shared confidence and enthusiasm for Cell Biology and for a better world.

Cells may communicate with each other by means of a variety of passageways. The communication may be effected by the transmission through complex pathways of regulatory ligands and active metabolites, produced and released by cells at various distances from target cells.

The plasma membrane of the latter cells possesses specific receptor sites for the informational signals, and a means to convey the signal from the exterior to the interior of the target cell. These interactions occur through extracellular spaces even when the two cells are in close proximity and imply that the plasma membranes operate as independent units for each individual cell.

Conversely, another form of passageway allows the rapid transfer of ions and active messengers directly from one cell to another (4—6). This type of horizontal cell-to-cell interaction requires contact and the presence of specialized junctions between closely apposed regions at the surface membranes.

The transmission of electrical potentials and tracer substances between "non-excitable" cells

When an electric current is passed into a single cell with a microelectrode, the measure of the electrical potential in the cell and the measured amount of the current that would pass between the cell interior and the exterior bathing medium indicate that the specific conductance of the plasma membrane is low. If two cells are in close contact, the electrical potentials measured simultaneously in the cell where the electric impulse has been initiated and in the adjacent cell, may be almost the same. The current, carried by small inorganic ions, may therefore proceed from one cell to the neighbour bidirectionally and without delay through low-resistance passageway joining adjacent plasma membranes at sites where they are in intimate contact.

This phenomenon, which shows the existence of a short-range intercellular transmission of electrical potentials, was in the past considered to be proper only of the excitable cells where electric or electrotonic synapses normally operate. In contrast with the latter assumption, the passage of small ions from one cell to the neighbour appears not to be confined to excitable cells, but it is extended to a variety of cell types of adult and embryonic, epithelial or connective tissues (cf. ref. 6).

The electrical potentials measured in the cell, where the inward current has been passed (I), and in the neighbour (II), reflect the partition of the electric pulse; the ratio of the two membrane voltages provides a convenient index of communication (V_{II}/V_I) (7).

The transmission of electrical potentials is not the only form of cell communication. Various types of dyes or fluorescent tracers injected into cells with a micropipette may pass from one cell interior to another. Consistent evidence has been provided that the upper limit of molecular weight to allow junctional passage is close to 1000 dalton. In fact neither cytochrome *c* (mol. wt. 12,000) nor myoglobin (mol. wt. 18,000) diffuse from one cell interior to another. On the contrary Procion yellow (mol. wt. 625), labelled sucrose (mol. wt. 342) and the dye Chicago sky blue (mol. wt. 897) were observed to pass between various types of cells (cf. ref. 6 and 8). In some instances the spread of small tracers (fluorescein, mol. wt. 320) could be followed from the injected cells through a series of many cells (5).

In all cases so far studied the transfer of larger molecules was always associated with the presence of low-resistance passageways. It is therefore tempting to conclude that small ions as well as larger substances proceed through the same passageway.

If the passage of large molecules is found to be constantly associated with the transmission of an electric current, the converse is not necessarily true. For instance electrical coupling is present between blastomers of the teleost *Fundulus*, but fluorescein does not cross the low-resistance pathway (9)*. Furthermore, fluorescent tracers do not pass between cells of early embryonic stages of *Xenopus*, although low-resistance junctions are functionally and morphologically detectable (10). Fluorescein may diffuse, only in reaggregated cells if they are originated from later embryonic stages (11). These experiments concerning the coupling between embryonic cells lead to the conclusion that the cell-to-cell passageways in embryonic tissues are probably more restrictive and/or selective than in adult cells (cf. ref. 12).

Cell-to-cell communication monitored by the passage of electrical potentials or tracers is not exclusive of cells of the same embryological origin. Communication can be established, between cells growing in vitro which are genetically different (13,14). The ratio between the resulting membrane potentials (index of communication) may vary within a given cell combination and between various combinations of cells. These differences in coupling ratio between heterogeneous cell combinations might reflect quantitative variations of the junctions, but the existence of a direct correlation between the efficiency of low-resistance junctions and their proper architecture has not yet been proved**.

The junctional efficiency relies in fact on many parameters, such as the ohmic resistance of junctional and non-junctional membranes, the shape, the size, the arrangement of the interconnected cells and the junctional area-to-cell volume ratio (cf. ref. 7). Hence the effective work of low-resistance junctions present in a complex cellular system cannot be fully evaluated

* Freeze-fracture experiments on developing *Fundulus* embryos have shown that the low-resistance pathway, impermeable to fluorescein, correlates with the presence of typical gap junctions between blastomers (M.V.L. Bennett and N.B. Gilula, J. Cell Biol. 63 (1974) 21a).

** It has been claimed that the increase in the junctional resistance, following the treatment with uncouplers is associated with a more regular geometrical packing of the intramembranous junctional particles (C. Peracchia and A. Dulhunty, J. Cell Biol. 63 (1974) 263a).

merely by the estimation of the coupling ratio and the rate of passage of tracers.

Other intracellular or extracellular factors appear to be involved in junctional permeability. Loewenstein (5) claimed that the passage through low-resistance junctions is maintained only if the intracellular Ca^{2+} concentration is kept below 10^{-5} M. Conversely, when the concentration of Ca^{2+} in the extracellular environment is lower than 10^{-3} M, then the junctional membrane permeability is rapidly reduced and approaches the low values of the non-junctional membrane. Likely an electrochemical equilibrium depending upon metabolic energy maintains these critical values within a physiological range. Hence, junctional coupling can be lowered or abolished by various physical or chemical metabolic inhibitors acting on the energy derived from the ATP hydrolysis (low temperature, dinitrophenol, cyanide, oligomycin, N-ethylmaleimide). In this line it has been claimed that intracellular injection of ATP counteracts the uncoupling by metabolic poisons (15).

Divalent cations may also stabilize the "perijunctional insulation" that prevents the leakiness of the junctional membrane from inside to the extracellular spaces. In addition, the concerted action of other types of intercellular junctions, not directly involved in cell communication, may either stabilize the cell attachment or form high resistance cross bridges between cells. Moreover, divalent cations concentration intervenes together with energy sources in cytoplasmic motility, influencing the assembly and contraction of microfilaments (16,17) and microtubules (18). One could therefore speculate that Ca^{2+} might also act on junctional permeability by affecting the cytoplasmic contractile devices which somehow exert an influence on the organization of the junctional sieve. Consistent with the latter assumption are the observations of Cox et al., showing that specific cell-to-cell interaction is affected by cytochalasin B which dissociates microfilaments and binds to plasma membrane (19).

Functional roles of low-resistance junctions. The metabolic coupling

The evidence that we have so far presented supports the postulate that the low-resistance junctions provide a direct passageway from cell to cell.

The coupling might therefore controls and buffers cellular concentration of metabolites and messengers which are essential for specific cellular functions. This private cell-to-cell passageway would also have the advantage to prevent loss, dilution or degradation of regulatory molecules by diffusion into the extracellular environment (12).

Some experimental observations on the coupling efficiency either during early embryogenesis or during organogenesis would also suggest that morphogenetic movement and differentiation may be controlled either by the apparition of low-resistance junctions or by the selective uncoupling between unlike cells (cf. ref. 12).

An active metabolic cooperation between cells in contact has been studied in several mixed cell experiments using genetic variants which are deficient of one of the enzymes of the nucleotide biosynthesis (19,20) (21—23). The variant cells cannot incorporate exogeneous hypoxanthine, guanine, thymidine or adenine into nucleic acid. However a correction of the mutant phenotype occurs when the variants are growing together with normal donor cells.

The metabolic cooperation between cells in heterospecific cell mixtures requires cell-to-cell contact and it is concomitant and/or associated with electrical coupling and free passage of tracers across low-resistance junctions (24,25).

The nature of the donor-recipient relationship is not yet fully understood. Consistent evidence has however been provided by Pitts (22,23) and Cox et al. (19,21) that specific nucleotides may be the transferred substances correcting the genetic defect, rather than enzymes or informational macromolecules.

Again from these experiments it appears that the electrical coupling, the passage of tracers and the metabolic cooperation, are not exclusive between cells of the same origin and may exist between cells that are genetically different. In addition, the ability to communicate appears to correlate with highly selective and specific factors. Specificity is exhibited in the nature of substances transferred and by the observation that not all cells are able to function as recipient or/and donors.

The possibility exists that the ability to form low-resistance passageways primarily depends upon the expression of phenotypic character(s) controlling the proper molecular configuration of the plasma membrane. This assumption is probably in keeping with the results of hybridization experiments between normal and cancerous cells in vitro. In mixed cultured the normal cells are coupled whereas the neoplastic derivatives are apparently unable to form the intercellular association necessary for cell-to-cell communication (26). Conversely, after cell hybridization the heterokaryons appear to be electrically and metabolically coupled both with each other and with the normal parental cells, as long as the hybrids have stable chromosome combination (27,28).

It is known from the hybridization experiments of Frye and Edidin (29) that, within a short period of time following cell fusion, specific antigens of both parental cell surfaces intermix at the heterokaryon plasma membrane. Taking in account the latter observations, it could be therefore assumed that communication between heterokaryons correlates with the circumstance that the hybrid, after the intermixing of both cell surfaces features, has predominantly retained the proper molecular configuration of the normal parental cell, which is needed for cell recognition and assembly of low-resistance junctions.

The structural differentiation of the plasma membrane associated with the low-resistance passageway. The nexus

The important question of whether the various functional aspects observed for the different types of cell-to-cell communication can be ascribed to a specific and unique structural arrangement of the plasma membranes cannot be easily answered.

Intercellular junctions present a variety of morphological forms. They include very elaborate types of desmosomes, intermediate junctions, septate desmosomes, "tight" and "gap" junctions (cf. ref. 30). Obviously, each of these forms of junctions or even more than one type might be purported as a candidate for the structural specialization responsible for the low-resistance passageway. Furthermore, very often animal cells are joined by an association of more than a single type of junction. Thus, the evidence implicating one specific type of junction as the basis of electrotonic or metabolic coupling is essentially inductive.

In spite of all these uncertainties many morphological observations have reinforced the postulate that, in almost all cell types which are electrically and metabolically coupled, "gap" junctions are present. In some instances, though very rarely, they are found to be the sole and unique junction type present between coupled cells (31).

The "gap" junction, also called *nexus*, belongs to the category of cell-to-cell junctions in which the plasma membrane of neighbouring cells appear to come into direct or close contact.

The "gap" junction in most types of tissues has the shape of a plaque of various extends, rather than the form of a continuous belt connecting adjacent cells. In thin sections of fixed and embedded tissue the two opposite outer layers of each junctional plasma membrane appear in close proximity. A minute (20 Å) "gap" is visible between the two junctional plasma membranes, which by standard staining techniques looks like an electron-transparent fence. Only with negative staining of isolated plasma membranes (cf. ref. 32) and by the introduction of colloidal electron opaque tracer (lanthanium hydroxide, potassium pyroantimonate) (cf. ref. 33) through the intercellular spaces, the architecture of the interlocking device between the junctional plasma membranes could be discovered. This structure consists of a bidimensional lattice of repeating subunits arranged in hexagonal array. The subunits are assumed to contain a small hydrophilic hole in the middle; the center-to-center distance of the subunits amounts of 80—90 Å. Negative staining and impregnation techniques, though they could outline the bidimensional organization in the plane of the occluding device of the "gap" junctions, were unable to provide information on the internal organization of junctional plasma membranes. Exclusively, the application of the new freeze-fracture technique could reveal that the structural organization of the "gap" junction involves a rearrangement of both the hydrophilic domain as well as of the hydrophobic core of the junctional membranes.

The freeze-etching technique has in fact the advantage of showing internal features of the membranes by fracturing and portraying these structures stabilized at very low temperature (—180° C) without the utilization of chemical fixation or drastic dehydration and staining of the specimens. Furthermore by the fracture and the sublimation (etching) of the vitreous ice enclosing the biological structures, multiple and large cleavage and etched surfaces are revealed. It is thus possible to collect informations useful for a

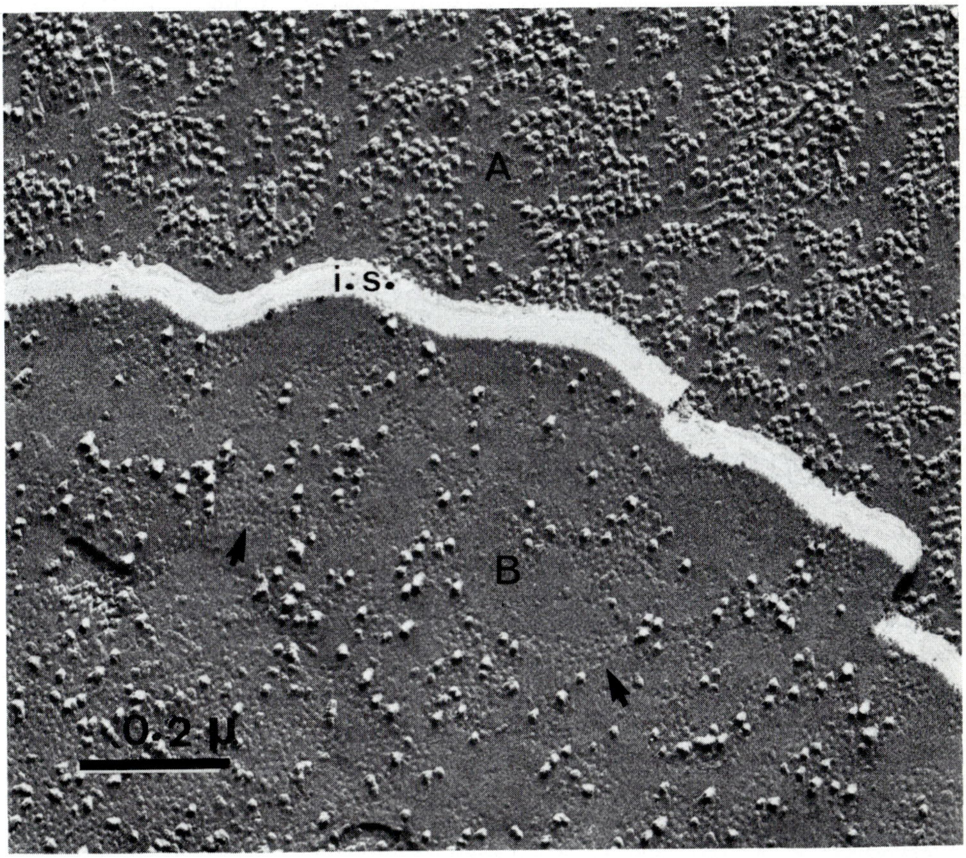

Fig. 1. Replica of eye lens fiber. The fracture has exposed large inner membrane surfaces of the plasma membranes of two adjoining lens fibers. The intercellular space (i.s.) separates the two fracture faces. The fracture face A is characterized by many inner membrane particles of heterogeneous shape and size, ranging between 40—90 Å in diameter. Note the clustered distribution of the particles on face A, probably due to a proper chemical feature of the lipid domain (cf. ref. 61). The fracture face B shows very few particles and many clusters of small pits and depressions corresponding to the emplacement of spanning particles which, during the cleavage, remain associated with the complementary fracture face A.

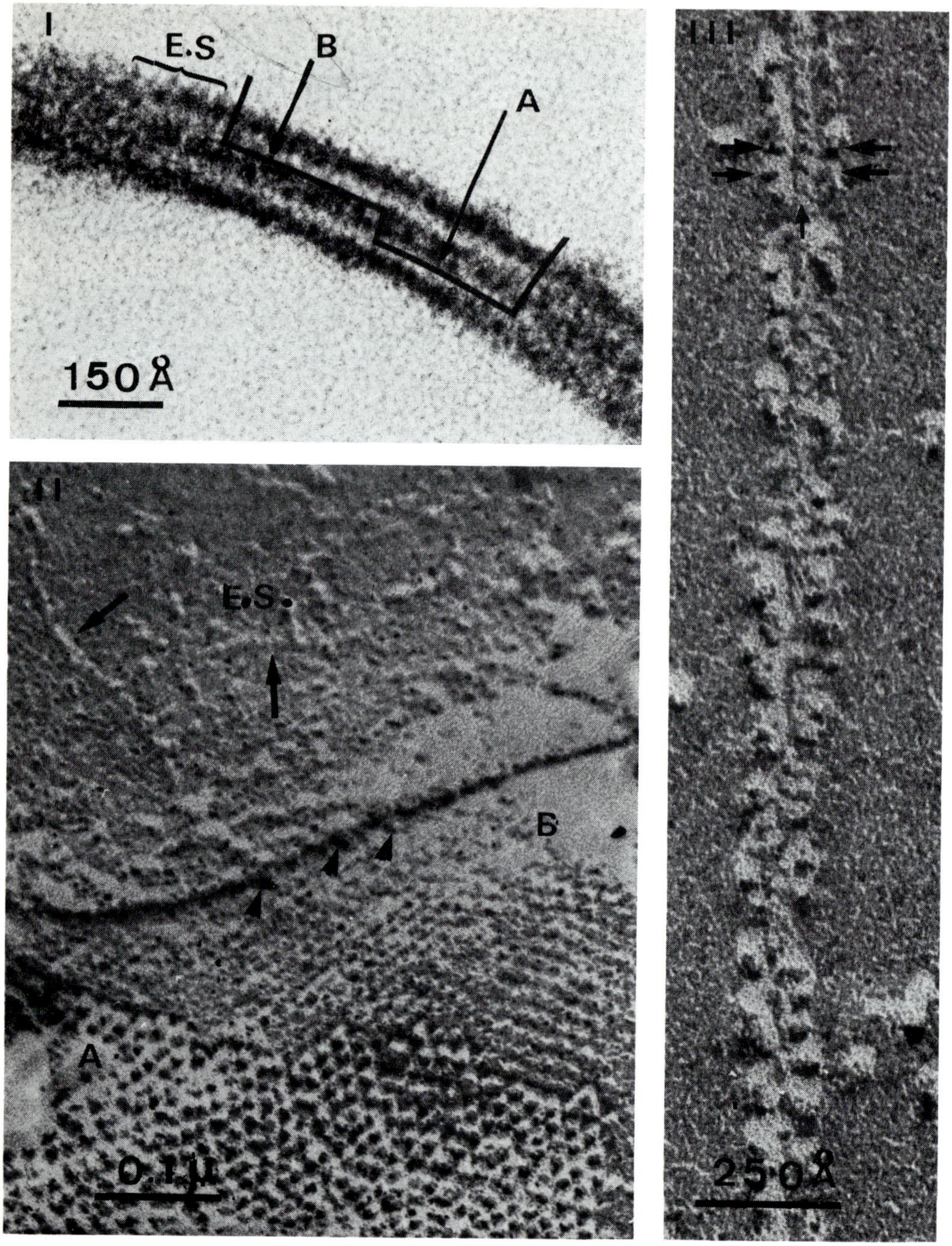

Fig. 2. I) Cross section of the nexus in isolated rat liver plasma membranes (thin section stained with uranyl acetate and lead citrate). The line depicts the path of the fracture exposing simultaneously the feature of the interlocking structure (face B) and the inner

tridimensional analysis of the membrane architecture. The interpretation of the images is based upon the principle that the membranes are fractured along their internal hydrophobic matrix, since at low temperature the hydrophobic bonding between the paraffin chains of the phospholipids are mostly weakened (34).

The fracture of the non-junctional plasma membrane produces two distinct types of cleavage faces characterized by a marked asymmetry of their structural features. In several types of plasma membranes in animal cells so far studied, the hydrophobic matrix exposed by the fracture is heterogeneous. It shows smooth areas, which correspond to membrane regions occupied almost exclusively by lipids, and particulate components consisting most probably of intercalated membrane proteins. The inner aspect of the leaflet adjacent to the cytoplasm (face A or (+) or outer fracture face (OFF)) is covered by a number of 80—90 Å particles randomly distributed. Conversely the inwardly directed fracture face of the outer leaflet (face B or (—) or inner fracture face (IFF)) is studded with fewer particles (cf. ref. 32) (fig. 1).

In the gap junctions, the freeze-fracture splits in a stepwise fashion simultaneously the two junctional plasma membranes. First the cleavage passes along one of the two plasma membranes, exposing the inner aspect of the leaflet adjacent to the cytoplasm (face A). Then the interior of the adjacent membrane is cleaved and this fracture surface corresponds to the inwardly directed face of the outer membrane leaflet (face B), or more precisely the inner aspect of the interlocking device (30,33) (fig. 2.I).

The junctional face A is characterized by regular arrays of 90 Å particles. The corresponding face B consists of a hexagonal lattice of regular facets showing in the best shadowing conditions a central pit or a hollow space with a center-to-center distance of 90 Å. Double replica and cross-fractures through the junctions clearly show that the hexagonal array of facets exposed on both faces B forms one single middle interlocking layer (32,35,36) (fig. 2.III).

---

aspect of the outer leaflet close to the cytoplasmic surface of the junction (face A). E.S. indicates the region where freeze-etching may reveal the true cytoplasmic surface of the junction. II) Freeze-etched replica of the same preparation as in (I). The true cytoplasmic surface of the nexus, revealed by the etching, has rather homogeneous appearance; the arrows point to filamentous structures probably corresponding to cytoplasmic components in close association with the junctional membrane. The edge between the etched surface and the fracture face B has a dotted appearance (▲) corresponding to the 90 Å junctional particles which are visualized on the fracture face A. The fracture face B is characterized by a distorted hexagonal array of pitted facets. III) Cross fracture of isolated nexus by deoxycholate treatment and density gradient centrifugation (cf. ref. 45). The structure of the nexus has a rather distorted appearance due to partial solubilization of membrane components. Nevertheless pairs of bimodal elements are visible (arrows). Apparently the bimodal element consists of two matching 90 Å particles interlocked by a device showing in some regions a periodicity (arrows).

The fracture does not expose the true cytoplasmic surface of the junction. The latter surface may be visualized by deep-etching on isolated plasma membranes. If freeze-fracturing and deep-etching are applied on isolated plasma membrane, then the etched cytoplasmic surfaces of the gap regions show a rather filamentous surface (fig. 2.II).

Although a number of experimental evidences implicate the gap junction as a site of intercellular communication, it would be unlikely to conclude that no other features of plasma membrane differentiation might be also associated with cell communication. In fact several variations of the nexus features have been described (cf. ref. 37). They involve the size and spacing of the junctional particles, the regularity of the hexagonal array both of the lattice of pitted facets and of the inner membrane particles.

In gap junctions of some vertebrate (38) and invertebrate (39) tissues, the membrane particles may appear to lie in rows which either anastomose forming an irregular net or interconnect aggregates of geometrically packed particles. Moreover, in various invertebrate tissues the fracture plane cleaves the junctional membranes in such a way that the inner membrane particles appear preferentially attached on fracture face B. Conversely, in this type of "inverted" junction, the fracture face A displays a hexagonal array of circular pits (40,41).

Chemical characterization of the nexus

The molecular architecture and chemical nature of the gap junction membranes are different from those of the general plasma membranes since the latter regions are almost completely solubilized by detergents, while the gap junctions are resistent to the same treatment.

Attempts of isolation of the gap junctions are in fact based upon the selective resistance of the junctional membrane to solubilization with sodium deoxycholate (42) or Sarkosyl (43,44). The result of biochemical investigations on the isolated gap junctions indicate that the gap junction-rich fraction, in contrast with the general plasma membrane, is characterized by high protein, very low phospholipid content, and by a preponderance of neutral lipid including free and esterified cholesterol, and triglyceride (cf. ref. 32).

By gel electrophoresis experiment in SDS the isolated junctions from eye-lens fibers are characterized by a rather simple protein pattern as compared to the general plasma membrane protein profile (45). One of the proteins, whose molecular weight varies in different experiments from 32 to 34,000 dalton, is always found as a main lens fiber plasma membrane constituent, and it is enriched in the membrane fraction isolated by detergents. Since the latter fraction mainly consists of lens fiber junctions, it has been proposed that this 34,000 dalton protein is a major and proper component of the fiber junctions. In accordance with these results, Goodenough (44) has reported the occurrence of a 34,000 dalton component as major constituent of the isolated gap junctions of mouse liver. It is not yet clear whether this 34,000 dalton component corresponds to an oligomeric association of low molecular weight polypeptides, or to a monomeric protein*.

The presence of a single predominant protein is not uncommon in biological membrane characterized by highly ordered geometrical assembly of repeating subunits. If further protein pattern analysis on gap junctions isolated from a greater variety of tissues will confirm the existence of a common junctional protein component, then the assumption that the process of junctional differentiation initiates with a convergent accumulation within the lipid membrane domain of preexisting particulate entities consisting of a single or very few quasi-equivalently related protein, will be greatly reinforced.

Moreover, the demonstration that in various tissues and in different animal species the gap junctions are characterized by a common protein profile may explain the occurrence of low-resistance junctions and metabolic cooperation between cells that are genetically different. One could then postulate that in the cell lines lacking gap junctions and cell-to-cell communication, either the phenotypic expression of the proper junctional protein pattern is repressed, or the assembly process of the junctional constituents is somehow altered. Obviously this view needs further investigations.

Molecular model of the gap junction

From the complex structural features and from some chemical properties of the gap junction an ideal molecular model may be derived. The coupling device at molecular level might consist of individual bimodal structures (cf. ref. 46) hexagonally packed in the plane of the junction. Each subunit could be either a single amphipatic protein or an oligomeric association of identical or quasi-equivalently related proteins, with one polar surface exposed to the aqueous cytoplasmic phase and the other, predominantly non-polar, intercalated into the hydrophobic junctional core. By a proper inwardly directed segregation of polar residues, the core of the bimodal element could be occupied by a hydrophilic channel connecting directly the two aqueous cytoplasmic phases.

The two opposite and complementary subunits could be interlocked by a single hexagonal lattice originated from the merging of both outer membrane leaflets. Selected species of lipid and/or polypeptides could be accommodated in a bidimensional geometrical configuration, displaying hydrophilic centers in register with the hydrophilic core of the 90 Å particles core. The observa-

* According to experiments of N.B. Gilula, the gap junctions isolated from rat liver contain two prominent low-molecular weight components of 10,000 and 20,000 daltons respectively. The molar ratio between the two components is 3:1 (J. Cell Biol. 63 (1974) 111a).

tion that hypertonic solutions split the nexus in two halves along the central hexagonal lattice is probably consistent with the presence of an individual structural device between the inner membrane array of repeating subunits (36).

Moreover, each hydrophilic channel through the bimodal junctional element must be completely insulated from the aqueous environment comprised within the extracellular spaces. Any leakage at the level of the interlocking device would connect directly the cell interior with the exterior and this circumstance would be incompatible with the proper function of the low-resistance passageway. The array of interlocking units may be longitudinally permeated by extracellular flow of matter circumventing each unit. Such a multi-units system therefore would bridge two opposite plasma membranes without sealing off the extracellular spaces, in contrast with other types of occluding junctions (30,37).

Assembly process of the nexus

There is some evidence to suggest that some membrane mediated functions, such as cell fusion (47), endo- and exocytosis (48) and cell contact (49), may be dependent or associated with the rearrangement, or perturbation, of the amphipatic protein/glycoprotein components forming the particles intercalated within the lipid domain.

In addition, it has been shown that several different actions affecting either the proteins (cf. ref. 50) or the lipids (51) of the plasma membrane may induce the rearrangement and the clustering of the inner membrane particles visualized by freeze-cleavage on the membrane fracture faces.

The clustering of the inner membrane particles in the plane of the plasma membrane core has been observed in mammalian cells agglutinated by concanavalin A (32,52) (figs. 3 and 4). The clustered distribution of the membrane particulate entities seems, however, not to be an inherent characteristic of the agglutinable cells, but rather an acquired feature dependent or associated with cell agglutination.

Other studies, in particular those of Hülser and Peters (53,54), have revealed the occurrence of electrical coupling between agglutinated lymphocytes by a mitogenic dose of phytohaemagglutinin. The electrical coupling, which rapidly follows the agglutination, appears to mediate the mitogenic effect. The view provided by these interesting experiments implies that cell activation by the lectin might be a multicellular event rather than a single cell process. One possible explanation of these observations could be that the low-resistance passageways between agglutinated lymphocytes correlate with the formation of small matching clusters of inner membrane particles interlocking the plasma membranes. In more general sense it might be postulated that cross-bridging between cell surface receptors or antigenic sites by multivalent ligands, and also "trans"-glycosylation of glycoprotein acceptors lo-

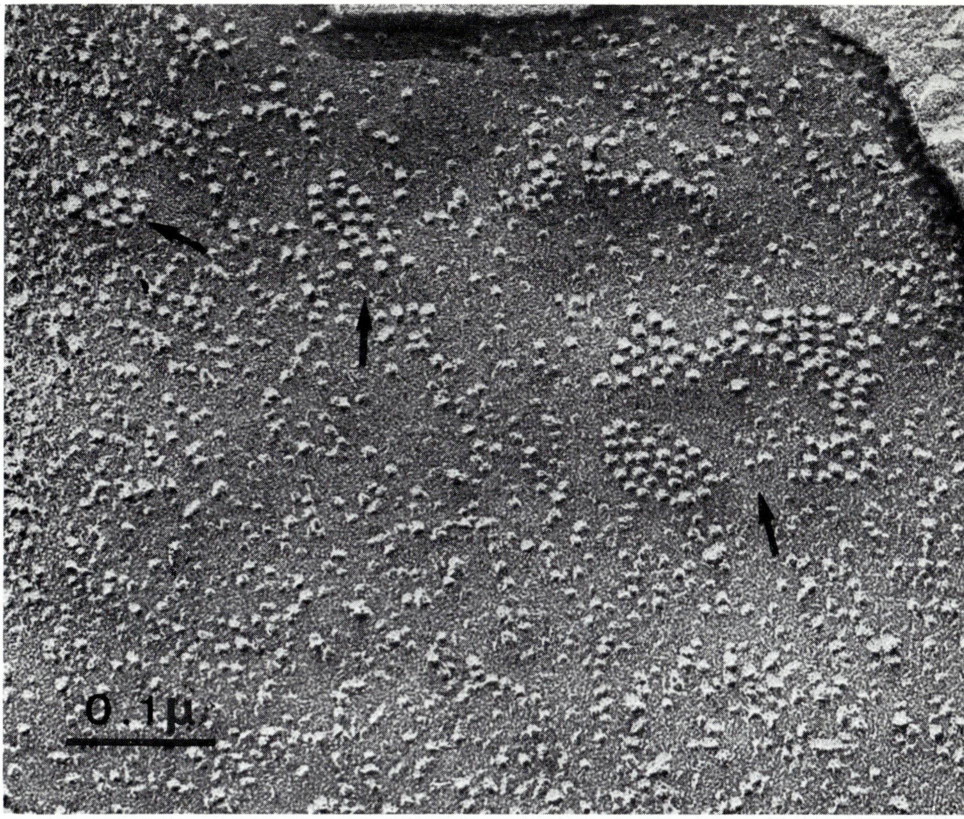

Fig. 3. Replica of fibroblast-like variant cell from BALB/C mouse plasmocytoma MOPC 173 treated by concanavalin A (50 μg/5.10^5 cells/ml). The fracture face A is characterized by the presence of inner membrane particles forming heterogeneous clusters or linear rows. Some larger clusters of identical 90 Å particles probably correspond to initial assembly of nexus. (Picture kindly provided by Mrs. M.A. Diawara).

calized on the opposite surfaces of adjoining cells (55,56), could trigger the rearrangement in the plane of the plasma membranes of preexisting or newly synthesized "junctional" particles. The accumulation of these identical or equivalently related protein subunits would lead to the formation of a bidimensional lattice or linear array of bimodal junctional units between the partner plasma membranes responsible for low-resistance pathways. This assumption probably accords well with the observations provided by freeze-fracturing technique applied to animal cells in culture (32,57) and to embryonic (58) or developing tissues (59—61), where low-resistance junctions are found. These studies, although incomplete, have shown that in areas of cell contact, the junctional assembly is characterized by progressive accumulation of structural identical particles in the plane of the membrane leaflets.

Fig. 4. Replica of epithelial-like variant cell from BALB/C mouse plasmocytoma MOPC 173 treated by concanavalin A (50 μg/5.10^5 cells/ml). The fracture face A shows heterogeneous inner membrane particles forming small clusters or short linear arrays. Note the round or oval small depressions of the fracture face surrounded by rows of particles. The latter features are suggestive for initial formation of endo- or exocytosis (cf. ref. 48). (Picture kindly provided by Mrs. M.A. Diawara).

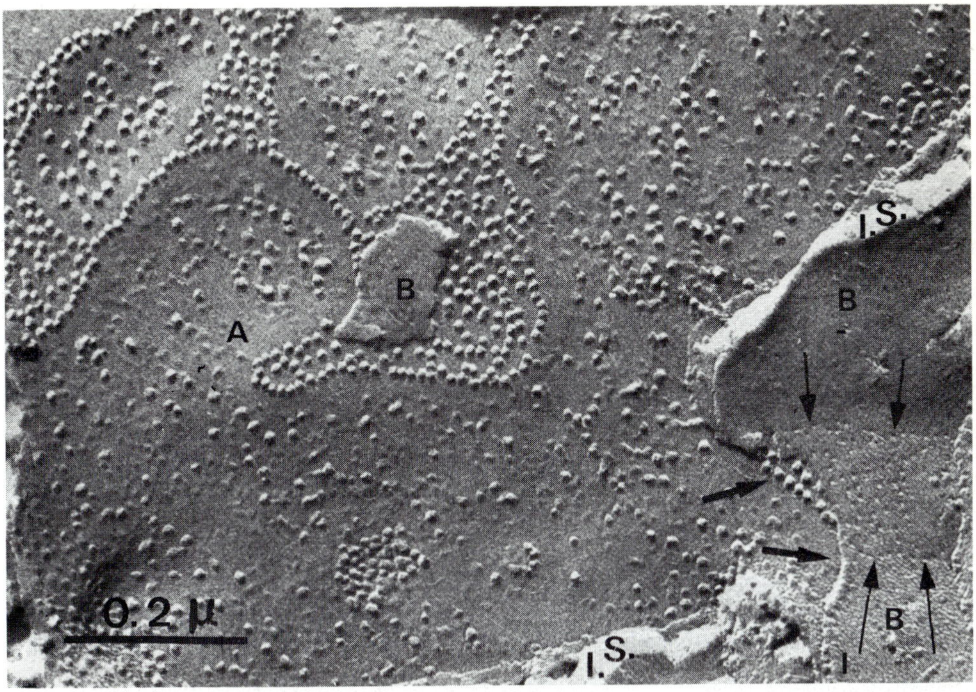

Fig. 5. Freeze-fracture replica of eye lens epithelial cell in the zone of cellular elongation. The fracture has exposed inner membrane aspects of the A and B faces in areas of junction formation. 90 Å junctional particles form linear rows or clusters of various dimensions. The two fracture faces are still separate by the intercellular space, except in regions of close apposition of two plasma membranes (arrows). These junctional particles are assembled on the face A and, on the matching face B, corresponding depressions are visible, forming either rows or clusters (arrows).

These particles may form either rows or polygonal clusters matching the two plasma membranes in close contact and obliterating the intercellular space (61) (fig. 5). As the differentiation of the junctions proceeds, the convergent accumulation of the bimodal junctional units forms more elaborate and extensive packed arrays (61) (fig. 6)*.

There is some evidence suggesting that in rapidly growing tissue low-resistance pathways may be formed, at any step of the cell cycle (cf. ref. 12), in a

* Freeze-fracture experiments of J.P. Revel at different times during reaggregation of hepatoma cells have demonstrated that the onset of the gap junction assembly is characterized by the presence of scattered large (100 Å) inner membrane particles on the fracture face A. It is not easily apparent whether these particles are cleaved into smaller junctional particulate entities or play a role in getting opposite cell surfaces in close contact (J.P. Revel, in A. Moscona (Ed.) The Cell Surface in Development, John Wiley, New York, London (1974) p. 51).

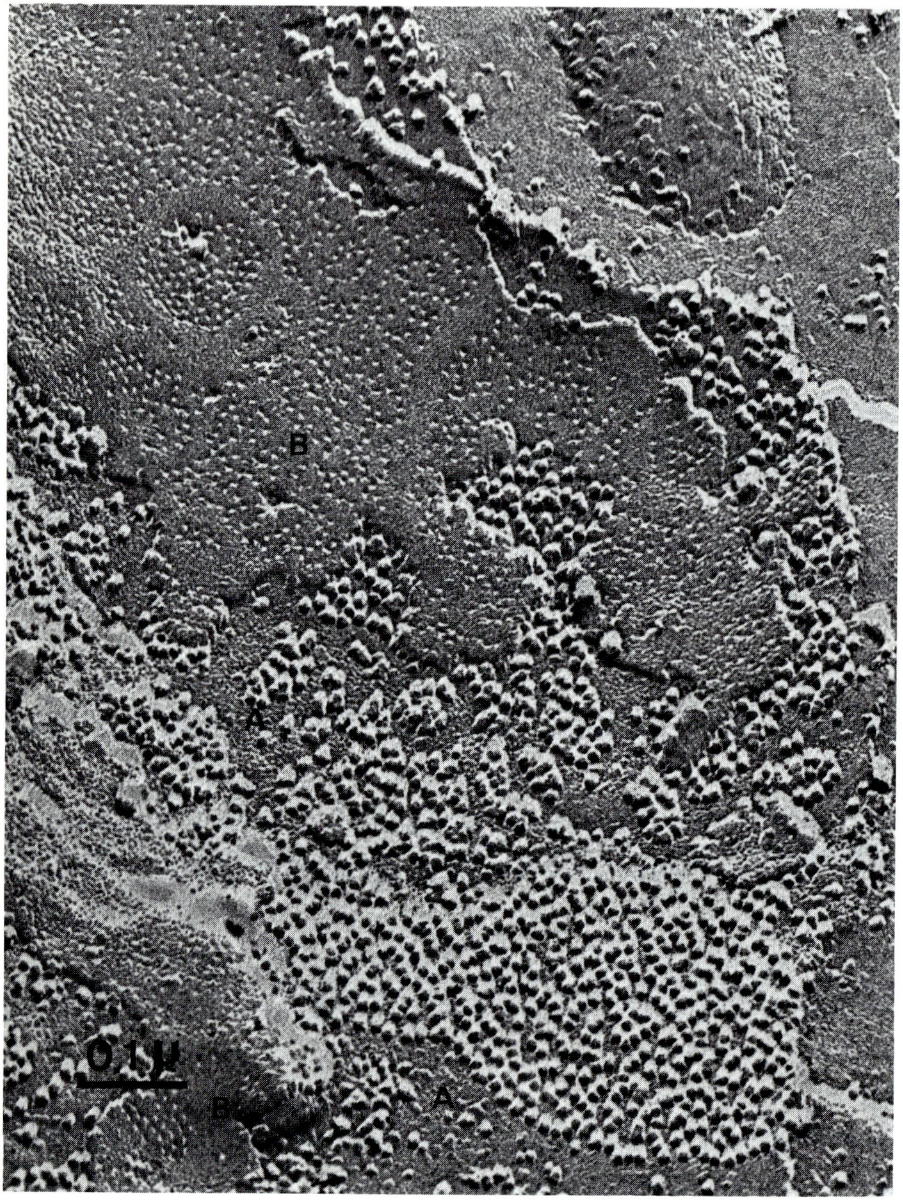

Fig. 6. Freeze-fracture replica of eye lens fiber in the zone of cellular elongation. The fracture has exposed the A and B faces of a developed junction. The A face is characterized by polygonal assembly of 90 Å particles and the B face by array and large cluster of depressions. These depressions correspond to the emplacement of the 90 Å particles which, during the cleavage, have been chipped off.

period of time which is probably too short to be compatible with a programme involving a new synthesis and insertion of junctional constituents (cf. ref. 62). In these cellular systems, therefore, the rapid and probably reversible assembly of low-resistance junctions may correlate with the lateral translation of a preexisting pool of "junctional" particles in the fluid lipid domain*.

In other slowly growing tissue, however, the process of junctional differentiation may not necessarily be the same, and the assembly process could even be favored by a rather rigid lipid core (61). The latter condition would in fact promote the segregation of newly synthesized protein particles. Consistently, from tissue culture experiments, evidence has been provided that the clustered distribution of inner membrane particles is associated with a rather rigid lipid domain and mediates cell-to-cell contact and density inhibition of growth (49,63).

Concluding remarks

Although incomplete and to some extend controversial, the present review, nevertheless, supports the conclusion that cell-to-cell communication associated with a specific type of junction may have profound effects on cell behaviour and tissue differentiation.

Yet the molecular mechanism whereby the low-resistance junctions ensure horizontal cell interactions and correlate with multicellular organization requires further explanations.

Metabolic cooperation between genetically different types of cells and the passage of inorganic ions or higher molecular weight substances seems to be dependent upon a proper structural differentiation of the junctional plasma membranes. Unfortunately we can only speculate on the manner by which the junctional sieve is controlled.

The presence of such a complex architecture of the junctional membranes could suggest a functional significance more elaborate than that of a merely passive mechanical sieve.

It is noteworthly that the hexagonal lattice of repeating subunits fulfils some topological principles of structuration postulated by Changeux et al. (64, 65) for a cooperative membrane regulatory function according to a model derived from the theory of regulatory allosteric enzymes.

It would be therefore extremely interesting to know whether the highly

* The times of coupling onset, measured in many different cellular systems, are rather short (1 min to 40 min). Yet the establishment of low-resistance pathways, monitored by the increasing junctional conductance, seems to be a progressive event, reflecting a gradual apposition of 'junctional units' (S. Ito, E. Sato and W.R. Loewenstein, J. Membrane Biol. 19 (1974) 305 and 355).

ordered array of junctional elements acts as a device regulating the selective and restrictive permeability by reversible conformational transitions following the interactions with ligands. But more experimental data on the chemical composition and on the enzymatic and/or antigenic properties of the junctional membranes are needed in order to understand whether this control is an inherent consequence of the organized complexity and proper architecture of the junctional structure or essentially relies upon the chemical specificity of the junctional components, in particular of the proteins. Furthermore it is not yet clear how the selective permeability or the uncoupling of the junction are related to critical intracellular and extracellular concentration of ligands including divalent cations, and to the presence of junctional receptors.

A new interesting approach to the problem of how divalent cations might mediate coupling has been reached by experiments showing that cell-to-cell communication can be induced by the action of lysolecithin on the plasma membrane of *Acanthamoeba castellanii.* A regional micelle formation dependent on lysophospholipid content would favour cell aggregation and membrane permeability. High Ca^{2+} concentration could on the contrary stabilize the lipid bilayer, thus counteracting the effect of lysophospholipid on junctional permeability (65). In addition, phospholipase(s) action on plasma membrane could also correlate with the formation of low-resistance pathways, since this treatment favors clustered redistribution of the inner membrane particles (51).

Other control mechanism appears to operate at the level of interactions between junctional components and cytoplasmic structural proteins (microfilaments) (19). A matter of further interest is whether this cytoplasmic-plasma membrane interactions correlate more with the mobility and the assembly process of the junctional units within the membrane lipid domain, rather than primarily with the control affecting the junctional permeability. The assembly of protein repeating subunits could account for a local tangential polarity of the plasma membrane coding for reciprocal recognition of the two membrane surfaces and specific cell-to-cell contact and triggering the formation of low-resistance pathways.

One of the most important questions yet unanswered on the function of the plasma membrane in animal cells concerns the molecular mechanism linking the interactions of the exposed receptors with various ligands, to the metabolic machinery controlling cell growth and differentiation.

On the face of the available evidence it may be postulated that one key-step of this control arises when plasma membranes of individual cells get mutually in contact and form low-resistance pathways (67,68). At this hierarchical order of multicellular organization it appears conceivable that the metabolic properties of each individual cell are not exclusively set by its own genotype but rather by the total gene pool of the cell community (20).

Acknowledgements

The authors would like to thank Dr. Hans Bloemendal (Department of Biochemistry, University of Nijmegen, The Netherlands) and Dr. Adolfo Martinez-Palomo (Departamento de Biologia Celular, Centro de Investigacion y de Estudios Avanzados del Istituto Politecnico, Mexico) for the useful and stimulating discussions during the redaction of the manuscript. They also gratefully acknowledge the help of Ms Michele Cornu for the preparation of the manuscript. The tissue culture experiments were realized in cooperation with Dr. A. Paraf (Laboratoire d'Immunodifferenciation, Institut de Biologie Moléculaire de l'Université Paris VII et du C.N.R.S., Paris).

The work has been supported by the Délégation Générale à la Recherche Scientifique et Technique (contract no. 74.7.0173).

References

1. A.D. Bangham, Ann. Rev. Biochem. 41 (1972) 753.
2. J.T. Bonner, Proc. Nat. Acad. Sci. U.S. 65 (1970) 110.
3. H.C. Aldrich and J.H. Gregg, Exp. Cell Res. 81 (1973) 407.
4. E.J. Furshpan and D.D. Potter in A.A. Moscona and A. Monroy (Eds.) Current Topics in Developmental Biology, Academic Press, New York (1968) ch. IV, p. 95.
5. W.R. Loewenstein, in M. Locke (Ed.) The Emergence of Order in Developing Systems, Academic Press, New York (1968).
6. M.V.L. Bennett, Fed. Proc. 32 (1973) 65.
7. J.D. Sheridan, Am. Zool. 13 (1973) 1119.
8. S.J. Socolar, Exp. Eye Res. 15 (1973) 693.
9. M.V.L. Bennett, Develop. Biol. 29 (1972) 419.
10. C. Slack and J.F. Palmer, Exp. Cell Res. 55 (1969) 416.
11. J.D. Sheridan, Develop. Biol. 26 (1971) 627.
12. R.L. De Haan and H.G. Sachs, in A.A. Moscona and A. Monroy (Eds.) Current Topics in Developmental Biology, Academic Press, New York, (1972) ch. V, p. 193.
13. W.R. Loewenstein, Fed. Proc. 32 (1973) 60.
14. W. Michalke and W.R. Loewenstein, Nature 232 (1971) 121.
15. A.L. Politoff, S.J. Socolar and W.R. Loewenstein, J. Gen. Physiol. 53 (1969) 498.
16. L.G. Tilney and M. Mooseker, Proc. Nat. Acad. Sci. U.S. 68 (1971) 2611.
17. J.F. Perdue, J. Cell Biol. 58 (1973) 265.
18. J.B. Olmsted and G.C. Borisy, Ann. Rev. Biochem. 42 (1973) 507.
19. R.P. Cox, M.R. Krauss, M.E. Balis and J. Dancis, J. Cell. Physiol. 84 (1974) 237.
20. H. Subak-Sharpe, R.R. Bürk and J.D. Pitts, J. Cell Sci. 4 (1969) 353.
21. R.P. Cox, M.R. Krauss, M.E. Balis and J. Dancis, Exp. Cell Res. 74 (1972) 251.
22. J.D. Pitts, in L.G. Silvestri (Ed.) Third Lepetit Colloquium on Cell Interactions, North-Holland, Amsterdam (1972) p. 277.
23. J.D. Pitts, in G.E.W. Wolstenholma and J. Knight (Eds.) Ciba Foundation Symposium on Growth Control in Cell Cultures, Churchill Livingstone, London (1971) p. 89.
24. N.B. Gilula, O.R. Reeves and A. Steinbach, Nature 235 (1972) 262.
25. R. Azarnia, W. Michalke and W.R. Loewenstein, J. Membrane Biol. 10 (1972) 247.
26. R. Azarnia and W.R. Loewenstein, J. Membrane Biol. 6 (1971) 368.

27. R. Azarnia and W.R. Loewenstein, Nature 241 (1973) 455.
28. R. Azarnia, W.J. Larsen and W.R. Loewenstein, Proc. Nat. Acad. Sci. U.S. 71 (1974) 880.
29. L.D. Frye and M. Edidin, J. Cell Sci. 7 (1970) 319.
30. N.S. McNutt and R.S. Weinstein, Prog. Biophys. Mol. Biol. 26 (1973) 45.
31. J.P. Revel, A.G. Yee and A.J. Hudspeth, Proc. Nat. Acad. Sci. U.S. 68 (1971) 2924.
32. E.L. Benedetti, I. Dunia and A. Diawara, Eur. J. Cancer 9 (1973) 263.
33. D.S. Friend and N.B. Gilula, J. Cell Biol. 53 (1972) 758.
34. D. Branton, in E.L. Benedetti and P. Favard (Eds.) Freeze-Etching: Techniques and Applications, Société Française de Microscopie Electronique, Paris (1973) p. 107.
35. J.P. Chalcroft and S. Bullivant, J. Cell Biol. 47 (1970) 49.
36. D.A. Goodenough and N.B. Gilula, J. Cell Biol. 61 (1974) 575.
37. P. Satir and N.B. Gilula, Ann. Rev. Entomol. 18 (1973) 143.
38. E. Raviola and N.B. Gilula, Proc. Nat. Acad. Sci. U.S. 70 (1973) 1677.
39. N.E. Flower, J. Cell Sci. 10 (1972) 683.
40. C. Peracchia, J. Cell Biol. 57 (1973) 54.
41. C. Peracchia, J. Cell Biol. 57 (1973) 66.
42. E.L. Benedetti and P. Emmelot, J. Cell Biol. 38 (1968) 15.
43. W.H. Evans and J.W. Gurd, Biochem. J. 128 (1972) 691.
44. D.A. Goodenough, J. Cell Biol. 61 (1974) 557.
45. I. Dunia, K. Sen Ghosh, E.L. Benedetti, A. Zweers and H. Bloemendal, FEBS Letters 45 (1974) 139.
46. D.E. Green, in D.E. Green (Ed.) Membrane Structure and its Biological Applications, New York Academy of Sciences, New York (1972) p. 150.
47. B. Satir, C. Schoolez and P. Satir, J. Cell Biol. 56 (1973) 153.
48. P. Satir and B. Satir, in B. Clarkson and R. Baserga (Eds.) Control of Cell Proliferation, Cold Spring Harbor, New York (1974) p. 33.
49. R.E. Scott, L.T. Furcht and J.H. Kersey, Proc. Nat. Acad. Sci. U.S. 73 (1973) 3631.
50. D. Branton, A. Elgsaeter and T. James, Eighth FEBS Meeting 28 (1972) 165.
51. A.J. Verkleij, R.F.A. Zwaal, B. Roelofsen, P. Comfurius, D. Kastelijn and L.L.M. Van Deenen, Biochim. Biophys. Acta 323 (1973) 178.
52. C. Guérin, A. Zachowski, B. Prigent, A. Paraf, I. Dunia, M.A. Diawara and E.L. Benedetti, Proc. Nat. Acad. Sci. U.S. 71 (1974) 114.
53. D.F. Hülser and J.H. Peters, Exp. Cell Res. 74 (1972) 319.
54. D.F. Hülser, in D.M. Prescott (Ed.) Methods in Cell Biology, Vol. VIII, Academic Press, New York (1974) p. 289.
55. S. Roth and D. White, Proc. Nat. Acad. Sci. U.S. 69 (1972) 485.
56. S. Roth, E.J. McGuire and S. Roseman, J. Cell Biol. 51 (1971) 536.
57. P. Pinto da Silva and N.B. Gilula, Exp. Cell Res. 71 (1972) 393.
58. J.P. Revel, P. Yip and L.L. Chang, Develop. Biol. 35 (1973) 302.
59. R.S. Decker and D.S. Friend, J. Cell Biol. 62 (1974) 32.
60. D.F. Albertini and E. Anderson, J. Cell Biol. 63 (1974) 234.
61. E.L. Benedetti, I. Dunia and H. Bloemendal, Proc. Nat. Acad. Sci. U.S. 71 (1974) 5073.
62. W.R. Loewenstein, in J.F. Woessner, Jr. and F. Huijing (Eds.), The Molecular Basis of Biological Transport, Academic Press, New York and London (1972) p. 85.
63. R.E. Barnett, L.T. Furcht and R.E. Scott, Proc. Nat. Acad. Sci. U.S. 71 (1974) 1992.
64. J.P. Changeux, J. Thiéry, Y. Tung and C. Kittel, Proc. Nat. Acad. Sci. U.S. 57 (1967) 335.
65. J. Cartaud, E.L. Benedetti, J.B. Cohen, J.C. Meunier and J.P. Changeux, FEBS Letters 33 (1973) 109.
66. W.M.A. Hax, G.E.P.M. van Venrooij, J.J. Denier van der Gon and P.F. Elbers, J. Membrane Biol. 13 (1973) 61.
67. J. Sheridau, in A.A. Moscona (Ed.) The Cell Surface in Development. John Wiley, New York, London (1974) p. 187.
68. R.P. Cox, M.R. Krauss, M.E. Balis and J. Dancig, in R.P. Cox (Ed.) Cell Communication, John Wiley, New York, London (1974) p. 67.

CHAPTER VII.2

NUCLEOCYTOPLASMIC RELATIONS IN MUTANTS AFFECTING CELL DIVISION AND CELL SHAPE IN *TETRAHYMENA PYRIFORMIS**

J. FRANKEL, F.P. DOERDER**, L.M. JENKINS, E.M. NELSEN and L.E. DEBAULT

Departments of Zoology and Psychiatry, University of Iowa, Iowa City, Iowa 52242, U.S.A.

Recessive temperature sensitive nuclear gene mutants affecting various stages of cell division of *Tetrahymena pyriformis* have been isolated at four gene loci. One non-temperature sensitive mutant affecting cell division, and one that alters cell shape from elliptical to conical, have also been obtained. In all division mutants, micronuclear divisions continue normally and synchronously for at least 3 or 4 cycles while cell division is arrested. Division of the macronucleus is, however, inhibited or greatly delayed in mutants arrested at early stages of the division process, and proceeds normally only in a mutant that is arrested in the cleavage process itself and produces linear chains with minimal distortion. In the "conical" cell shape mutant, the cleavage furrow is displaced posteriorly while the macronucleus divides normally, initiating its division at the usual central position. The cell cycle is prolonged in the smaller posterior cell division product.

The ciliated protozoan *Tetrahymena pyriformis* is highly amenable to application of both physiological and genetic approaches to problems of nucleocytoplasmic relations in the cell cycle. What follows is an initial brief report of results of the application of a genetic approach. We mutagenize exponentially growing cultures with N-methyl-N′-nitro-N-nitrosoguanidine (Aldrich Chemical Co.) at a concentration of 10 μg/ml, and then cross mutagenized cells of opposite mating type with one another to bring mutants into the macronucleus. Time is then allowed for expression of recessive mutant genes through macronuclear assortment — an intramacronuclear process that appears to bring about random repression of one of the two alleles at each gene locus, and thus permits expression of recessive alleles (1). Selection of mutants is accomplished by an adaptation of the replica plating technique, with a search for morphological abnormalities in replicated clones tested at

* Research supported by Biological Sciences Development Award NSF-GU-2591 to the University of Iowa, NSF grant GM 32408 and NIH grant HD 08485-1 to J. Frankel, and NIH grant GM-18966-02 to L.E. DeBault.

** Present address: Department of Biology, University of Pittsburgh, Pittsburgh, Pa. 15260, U.S.A.

40°C. Clones showing such abnormalities are outcrossed to normal cells, and their heterozygous progenies then subjected to a conventional Mendelian analysis.

Using these techniques, recessive nuclear gene mutations affecting various stages of cell division have been obtained. The genetic and phenotypic properties of these mutants, as well as of an additional mutant that affects cell form, are summarized in table 1. Salient features include the following:

1) Inbreeding of the original phenotypic "assortants" resulted in pure-breeding homozygous stocks that differ from wild-type cells by recessive alleles at a single gene locus as defined by conventional Mendelian criteria. Certain complications, suggestive of unstable suppressors, appeared early in the course of inbreeding of two of the mutants; these will be considered in more detail elsewhere.

2) Results of complementation tests revealed two alleles at each of two gene loci (mo-1, mo-6) and four at a third locus (mo-3). The phenotypes of the two mo-1 alleles differ somewhat in severity of expression (see below) while the four mo-3 alleles display an essentially identical phenotype under restrictive conditions.

3) Most of the mutants are straightforward high-temperature sensitives.

TABLE 1

Summary of properties of mutants affecting cell division.

Mutant designation	Genetic basis	Temperature at which expressed	Nature of effect	Division of:	
				Micronucleus	Macronucleus
mo-1^a mo-1^b	single-gene recessive	high (40°C)	failure of fission line establishment	continues normally	arrested or severely retarded
mo-2	single-gene recessive	all	blockage of furrowing process		
mo-3^a mo-3^b mo-3^c mo-3^d	single-gene almost complete recessives	3^a: high (40°C) and low (15°C) 3^b and 3^d: high 3^c: low	blockage of pre-division shape changes and furrowing process	continues nearly normally	continues nearly normally
mo-6^a mo-6^b	single-gene recessive	high (40°C)	instability of division furrow	continues nearly normally	sometimes arrested
mo-8	single-gene recessive	high (40°C)	see text		
co	single-gene recessive	all	conical shape; posterior displacement of fission line	normal	normal, without displacement

However, certain mo-3 alleles are also (mo-3^{a}) or predominantly (mo-3^{c}) low temperature sensitive. The mo-3 data establish that the type and degree of thermosensitivity is a property of the allele, not the locus. The mo-2 heterozygous assortant was high-temperature sensitive when originally selected, but eventually it, like the homozygote derived from it, expressed the abnormal phenotype at room temperature.

4) At 40°C mo-1 fails to carry out the equatorial subdivision of the longitudinal ciliary meridians, a process which in normal cells helps establish the site of future division furrowing. This arrest is nearly total in mo-1^{a} homozygotes, and partial in mo-1^{b} homozygotes (and also in some of the original mo-1^{a} heterozygous assortants). Furrowing does not take place where the subdivision of the ciliary meridians has not occurred, and large irregular "monsters" eventually result at 40°C. The other division mutants (with the exception of mo-8) can subdivide their ciliary meridians (apparently) normally, and produce an equatorial fission line. Cell division, however, is not completed, apparently for different reasons in different mutants. In all mo-3 mutant alleles the shape changes that normally precede division do not occur under restrictive conditions, and cleavage is aborted with the division products connected as stiff linear "chains" that sometimes secondarily pull apart. In some cases 4 or even 8 cellular units may be linearly connected, each complete with macronucleus, micronucleus, and (generally) oral apparatus. In mo-6 the fission line appears to be "weak" at 40°C so that the prospective division products rotate around the common axis, and the posterior component then slides forward to produce an irregular homopolar doublet which then attempts to divide again, so that an irregular monster results. Finally, mo-8 is somewhat enigmatic, in that both the pre-division shape changes and the furrowing process appear relatively normal, except that one or two ciliary rows remain unbroken and the oral apparatus of the posterior component may be abnormal and come to be wedged within the advancing division furrow. The correlated oral and division abnormalities of this mutant suggest a disturbance of a morphogenetic field that underlies both oral development and the subdivision of one cell unit into two.

5) Examination of mo-1^{a}; mo-3^{a} double homozygotes after shifting to 40°C has confirmed that the subdivision step mediated by the normal allele at the mo-1 locus precedes the furrowing process controlled by the normal allele at the mo-3 locus. However, these double homozygotes develop an elongated, "snaky" shape upon prolonged maintenance at 40°C that is not found in simple mo-1 homozygotes, suggesting that the mo-3 gene must act on cell shape at times other than the point of final division arrest in cells bearing mutations at this locus. mo-3^{a}; mo-8 homozygotes display the characteristic terminal configuration of *both* mutants after shifting to 40°C, suggesting that the normal alleles at these two loci mediate two parallel pathways that are both necessary for normal division. mo-6 may well affect a fourth pathway, though this has not yet been established.

6) In normal *T. pyriformis*, the diploid micronucleus completes its mitotic division prior to the subdivision of the ciliary meridians. In all investigated cell division mutants micronuclear division continues normally at 40°C for at least 3 or 4 cycles, with a timing that is unaffected or minimally affected by the concurrent blockage of the cell fission process. Single large cells with 16 or more micronuclei (fig. 1) may result. Micronuclei divide synchronously in such cells.

7) The compound macronucleus normally begins its elongation as the ciliary meridians are becoming subdivided, and divides non-mitotically during the early part of the cell furrowing process. Macronuclear division is initially arrested at 40°C in mo-1 (fig. 1) and mo-6, with occasional and often unequal later division during prolonged maintenance of cells at 40°C. The macronucleus divides normally in the chain-forming mo-3 mutants suggesting an association of macronuclear division with maintenance of the normal linear arrangement of cell fission products.

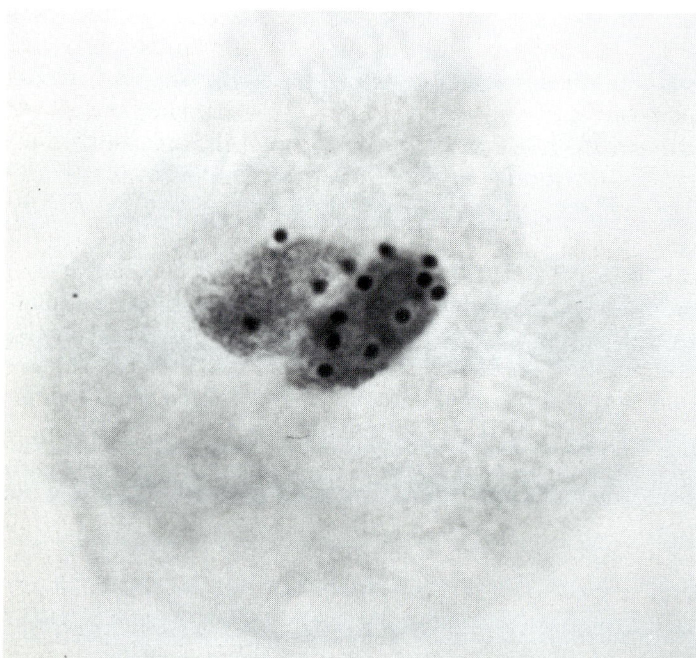

Fig. 1. A Feulgen-fast green preparation of a mo-1^{a} cell fixed after 10 1/2 hr at 40°C. The small densely staining nuclei are micronuclei, the larger and lighter one is the macronucleus. The generation time of normal *Tetrahymena* cells under these conditions is about 3 hr; micronuclear division in this mo-1 cell has thus continued with normal periodicity while cell and macronuclear divisions were blocked. 925 ×.

8) The "conical" cell-shape mutant cleaves normally, but at a site posterior to the geometric cell equator, resulting in two unequal cell fission products. The site at which the macronucleus begins its division *is* equatorial, anterior to the posteriorly shifted cell division furrow. Microfluorimetric measurements of Feulgen stained cells, following the procedures of Bohm and Sprenger (2), have revealed that the macronuclear DNA of co cells is divided no more unequally than that of wild-type cells (cf. 3), and that the macronuclear division product that has the greater amount of DNA is equally likely to enter either the larger anterior or the smaller posterior cell division product. The small posterior division product almost invariably has a longer cell generation time than the larger anterior division product.

Several conclusions can be drawn from the information summarized above, of which the following have special pertinence to nucleo-cytoplasmic relations: a) division of the macronucleus and micronucleus are subjected to controls that can be readily dissociated from each other (the same dissociation had been established earlier (4) for DNA synthesis in these organelles); b) the running of a cyclic process that cues micronuclear division continues even in mutants that fail to complete the cortical subdivision that normally precedes cell division; c) the *timing* of macronuclear division is intimately related to the onset of a geometrically fairly normal cell division, while the *location* of the macronuclear division furrow is dissociable from that of the cell division furrow. Overall, the results suggest parallel pathways of preparation for cell division and considerable, though not complete, dissociability of these pathways.

References

1. Nanney, D.L., in M. Locke (Ed.) The Role of Chromosomes in Development, Academic Press, New York (1964) p. 253.
2. H. Bohm and E. Sprenger, Histochemie 16 (1968) 100.
3. G. Cleffmann, Exp. Cell Res. 50 (1968) 193.
4. M. Jerka-Dziadosz and J. Frankel, J. Exp. Zool. 173 (1970) 1.

CHAPTER VII.3

MOVEMENTS AND POSITIONING OF ORGANELLES IN *PARAMECIUM AURELIA**

J. BEISSON and M. ROSSIGNOL

Centre de Génétique Moléculaire du C.N.R.S. 91190 Gif-sur-Yvette, France

The mechanisms responsible for the movements and precise positioning of organelles within a cell can be studied in *Paramecium aurelia* for two sets of organelles: the nuclei and the trichocysts. The normal movements of these organelles during growth and division can be altered by mutation or by drugs. Two monogenic mutations (genetically independent) have been studied which determine both abnormal macronuclear division (due to a failure of the macronucleus to migrate to its normal site) and failure of trichocysts to attach to their specific membranar site. In wild type cells, vinblastine and cytochalasin B induce effects similar to those of the mutations. Such mutants may help in understanding the role of microfilaments and microtubules in the movements of organelles and in the processes of "recognition" between organelles and membranar sites.

During cell growth and division, intracellular organelles undergo series of movements, such as changes in shape and location or precise positioning at specific sites. *Paramecia* are quite suitable to study the mechanisms of such movements, because of their large size, the variety of their organelles and the complexity of their structural organization. What follows is a preliminary report on genetic and physiological alterations of the movements of two sets of organelles, the nuclei and the trichocysts, in *Paramecium aurelia* (species 4, stock d4-2).

Movements of nuclei and trichocysts in normal cells

The nuclei

In a non-dividing cell, the highly polyploid macronucleus is anchored in a central position, behind the gullet, and the two diploid micronuclei lie very close to the macronucleus. At fission, the macronucleus leaves its central location and stretches beneath the dorsal surface, in close contact with the cortex; as fission proceeds, it keeps it subcortical position while elongating and finally constricts at the level of the fission furrow; after completion of

* This work was supported by grants from the Action Thématique Programmée of the C.N.R.S. 'Differenciation Cellulaire' (contract No. 4317).

division, each daughter macronucleus rounds up and moves back to a central location. The two micronuclei move away from the macronucleus, undergo mitosis and develop two very long spindles running across the fission furrow. The equal repartition of macronuclei and micronuclei between the two daughter cells is thus achieved mainly by a proper location and elongation of macronucleus and micronuclei spindles in relation to the plane of the fission furrow. As is the case in *Stentor* (1) the macronuclear elongation and constriction seem to be guided by some interaction between cortical membranes and macronucleus.

The trichocysts

These special organelles develop and reach their mature size (as a membrane bound structure) in the cytoplasm, then move to the plasma membrane where they become attached by their tip at very precise sites: trichocysts are inserted all over the cell surface in regular alternation with cilia; furthermore, each site of trichocyst insertion is marked by a differentiation in the molecular organization of the plasma membrane, as can be seen by freeze fracturing this membrane (2,3). Attached trichocysts can be "excreted" by the cell by a process of exocytosis involving fusion between plasma membrane and trichocysts membrane, similar to the process described in *Tetrahymena* for mucocysts (4).

Mutations affecting nuclear and trichocyst movements

Various monogenic recessive nuclear mutations are known which alter simultaneously attachment of trichocysts to the membrane or trichocyst extrusion and normal movements of nuclei at fission.

Two such mutations have been studied: tam_6 and tam_8, both induced by nitrosoguanidine. The two mutations are genetically independant. The most striking features of the two rather similar mutant phenotypes are the following. 1) A normal complement of normally shaped mature trichocysts are randomly distributed throughout the cytoplasm. In tam_8 cells no trichocyst is ever attached to the membrane; in tam_6, a variable number of trichocysts can attach and be normally excreted: very few can attach at 18° C, more at 27° C, but in any case much fewer than in wild type cells; 2) Macronuclear division is generally abnormal. Stained (5) dividers show a variety of abnormalities: defective location (completely anterior or posterior, central instead of subcortical) defective elongation, defective place of constriction. Fission yields in general an unequal distribution of macronuclear volumes between the two daughter cells and a certain percentage of amacronucleate cells, which in the case of tam_6 is temperature dependent (more abnormalities at lower temperature). In tam_8 cells, this defective macronuclear division is clearly correlated with the fact that the macronucleus remains in the center of the cell and never reaches its normal dorsal subcortical location.

Besides these two main abnormalities, in tam_6 or tam_8 populations, a certain percentage of cells with 1 or 3 (less frequently with 0 or 4) micronuclei instead of two are found. This fact suggests that either the localization of the micronuclei during fission or the elongation of their spindles or both are also affected. Finally, in non-dividing mutant cells, the shape and anchorage of the macronucleus behind the gullet appear more loosely controlled than in wild type cells.

Experimental alterations of nuclei and trichocysts movements

As previously described, the nuclei undergo a regular sequence of movements during the cell cycle involving precise changes in location, interactions with the cell cortex and changes in form: elongation, constriction, contraction. Trichocyst movements involve a specific recognition of their membranar site of attachment and membrane fusion and rearrangement when they are excreted.

The fact that single nuclear mutations can affect the movements of nuclei *and* trichocysts suggest the existence of a common molecular basis in the mechanisms responsible for the movements of both sets of organelles. Microtubules and microfilaments could reasonably be considered as the common basis and the effect of vinblastine (known to inhibit the development of microtubular systems) and of cytochalasin B (known to inhibit the function of microfilaments) was investigated. It was found that both drugs, applied to wild type cells, could mimic the effects of the two studied mutations.

Effects of vinblastine

Dividing cells picked up from an exponentially growing population, 5—30 min after exposure to 25 γ/ml of vinblastine, show a variety of macronuclear abnormalities: defective positioning, elongation or constriction, quite similar to those observed in untreated dividing tam_6 or tam_8 cells. The sister cells derived from such disturbed divisions yield some pairs of clones with 0 : 4 or 1 : 3 instead of 2 : 2 micronuclei.

Effects of cytochalasin B

Dividing cells picked up from an exponentially growing population show, within 1—2 hr of exposure to cytochalasin B (200 γ/ml) less conspicuous but definite alterations in macronucleus localization and elongation. In all cells, trichocysts extrusion, provoked by the routine picric acid assay, is partly inhibited. Furthermore, if cytochalasin does not displace attached trichocysts, it prevents new attachment of mature trichocysts: cells which have been induced to expel most of their trichocysts (by osmotic shock or exposure to alcohol for instance) and then exposed to cytochalasin, are very similar to tam_6 or tam_8 cells, i.e. full of unattached mature trichocysts randomly located in the cytoplasm.

Conclusions

The similarities between the effects of the studied mutations and the treatments by vinblastin or cytocholasin B suggest that both microtubules and microfilaments might be involved in the movements of nuclei and trichocysts. However, both drugs have been reported to affect membranes properties (6,7,8). It seems also unlikely that the mutation tam_6 or tam_8 affect the structural proteins of either microtubules or microfilaments, which are involved in other structures and functions (cilia, excretory processes etc.) that remain normal in tam_6 or tam_8 cells. A more likely hypothesis is that these mutations affect some membrane proteins interacting with a complex network of microtubules and microfilaments. The functional relationships between these membrane proteins and microtubules and microfilaments might be involved in guiding or triggering organelle movements. Further study of these and other mutants with similar effects might help in dissecting this network of interactions.

References

1. N. De Terra, Dev. Biol. 32 (1973) 129.
2. R. Janisch, J. Protozool. 19 (1972) 470.
3. H. Plattner, F. Miller and L. Bachmann, J. Cell Sci. 13 (1973) 687.
4. B. Satir, C. Schooley and P. Satir, J. Cell Biol. 56 (1973) 153.
5. R.V. Dippell, Stain Technol. 30 (1955) 69.
6. L.P. Everhart and R.W. Rubin, J. Cell Biol. 60 (1974) 442.
7. R.D. Estensen and P.G.W. Plagemann, Proc. Nat. Acad. Sci. U.S. 69 (1972) 1430.
8. A.W. Waddell, R.T. Robson and J.G. Edwards, Nature 248 (1974) 239.

CHAPTER VII.4

EFFECTS OF CYCLIC AMP ON YEAST PLASMA MEMBRANE FUNCTIONS

F. FOURY

Laboratoire d'Enzymologie, Université de Louvain, Place Croix du Sud, 1, Science 14 B, 1348 Louvain-la-Neuve, Belgium

In the glycerol-grown yeast *Schizosaccharomyces pombe* subjected to carbon deprivation, the addition of 1 mM cAMP stimulates within 30 sec the active uptake of L-amino acids and nucleosides by a factor of 2. Subsequent stimulations of RNA and protein syntheses occur. Adenosine 3′,5′-monophosphate and $N^6O^{2'}$-dibutyryl cAMP increase the maximum velocity of uptake of L-leucine, L-aspartic acid, L-valine, L-tryptophan, L-glycine, α-amino isobutyric acid, uridine and adenosine, but do not modify their K_M. The cyclic nucleotide is without effect on guanosine and 2-D-deoxyglucose uptakes. Although the stimulatory effects by cAMP are observed only under glucose limiting conditions, metabolic energy is strictly required. The stimulation can occur in the absence of respiration, but is inhibited by an ATPase inhibitor, Dio-9. Carbon deprivation decreases the endogenous cAMP cellular content by about 30%.

In normal fibroblasts, cycloheximide reverses the pleiotypic control exerted by serum deprivation on a number of parameters including cellular transport and RNA synthesis (1). The antibiotic also prevents the production of cAMP which is a negative mediator of transport in normal fibroblasts (2). We have shown previously (3) that in the yeast *Schizosaccharomyces pombe* COB5 submitted to glucose starvation, cAMP as well as cycloheximide stimulate RNA synthesis. We show here however that cAMP seems to act primarily on the uptake of metabolites across the plasma membrane of yeast.

In glycerol-grown cells of *S. pombe* COB5 (3) transferred into 20 mM Tris adjusted to pH 4.5 with citric acid (starvation medium), cAMP stimulates RNA synthesis just as cycloheximide does in the same conditions (4). The cyclic nucleotide also stimulates protein synthesis, but to a lower extent. The stimulations of incorporation of uridine into RNA by cycloheximide and by cAMP were found to be additive, suggesting that the two stimulations are at least partly independent processes. In fact, cAMP stimulates also the uptake of uridine across the plasma membrane, and this stimulation precedes the stimulation of uridine incorporation into RNA by about 3 min. Furthermore, cAMP stimulates the uptake of chemically unrelated compounds such as metabolized and non-metabolized amino acids, nucleosides, sucrose and succinate. These results strongly suggest that the stimulation of uptake by cAMP is not the consequence of the stimulation of the incorporation of uridine into RNA, and that cAMP effects primarily the metabolite transport across the plasma membrane.

Cyclic AMP increases the maximum velocity of the uptake of all the studied metabolites by a factor of two and does not change their apparent K_M (5). The uptake of guanosine which does not seem to be carrier-mediated is not stimulated by the cyclic nucleotide.

The stimulation of uptake by cAMP can be observed only in starvation conditions eliciting low uptake of the metabolites. A close relationship seems to exist between the cellular ATP content and the active uptake in yeast. cAMP however does not stimulate the uptake by increasing the cellular ATP content. Furthermore the stimulation requires high cellular ATP levels and concerns only active transport against concentration gradient. The addition of antimycin A, a respiratory inhibitor, still allows some transport but inhibits completely the stimulation of uptake by cAMP in glucose deprived cells. These results do not favor the hypothesis that cAMP stimulates the degradation of glycogen and trehalose and increases glycolysis in *S. pombe* COB5. The uptake of 2-D-deoxyglucose which is carrier-mediated but seems to be a passive transport down its concentration gradient is not enhanced by the cyclic nucleotide.

The uptake of metabolites in yeast might be driven by a potential gradient generated across the plasma membrane by the active extrusion of protons into the external medium produced by the hydrolysis of ATP by a plasma membrane ATPase (6,7). Indeed, cAMP stimulates slightly but significantly the extrusion of protons. The role of the cyclic nucleotide however is not to generate this electrochemical potential since the addition of a proton-conducting agent, such as carbonylcyanide m-chlorophenylhydrazone inhibits the uptake but does not specifically decrease the stimulation of uptake by cAMP.

Dio-9, a potent inhibitor of the ATPases of chloroplasts, bacteria and yeast mitochondria, inhibits both proton extrusion and uptake in antimycin A-treated cells. In the same conditions, other mitochondrial inhibitors such as dicyclohexylcarbodiimide, oligomycin or venturicidin A are inefficient, suggesting that Dio-9 does not act through the mitochondrial ATPase. Furthermore the concentrations required to inhibit the production of mitochondrial ATP in *S. pombe* COB5 are much higher than those required to inhibit transport. The stimulation of uptake by cAMP is strongly inhibited by Dio-9, suggesting that cAMP might control a Dio-9 sensitive function involved in active transport, possibly at the level of the plasma membrane ATPase.

References

1. R. Kram, P. Mamont and G.M. Tomkins, Proc. Nat. Acad. Sci. U.S. 70 (1973) 1432.
2. R. Kram and G.M. Tomkins, Proc. Nat. Acad. Sci. U.S. 70 (1973) 1659.
3. F. Foury and A. Goffeau, Biochem. Biophys. Res. Commun. 48 (1972) 153.
4. F. Foury and A. Goffeau, Nature New Biol. 245 (1973) 44.
5. F. Foury and A. Goffeau, J. Biol. Chem. 250 (1975) 2354.
6. P. Matile, H. Moor and K. Muhlethaler, Arch. Mikrobiol. 58 (1967) 201.
7. A. Pena, G. Cinco, A. Gomez-Puyou and M. Tuena, Arch. Biochem. Biophys. 153 (1971) 413.

CHAPTER VII.5

NUCLEAR INFLUENCE ON THE MEMBRANE FIXATION OF INSULIN. STUDIES ON *ACETABULARIA MEDITERRANEA*

F. LEGROS, B. HANSON, I. DUMON, J. JEANMART and V. CONARD

Laboratoire de Physiologie et de Physiopathologie, 2, rue Evers, 1000 Brussels, Belgium

Insulin binding to the membrane fraction from *Acetabularia mediterranea* increases during incubation in darkness. This process is nucleus dependent since it is observed on nucleate but not on anucleate fragments of algae.

Incorporation of insulin into the cytosol is not influenced by passage from the circadian rhythm to continuous dark culture conditions.

Introduction

Binding to specific receptors located at the level of plasma cell membranes is the first step for insulin to exert its metabolic effects. This fact is very well documented for animal cells (1—4).

The number of insulin binding sites may be influenced by genotypic and/or metabolic factors. Insulin fixation to adipose cells of genotypically obese mice is inferior to hormonal sensitivity of normal mice adipocytes (5—8). In the normal rats, the capacity of purified hepatic plasma membranes to bind insulin decreases by 60% when animals age from 2 to 24 months (9).

On an unicellular alga, *Acetabularia mediterranea*, Legros et al. have observed an oxygen uptake enhancement (10—12), induced by a specific binding of insulin (13,14). Algae were kept in darkness during several days in order to prevent the interference of photosynthesis. As proved by crude membrane and cytosol extracts isolation and by immunocytochemical methods (14), the binding of insulin preferentially occurs at the membrane level, but significant amounts of hormone molecules also penetrate into the cytosol.

It is known that *Acetabularia* may be cut into nucleate and anucleate fragments which are able to survive for several weeks, thanks to the presence of nucleic acid material in the cytosol (15,16).

The aim of the present work is to study the influence of the cell nucleus on the specific binding of insulin to *Acetabularia*, comparing the sensitivity of nucleate and anucleate fragments cultured in darkness during various time periods.

Materials and methods

Acetabularia cells without cap were grown in darkness for 1 to 28 days. In the course of dark culture, samples of 25 cells having approximately the same size were incubated in ^{125}I-insulin (^{125}I-iodinated insulin in HCl solution, pH 3; Sorin) dissolved in 120 ml of sea water. Final hormone concentration was 1 nM.

As incubation lasted 30 min, it may be assumed that maximal labelling was obtained (14). Three successive rinsings of 10 min were performed in solutions of 100 ml plain sea water in order to release radioactive material unspecifically adsorbed to the cells. Measurements of fixed radioactivity were determined either on intact cells or on homogenized and centrifuged fractions (70 g for 2 min). This method crudely isolates cytosol and membrane components (17).

Enucleation has been performed on cells measuring about 4 cm; the stalk was cut approximately at one third from the rhizoid. Nucleate and anucleate fragments were kept for 5 days in 12 h Light—12 h Dark conditions. After that time, they were cultured in constant darkness. In the course of dark incubation, samples of 25 nucleate and anucleate parts were labelled for 30 min in ^{125}I-insulin as described above.

Results are expressed as cpm/25 cells, membrane or cytosol fractions from 25 cells samples. Radioactivity may be referred to cpm/1 ml of the incubation solution (% of total radioactivity). In the course of increasing dark culture periods, membrane and cytosol labellings may be compared to radioactivity fixed to intact cells (% of whole cells labelling). Thus, according to the following relationships, the influence of darkness on insulin binding may be studied.

Results

Fig. 1 shows the progressive increase of insulin binding to the intact cells and the membrane fraction during incubation of *Acetabularia* in darkness. On the contrary, incorporation into the cytosol is not significantly modified. Same results are illustrated by table 1, demonstrating a net increase of insulin fixation to the membrane fraction after dark culture periods ranging from 7 to 10 days. Prolongation of this dark treatment during 3 to 4 weeks severely depresses insulin binding.

As indicated on fig. 2, the insulin binding to nucleate fragments follows the same evolution as fixation to the intact cells: increase of labelling of the nucleate parts and of their membrane fractions, no modification of the cytosol incorporation. On the same figure, it may be observed that insulin binding to anucleate fragments, their membrane or cytosol extracts is not modified during periods of 14 days dark culture.

Fig. 3 indicates that percent of radioactivity bound to cell membranes

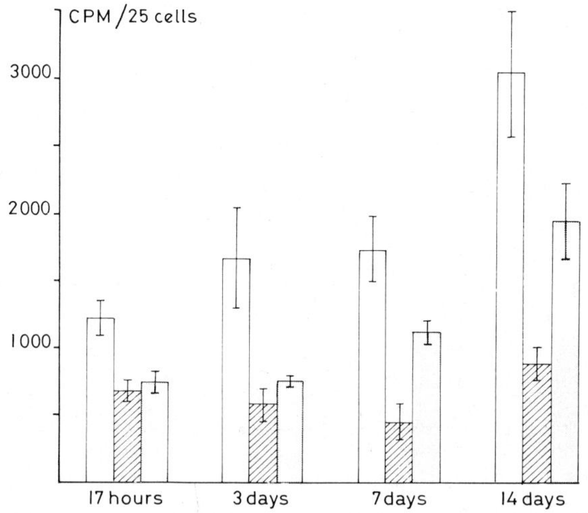

Fig. 1. Evolution of intact cells, cytosol and membrane crude extracts of 5 cm long *Acetabularia* cells during incubation in darkness. Time periods of dark culture are plotted against cpm/25 cells or cytosol and membrane extracts from 25 cells samples. n = 8 Columns ▭ represent radioactivity bound to intact cells. Columns ▨ represent insulin incorporation into cytosol extracts. Columns ▩ show the increase of membrane sensitivity to insulin.

becomes more and more important in the total amount of radioactivity fixed to *nucleate* fragments, while cytosol and membrane labelling stay in the same ratio when compared to intact *anucleate* fragments radioactivity.

TABLE 1

Percent of radioactivity fixed to membrane fraction as compared to whole cells labeling during increasing periods in darkness of *Acetabularia mediterranea*. n = 6.

Time in darkness	3.5 long cells	5.5 long cells
12 hr		41.2 ± 7.3
1 day	22.5 ± 2.5	48.8 ± 6.4
2 days	26.9 ± 2.6	49 ± 5.3
4 days		56.2 ± 3.8
6 days		57.6 ± 0.5
7 days	38.4 ± 5.5	65.2 ± 5.3
10 days	56 ± 3.8	
14 days	57.9 ± 2.7	64.9 ± 6
28 days		33.2 ± 4.7

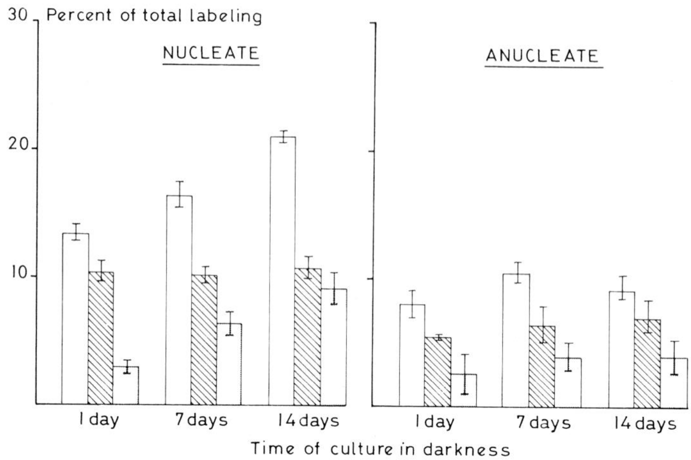

Fig. 2. Evolution of intact cells, cytosol and membrane crude extracts radioactivity of nucleate and anucleate fragments of *Acetabularia* during incubation in darkness. Labelling is expressed as percent of total radioactivity. n = 8. Symbols of the columns are the same as in fig. 1.

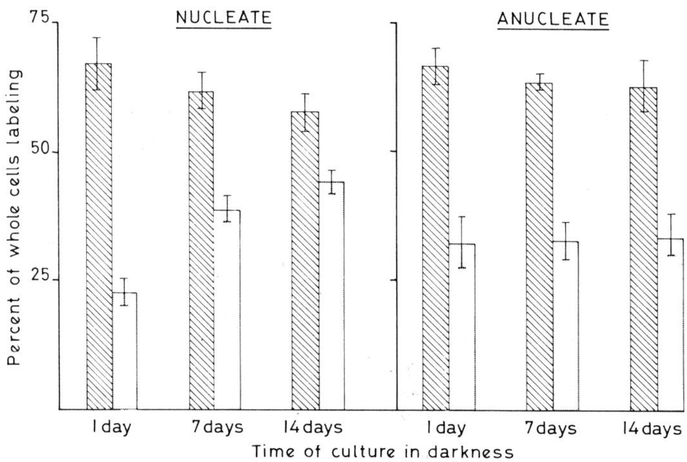

Fig. 3. Comparison between cytosol ▨ and membrane ▩ extracts labelling of nucleate and anucleate fragments of *Acetabularia* during incubation in darkness. cpm fixed to crude extracts are referred to labelling of corresponding intact cells (4.5 cm long). n = 10.

Discussion

Passage from the circadian rhythm of photosynthesis to continuous darkness culture conditions induces a progressive increase in the number of membrane receptors to insulin. This increase is nucleus dependent since it is observed on nucleate but not on anucleate fragments of algal cells.

In previous studies, we have shown that insulin stimulates oxygen uptake of nucleate as well as anucleate parts of *Acetabularia* (18). This discrepancy may be explained by the fact that insulin binding to a limited number of receptors is sufficient to modify oxygen uptake of both types of cell fragments. This assumption is supported by demonstration of Kono and Barham (19) that metabolic and permeative effects of insulin (measured by glucose metabolism of adipose cells) are maximally stimulated when 2.5% of the total insulin receptors are occupied by hormone molecules.

In conclusion, change of culture condition which stops the photosynthetic pathway of an autotrophic cell induces an increase in the number of membrane receptor sites to insulin. This process involves a nuclear participation.

Acknowledgments

We wish to thank Professor Jean Brachet who allowed us to use the *Acetabularia* culture of his laboratory.

^{125}I-Insulin has been provided by Professor J.R.M. Franckson and Dr. H.A. Ooms.

This work has been supported by grant from the Fonds de la Recherche Scientifique Médicale (Contrat No. 20.059).

References

1. I. Pastan, J. Roth and V. Macchia, Proc. Nat. Acad. Sci. U.S. 56 (1966) 1802.
2. P. Cuatrecasas, Proc. Nat. Acad. Sci. U.S. 63 (1969) 450.
3. R.J. Lefkowitz, J. Roth, W. Pricer and I. Pastan, Proc. Nat. Acad. Sci. U.S. 65 (1970) 745.
4. P. Freychet, J. Roth and D.M. Neville, Jr, Proc. Nat. Acad. Sci. U.S. 68 (1971) 1833.
5. P. Freychet, M.H. Laudat, P. Laudat, G. Rosselin, C.R. Kahn, P. Gorden and J. Roth, FEBS Letters, 25 (1972) 339.
6. C.R. Kahn, D.M. Neville, Jr, and J. Roth, J. Biol. Chem. 249 (1973) 244.
7. C.R. Kahn, D.M. Neville, Jr, P. Gorden, P. Freychet and J. Roth, Biochem. Biophys. Res. Commun. 48 (1972) 135.
8. P. Freychet and P. Laudat, Le Diabète 4 (1973) 185.
9. C. Freeman, K. Karoly and R.C. Adelman, Biochem. Biophys. Res. Commun. 54 (1973) 1573.
10. F. Legros and V. Conard, Hormone Res. 4 (1973) 107.
11. F. Legros, C. Rogister and V. Conard, in M. Margoulies and F.C. Greenwood (Eds), Structure-Activity Relationship of Protein and Polypeptide Hormones, Excerpta Medica I.C.S. 241 (1972) 542.
12. F. Legros, M. Saines and V. Conard, Arch. Intern. Physiol. Biochim. 81 (1973) 745.
13. F. Legros and V. Conard, Arch. Intern. Physiol. Biochim. 82 (1974) 353.

14. F. Legros, P. Uytdenhoef, I. Dumon, B. Hanson, J. Jeanmart, B. Massant and V. Conard, Protoplasma (submitted).
15. J. Hammerling, Int. Rev. Cytol. 2 (1953) 475.
16. J. Brachet, Biochemical Cytology, Academic Press, New York (1957).
17. A. Goffeau and J. Brachet, Biochim. Biophys. Acta 95 (1965) 302.
18. F. Legros, M. Saines, M. Renard and V. Conard, Plant Sci. Letters 2 (1974) 339.
19. T. Kono and F.W. Barham, J. Biol. Chem. 246 (1971) 6210.

CHAPTER VII.6

MIGRATION OF CYTOPROTEINS INTO THE NUCLEUS OF *ACETABULARIA*

P. CLÉRIN, T. VANDEN DRIESSCHE and J. BRACHET

Laboratoire de Cytologie et Embryologie moléculaires, Département de Biologie moléculaire, Université libre de Bruxelles, 67, rue des Chevaux, 1640 Rhode St. Genese, Belgium

It is shown that cytoplasmic proteins enter the nucleus of *Acetabularia* by an autoradiographic study of nuclei present in normal rhizoids grafted on labelled stalks. Preliminary evidence is presented that this entry of proteins into the nucleus differs according to the amino acid (lysine, leucine or tryptophane) used for labelling the stalks.

Introduction

Transcriptional control is thought to be of crucial importance in cell differentiation, but the way in which this control is exerted is poorly understood. The fact that cytoplasmic proteins can enter the nucleus suggests the possibility that they might exert a regulatory control on gene expression and indirect evidence along this line has accumulated (1—11).

Acetabularia might be a favorable material for investigating this problem, since cytoplasmic signals must enter the nucleus in at least the two following instances. When a young cap is removed from an alga which is on the verge of initiating nuclear division, this process is immediately halted and delayed until a new cap has been formed. On the contrary, the nucleus is induced to divide (although it would normally have remained intact during several months) when a mature cap is grafted onto a young rhizoid (12).

It seemed of interest to investigate whether proteins formed in the cytoplasm enter the nucleus of *Acetabularia* and, if so, whether a special class of proteins is concerned since it is thought that the non-histone chromosomal proteins control in a more specific way RNA synthesis than the histones (see McClure and Hnlica (13) for a review). The histones are known to be synthesized in the cytoplasm. They are basic and rich in lysine. The site of synthesis of the acidic proteins is not known with certainty; these proteins, in contrast with the histones, contain tryptophan. Other amino acids, leucine for instance, are present in all kinds of proteins.

In the present paper, we show that cytoplasmic proteins enter the nucleus of *Acetabularia* by an autoradiography study of nuclei present in normal rhizoids grafted on labelled stalks. Preliminary evidence that this entry of proteins into the nucleus differs according to the amino acid (lysine, leucine or tryptophan) used for labelling the stalks, is also presented.

Material and methods

The algae (*Acetabularia mediterranea*) were cultivated in the standard medium (14) and maintained in alternating light and dark periods of 12 h : 12 h; the light intensity of 1000 to 1500 lux was given by "daylight" fluorescent tubes (Phytor, ACEC).

The algae used in the experiments were 20—30 mm long, but had not yet formed a cap. Since it is much easier to graft algae which have a slightly different stalk diameter, the grafts were performed with *Acetabularia* of two different batches.

All the radioactive amino acids were tritiated (Amersham). The incubations were performed during 1 h in sea-water + the following radioactive amino acids (50 μCi/ml), ^{3}H-leucine (specific radioactivity 1 Ci/mM or 38 Ci/mM, concentrations 50 $\times$ 10^{-8} or 1, 31 $\times$ 10^{-6} mM/ml). ^{3}H-lysine (spec. act. 5.58 Ci/mM, concentration of 9.09 $\times 10^{-6}$ mM/ml); ^{3}H-tryptophane (spec. act. of 5.58 Ci/ml, a concentration of 8, 9 $\times$ 10^{-6} mM/ml). It has been ascertained, in preliminary experiments that incorporation in the afternoon and in the morning gave similar results and that a 2 hr incubation (instead of 1 hr) did not improve variability of the grain counts.

The grafts and chases were performed in sea-water containing the corresponding amino acids at a concentration 2000 times higher than that of the radioactive precursors; in addition, puromycin (20 μg/ml), which prevents protein synthesis both in the cytoplasm (15) and in the nucleus (16 and 17) was added in several experiments.

The treated algae were fixed by freeze-substitution, embedded in paraffin, sectioned at 10 μm with a refrigerated microtome and submitted to autoradiography according to Ficq (18). The grain counts were made on a surface of 60.5 μm^2 (i.e. 605 μm^3) on 10 sections of a given sample of algae (exceptionally on 5 sections, when indicated). The standard deviation was always calculated.

Results

In two experiments, stalks labelled with *lysine* were grafted on "cold" rhizoids and maintained in puromycin. In the first experiment, the chases lasted 3 hr, 2, 4 and 6 days; in the second experiment, they lasted 1, 6, 8 and 15 days.

A fast labelling of both "cold" cytoplasm and nucleus was observed. The number of grain counts over the cytoplasm increases until the 2nd day; it then levels off and decreases after a long chase of 15 days. In the nucleus also, the radioactivity increases rapidly. A well delineated hightly significant peak appears at the second day in both the nucleoplasm and the nucleolus. A rather high nuclear radioactivity is maintained until the 6th day and then decreases (fig. 1). In the absence of puromycin, the variability is much greater than in its presence and the evolution of labelling in the nucleus is less apparent: it more or less follows that of cytoplasm. The radioactivity of the nucleus is lower in the absence of puromycin; this might be due to a larger utilization of the precursor molecules for the building up of cytoplasmic proteins.

After 1 hr of incubation with radioactive *leucine* and chases of increasing lengths of time in the presence of puromycin, the distribution of radioactivity is strikingly different from that observed with lysine. Almost no radioactivity can be detected in either compartment after 2 hr, and labelling is

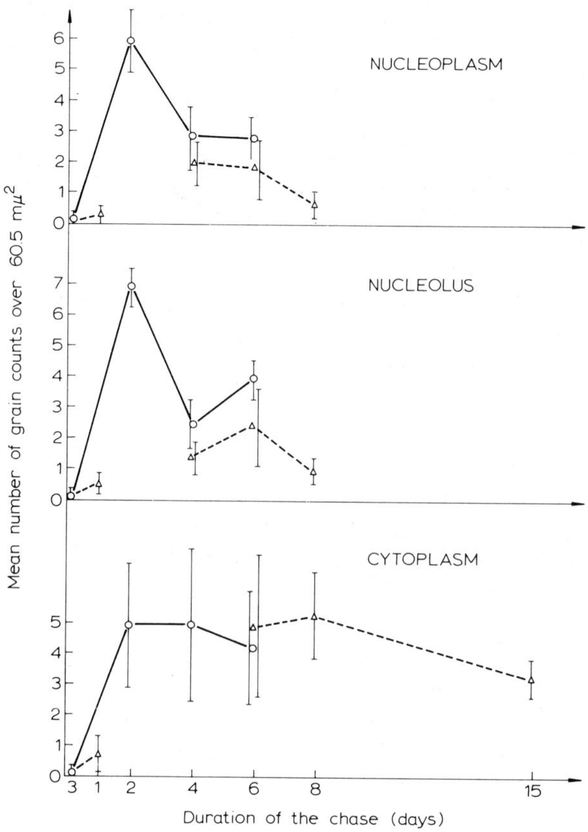

Fig. 1. Kinetics of labelling after 1 hr incubation with ^{3}H-lysine and chase in the presence of puromycin. Mean number of grain counts over 60.5 μm^2. Experiment I, solid line. Experiment II, dotted line.

still very low after 2 days (in sharp contrast with lysine labelling). But the number of grain counts increases gradually with time (fig. 2). In experiment II, the three curves showing respectively the labelling of the cytoplasm, the nucleoplasm and the nucleolus as a function of time are smooth; in experiment I, large fluctuations were observed. However, taking into account the standard deviation, a smooth curve could be drawn, which is very similar to the one obtained in experiment II. Without puromycin, in experiment II, the increase in labelling was observed until the 8th day, in all three compartments; it was followed by decrease.

When *tryptophan* is used as a precursor of proteins and when incubation is followed by a chase in the presence of puromycin, the number of grain counts increases until the 4—6 day in the cytoplasm, the nucleoplasm and the nucleolus. Only the cytoplasm has been analyzed after 15 days: its radioactivity did not change significantly from the 4th to the 15th day

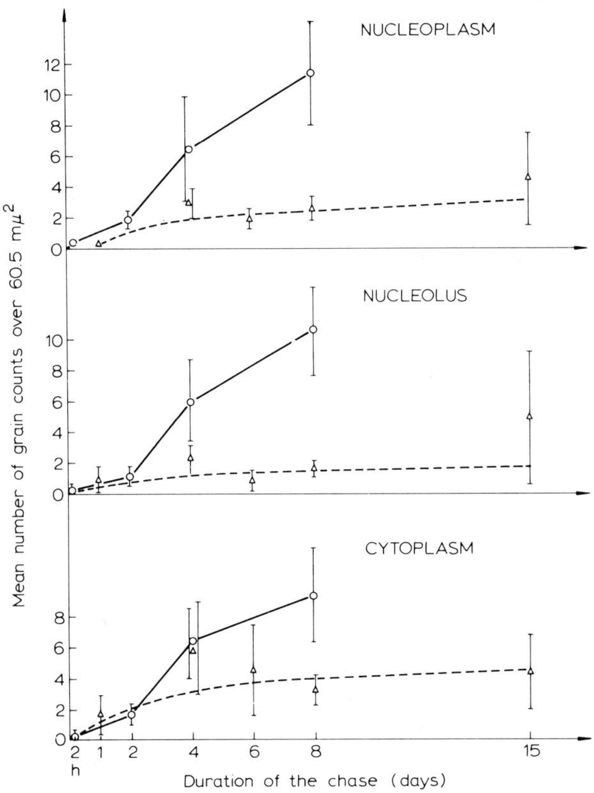

Fig. 2. Kinetics of labelling after 1 hr incubation with ^{3}H-leucine and chase in the presence of puromycin. Mean number of grain counts over 60.5 μm^2. Experiment I, solid line. Experiment II, dotted line.

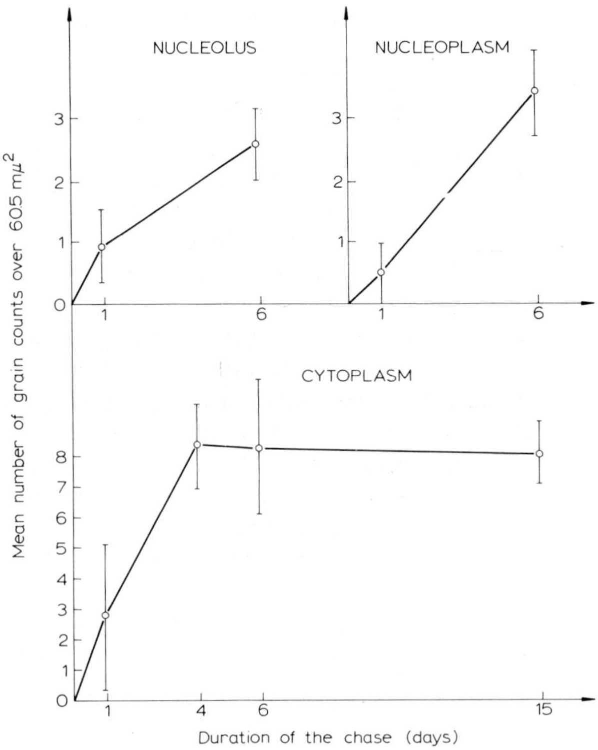

Fig. 3. Kinetics of labelling after 1 hr incubation with ^{3}H-tryptophan and chase in the presence of puromycin. Mean number of grain counts over 60.5 μm^2. Experiment II.

(fig. 3). Again, in the absence of puromycin, a much greater variability was observed. It is interesting that, after 6 days, there is more radioactivity in both the nucleoplasm and the nucleolus than in the algae which were treated with puromycin.

Discussion and conclusions

In all experiments made in order to establish whether proteins migrate from the cytoplasm into the nucleus, the results have been positive: cytoplasmic proteins penetrate into the nucleus.

In the absence of puromycin, the radioactivity of the cytoplasm is much greater than in its presence. This reflects the utilization of the labelled amino acids in building up new cytoplasmic proteins after the incubation period. In addition it seems that the nuclear labelling increases in parallel with that of the cytoplasm, when the precursor is lysine or leucine. This is not, however, the case for tryptophan: there is a greater difference in the labelling between cytoplasm and nucleus.

In the presence of puromycin where no protein synthesis can take place, the proteins which were synthesized during the one hour incubation with the radioactive amino acid penetrate into the nucleus of *Acetabularia*. Since significantly more radioactivity is found, after certain periods of time within the nucleus than in the cytoplasm, in the case of lysine, a concentration of proteins of cytoplasmic origin must take place inside the nucleus. The ulterior decrease is probably due to the turn-over of the radioactive proteins. Nuclear labelling after incubation with leucine is always delayed as compared to lysine. The nuclear accumulation of proteins labelled with tryptophan seems to be increased in the presence of puromycin.

All the data have been statistically analyzed in order to ascertain the confidence limits of the results. However, owing to the great variability of some of the available data and the usual variations observed between batches of *Acetabularia*, the kinetics of the uptake of the proteins labelled with the various amino acids must be regarded as preliminary. However, the highly significant differences observed both in the rate of penetration and the evolution of the radioactivity of the proteins labelled with the various amino acids indicate that penetration into the nucleus does not occur at random and depends on the nature of the protein.

Similar conclusions have been drawn by Gurdon (5), who studied the penetration into the nucleus of labelled proteins injected in the cytoplasm of *Xenopus laevis* oocytes. It is likely that, in our experiments as well as in those of Gurdon, both the size and the chemical nature of the cytoplasmic proteins play an important role for their uptake into the nucleus.

References

1. K. Arms, J. Embryol. Exp. Morphol. 20 (1968) 367.
2. S.A. Carlsson, G.P.M. Moore and N.R. Ringertz, Exp. Cell. Res. 76 (1973) 234.
3. A. Ficq, Exp. Cell Res. 76 (1972) 242.
4. J.B. Gurdon, Proc. Nat. Acad. Sci. U.S. 58 (1968) 545.
5. J.B. Gurdon, Proc. Roy. Soc. 13 (1970) 303.
6. L. Goldstein and D. Prescott, J. Cell Biol. 33 (1967) 637.
7. L. Goldstein, M.V. Rao and D.M. Prescott, Ann. Embryol. Morphol. Suppl. 1 (1969) 189.
8. L. Goldstein, G.E. Wise and M. Besson, Exp. Cell Res. 76 (1973) 281.
9. H. Harris, J. Cell. Sci. 2 (1967) 23.
10. R.W. Merriam, J. Cell Sci. 5 (1969) 333.
11. P.L. Paine and C.M. Feldherr, Exp. Cell Res. 74 (1972) 81.
12. J. Hämmerling, Int. Rev. Cytol. 2 (1953) 475.
13. M.E. Mc. Clure and L.S. Hnlica, Suc. Cell Biochem. 1 (1972) 311.
14. L. Lateur, Rev. Alg. 1 (1963) 26.
15. P. Lengyel and D. Söll, Bacterial Rev. 33 (1969) 264.
16. Fleischer, H. Lambropoulos and I. Reinsch, Hoppe Seyler's Z. Physiol. Chem. 352 (1971) 593.
17. M. Laval and M. Bouteille, Exp. Cell Res. 79 (1973) 391.

Dr. SCHATZ:

Does lysine incorporated into the nucleus reside mainly in protein?

Dr. VANDEN DRIESSCHE:

It is assumed that lysine incorporated into macromolecules (not washed out during the successive baths in the preparation of sections) is mainly in the form of proteins.

Dr. MARGULIES:

How do you know that protein labelled in cytoplasm is the same as that found later in the nucleus?

Dr. VANDEN DRIESSCHE:

The aim of the work was to determine whether or not proteins migrate from the cytoplasm into the nucleus.

Dr. MARGULIES:

Why is there a continued increase in label in cytoplasm even in the presence of puromycin?

Dr. VANDEN DRIESSCHE:

There is a continuous increase in label in the presence of puromycin in the cytoplasm of the initially cold rhizoid grafted into the stalk of a labelled alga.

SECTION VIII

AN OUTSIDER'S VIEW OF CELL PHYSIOLOGY AND DIFFERENTIATION

CHAPTER VIII.1

SOME REMARKS ON THERMODYNAMICS AND KINETICS OF SELF-ORGANIZATION

A. BABLOYANTZ, J. HIERNAUX and I. PRIGOGINE*

Université Libre de Bruxelles, Faculté des Sciences, Brussels, Belgium

Introduction

Theoretical studies of open chemical systems subject to non-linear kinetics have shown that the homogeneous state of such systems may become unstable far from thermodynamic equilibrium and evolve to time or space dependent regime. Non-linear evolution equations may also allow far from thermodynamic equilibrium for multiple steady states. Such situations have been called by Prigogine *dissipative structures* to indicate that they are created and maintained by the dissipative entropy producing processes inside the system. A generalized thermodynamic has also been developed for these systems by Prigogine and Glansdorff (1,2).

It seems that these ordering phenomena may provide a physico-chemical basis for the understanding of certain biological systems. Indeed the latter are highly complex and ordered objects, operating far from thermodynamic equilibrium and they require metabolism, synthesis and regulation for their maintenance. These functions are essentially characterized by induction, repression, cross-feedback, and cooperative mechanisms, which obey to non-linear kinetics. Moreover the morphological patterns of living systems are characterized by highly heterogeneous distribution of matter which results from the process of embryological development. The simplest term to describe these characteristic features of biological systems seems to be *coherent behavior.*

In this paper, we shall first describe briefly the properties of dissipative structures for a simple model system and their relevancy to the actual biological problems. In another section, we shall discuss in more detail the possible role of dissipative structures for the *generation* of *polarity* in an initially homogeneous morphogenetic field as well as the resulting spatial differentiation which give rise to developmental patterns.

---

* Also the Center of Statistical Mechanics and Thermodynamics University of Texas, Austin, Texas, U.S.A.

Model dissipative system

We shall consider the following scheme

$$A \rightarrow X$$
$$B + X \rightarrow Y + D$$
$$2X + Y \rightarrow 3X$$
$$X \rightarrow E$$

Scheme 1

The system is open and it is subject to the flow of chemicals A, B, D and E which are converted within the reaction volume V into the intermediates X and Y.

Moreover, we assume that the system is isothermal, at mechanical equilibrium and subject to time independent boundary conditions. It is assumed that the diffusion of X and Y within V obeys Fick's law, and only the diagonal elements of the diffusion coefficient matrix are taken into account.

Let X and Y denote the average mole fractions or the average distributions of the two intermediate products. The time evolution of these variables are given by the well known conservation mass equations:

$$\frac{\partial X}{\partial t} = A - (B + 1)X + X^2 Y + D_X \frac{\partial^2 X}{\partial r^2}$$

$$\frac{\partial Y}{\partial t} = BX - X^2 Y + D_Y \frac{\partial^2 X}{\partial r^2}$$

D_X and D_Y are the constant diffusion coefficients of X and Y. All the kinetic constants are taken equal to 1. One has neglected the inverse reactions. Thus the system operates automatically far from thermodynamic equilibrium.

An analytical study of this system has been performed recently by Nicolis and Auchmuty (3,4). We shall report here the main properties of the model as obtained by analytical and computer calculations for fixed boundary conditions (1,5):

a) At or near equilibrium, there is a unique steady state solution which is stable towards arbitrary perturbations.

b) When structure formation is possible, the steady state corresponding to the extrapolation of the close-to-equilibrium behavior becomes unstable. Then the homogeneous system evolves to a new regime corresponding to

1) inhomogeneous distribution of chemical substances in space, as seen in fig. 1.

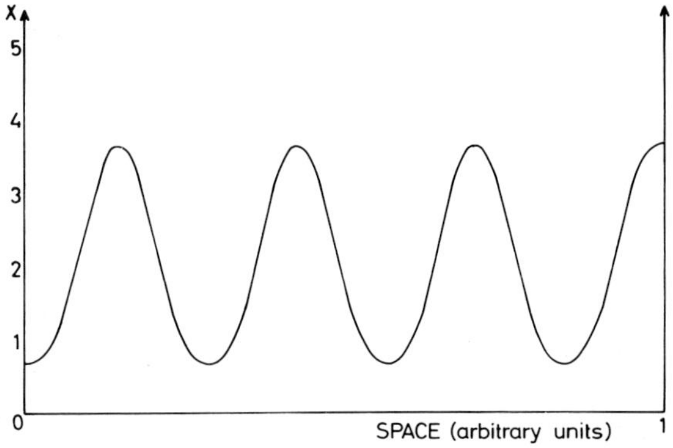

Fig. 1. Spatial dissipative structure. This presents the inhomogeneous distribution of the intermediate for a suitable choice of the values of the parameters.

2) under different conditions, the concentration of the chemicals may show sustained oscillations (6) as seen in fig. 2. Such oscillations have been found in glycolytic cycle for which a model has been developed by Goldbeter and Lefever (7).

3) one can also obtain a regime depending on time and space which is characterized by propagating concentration waves (8) as seen in fig. 3. This may be interesting for the description of information propagation as in the case of aggregating slime molds (9).

4) multiple steady states and hysteresis may appear for certain type of

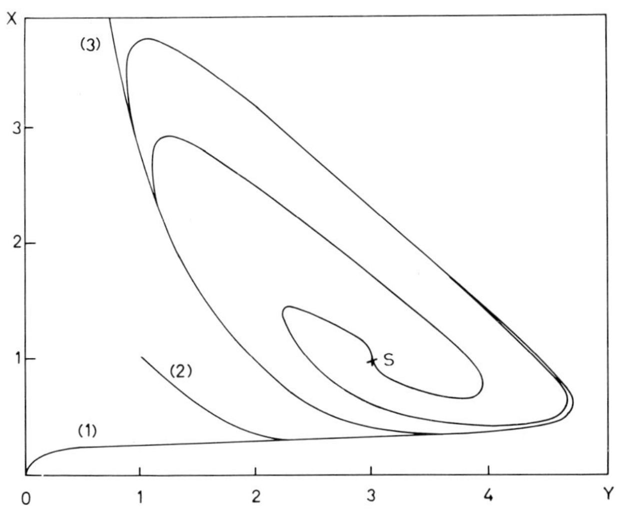

Fig. 2. Sustained oscillations in the homogeneous case for A = 1 and B = 3.

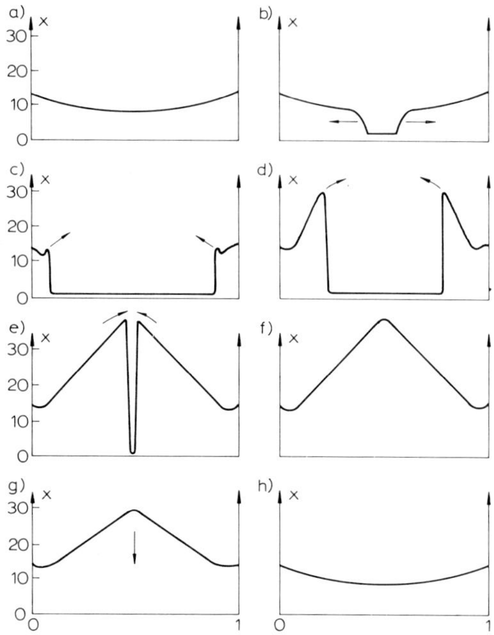

Fig. 3. Propagating concentration waves of the intermediate X. The value of the parameters are $D_X = 1.05 \times 10^{-3}$, $D_Y = 0.60 \times 10^{-3}$, $B = 77$; $D_A = 1.95 \times 10^{-1}$ for the diffusion of the chemical A.

kinetics. This kind of behavior can be used for the description of biological phenomena exhibiting "all-or-none" type models for genetic regulation (10), β-galactosidase induction in *E. coli* (11), excitable systems (12) and the synthesis of prebiotic polymers (13—15).

Spatial dissipative structures with fixed boundary conditions cannot generate spontaneously dissymetrical distribution of matter in space. Recent-

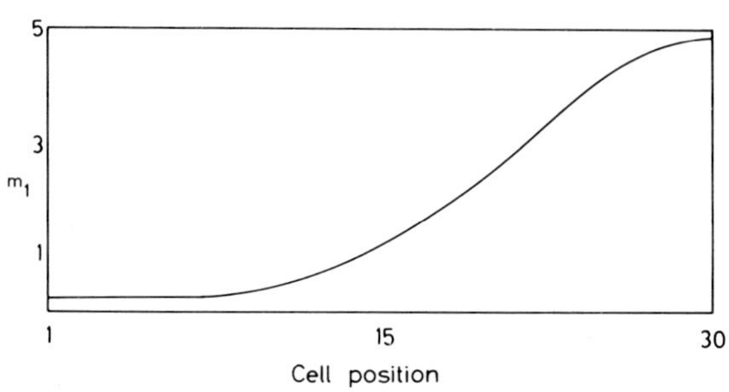

Fig. 4. Concentration profile of a morphogen m_1 in a morphogenetic field of 30 cells, which presents a polarity.

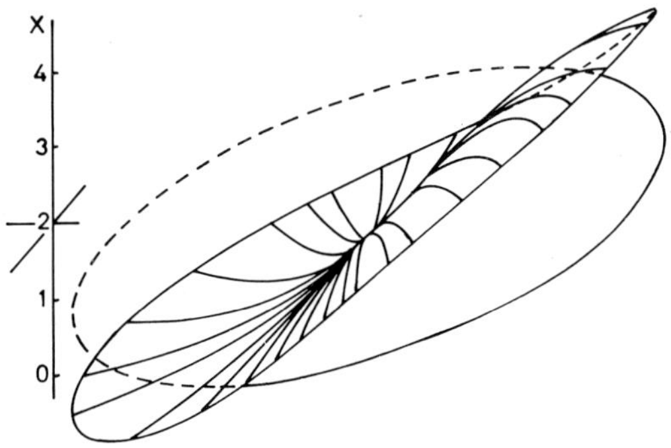

Fig. 5. Concentration profile of a morphogen X in a circular morphogenetic field, which presents a polarity.

ly, it has been shown that a dissipative structure may generate polarity in an initially homogeneous morphogenetic field provided the flux of diffusing substances are supposed to be zero across the boundaries of the system (16) as seen in fig. 4. This problem will be treated in detail in the next part. Similar structure can also be obtained for the scheme 1 (17). Fig. 5 shows a polar structure for the particular case of the circle.

Generation of polarity and positional information

One of the major problems of developmental biology is that of reproducible pattern formation of differentiated cells in a multicellular organism, in spite of the fact that cells forming the system possess the same genetic potentialities.

The existing theories of pattern formation postulate a universal two-step process with two distinct time scales (18,19). In a first stage positional information is assigned to the cells in a coordinated system by a gradient generating mechanism. In a second stage the action of positional information at the genetic level gives rise to molecular differentiation.

If one assumes the existence of a gradient in the morphogenetic field one can construct models which interrelates the establishment of the gradient of positional information and its subsequent translation by the cell's genome into positional differentiation and pattern formation. The establishment of the gradient can be assumed to result from active transport (20) or diffusion of morphogen (21) between source and sink.

However, a mechanism giving rise to the spontaneous onset of polarity within an initially uniform tissue seems to be necessary in any coherent theory of pattern formation. This question constitutes a special case of a more general one, related to the problem of spontaneous pattern formation in chemical systems as treated in the preceding part.

Our aim here is to present a consistent molecular model for pattern formation which, starting from a completely homogeneous tissue, may generate polarity under the action of a disturbance, thus giving rise to the establishment of positional information in the morphogenetic field. Then as a result of the interaction of the gradient at the genetic level, developmental patterns of differentiated cells appear.

We consider that each cell is subject to chemical reactions described by the scheme 2

$$
\begin{aligned}
a &\xrightarrow{k_1} m_{1i} \\
c + 2m_{1i} &\xrightarrow{k_2} 2m_{1i} + m_{2i} \\
m_{1i} &\xrightarrow{k_3} f \\
m_{2i} &\xrightarrow{k_4} g \\
m_{2i} + p &\underset{k_6}{\overset{k_5}{\rightleftharpoons}} m_{1i} \\
b + m_{1i} + p &\xrightarrow{k_7} p + 2m_{1i}
\end{aligned}
\qquad (2a)
$$

$$
\begin{aligned}
m_{2i} &\xrightarrow{k_8} I_i \\
R' &\underset{k_{10}}{\overset{k_9}{\rightleftharpoons}} R_i \\
R_i + O_i^+ &\underset{k_{12}}{\overset{k_{11}}{\rightleftharpoons}} O_i^- \\
R_i + 2I_i &\underset{k_{14}}{\overset{k_{13}}{\rightleftharpoons}} F_1 \\
\alpha + O_i^+ &\xrightarrow{k_{15}} O_i^+ + E_i + M_i \\
I_{i-1} + M_i &\underset{k_{17}}{\overset{k_{16}}{\rightleftharpoons}} M_i + I_i \\
I_{i+1} + M_i &\underset{k_{17}}{\overset{k_{16}}{\rightleftharpoons}} M_i + I_i \\
E_i &\xrightarrow{k_{18}} F_2 \\
M_i &\xrightarrow{k_{19}} F_3
\end{aligned}
\qquad (2b)
$$

for the ith cell

Scheme 2.

The part (2a) of the scheme involves the interaction of different morphogens responsible for gradient formation. Precursor a produces a morphogen m_1, which in turn catalyzes the formation of another morphogen m_2. The latter will combine with a third molecule p to give m_1. m_1 and m_2 diffuse throughout the system and may also decay. $k_1 ... k_7$ are the kinetic constants a, b, c, f and g are maintained space and time independent.

The part (2b) of the scheme describes the interaction of the morphogen m_2 with cell's genome. A detailed description of this part of the model and its justification may be found elsewhere (20,21,11). We shall only remind the essential features of the model. The concentration profile of I is determined by the diffusion gradient of m_2, which establishes positional information. If the concentration of I is less than a critical value I_c, the cells will remain undifferentiated. On the other hand, if I is greater than I_c, a luxury protein E is synthesized in great quantities and the cells are considered to be differentiated.

We shall consider successively a morphogenetic field composed of a single array of 30 cells and a two dimensional sheet of 15 × 5 cells. The value of the fluxes across the boundaries of the system are zero. The kinetic equations of the system describe the time evolution of m_{1i}, m_{2i}, E_i, M_i and I_i. The other variables are assumed to be at quasi steady state.

If one follows the time evolution of the various concentrations in the system, it appears that first E and I remain at their initial value, and, m_1 and m_2 only vary and reach a stable steady state. Later on, E and I vary and reach their steady state values. Thus, at the first stage, the part 2a of the scheme 2 behaves as an independent set of chemical reactions and it can be shown that it constitutes an example of a dissipative system. That is for a given range of values of the parameters and diffusion coefficients, the homogeneous steady state concentrations may become unstable, due to the action

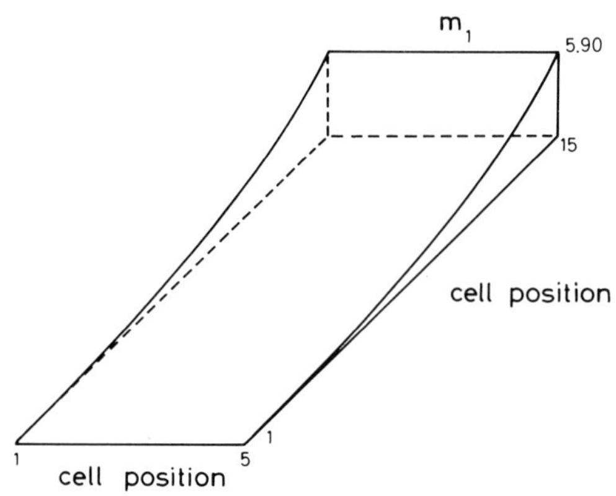

Fig. 6. Concentration profile of a morphogen m_1 in a rectangular field of 15 × 5 cells, which presents a polarity.

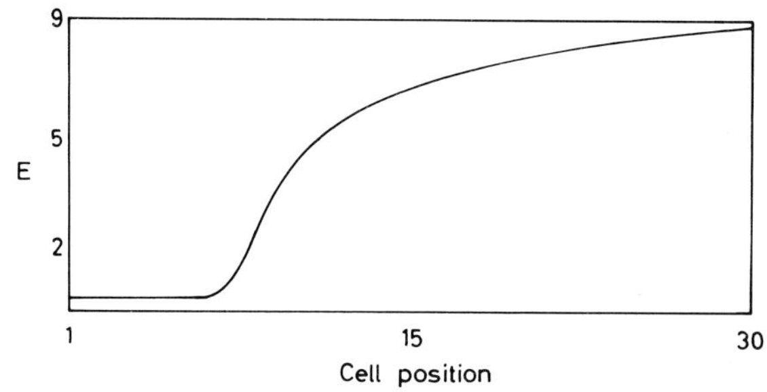

Fig. 7. This represents the spatial concentration profile of enzyme E, corresponding to the gradient of fig. 4.

of random disturbances. The system then evolves towards an inhomogeneous state in concentration of m_1 and m_2. As a result, all the cells of the field do not possess the same amount of m_1 and m_2. For given values of parameters, the profiles of the concentrations in the field are critically dependent on the geometry of the system. There is a length L (or $L_1 \times L_2$ for the two-dimensional case) such as a unique concentration profile appears in the system as seen in figs. 4 and 6 and one sees that the boundary cells contain high and low concentration of m_2, thus there appears a gradient of morphogen between source and sink. So, a dissipative structure can generate and maintain polarity in a morphogenetic field.

The part of kinetic equations describing part (2a) of the scheme is the same as the one used by Gierer and Meinhardt (22,23). However, they had assumed unnecessarily the presence of a shallow gradient in their system. Therefore their structures cannot be considered as giving rise to the spontaneous generation of polarity in the morphogenetic field.

The choice of scheme 2 should not be considered as a limitation for the generation of polarity in developmental fields. Various schemes will exhibit the same properties if they enter into the class of systems that can generate dissipative structures.

When the dimensions of the field are increased, more and more complex patterns may appear and a degeneracy in the number of profiles can be seen; the form of the patterns is then critically dependent on the magnitude and location of the perturbations.

Once the gradient of positional information is established, the cells differentiate in the morphogenetic field, according to the scheme (2b), and a developmental pattern of differentiated cells appear as seen in fig. 7.

Concluding remarks

Dissipative structures, as described in the two preceeding parts, can be considered as a new state of matter, induced by the flow of free energy under non-equilibrium conditions and provided that these systems undergo non-linear kinetics.

This new state of matter gives rise to a new type of physical chemistry on the macroscopic level although the usual laws of quantum mechanics remain obviously valid on the microscopic level.

These developments seem to be particularly suitable for description of a wide variety of biological problems related to self-organization, to macroscopic aspects such as polarity, form... These can be brought into direct relation with the functioning of the system in terms of chemical reactions and transport processes.

Acknowledgments

We express our thanks to the members of our groups in Brussels and Austin, who have always collaborated in this research.

This work has been supported by the Welch Foundation, Houston and the Fonds de la Recherche Fondamentale Collective of Belgium. J.H. acknowledges financial support from the Institut pour l'Encouragement de la Recherche Scientifique dans l'Industrie et l'Agriculture.

References

1. P. Glansdorff and I. Prigogine, Thermodynamic Theory of Structure, Stability and Fluctuations, Wiley-Interscience, New York (1971).
2. I. Prigogine and G. Nicolis, Quart. Rev. Biophys. 4 (1971) 107.
3. G. Nicolis and J.F.G. Auchmuty, Proc. Nat. Acad. Sci. U.S. 71 (1974) 2748.
4. J.F.G. Auchmuty and G. Nicolis (submitted to Bull. Math. Biol.).
5. G. Nicolis, Adv. Chem. Phys. 19 (1971) 209.
6. for a recent review, see G. Nicolis and J. Portnow, Chem. Rev. 73 (1973) 365.
7. A. Goldbeter and R. Lefever, Biophys. J. 12 (1972) 1302.
8. M. Herschkowitz-Kaufman and G. Nicolis, J. Chem. Phys. 59 (1972) 1890.
9. See for example Robertson, A. and Cohen, M.H., Ann. Rev. Biophys. Bioengineering 1 (1972) 409.
10. A. Babloyantz and G. Nicolis, J. Theor. Biol. 34 (1972) 185.
11. A. Babloyantz and M. Sanglier, FEBS Letters 23 (1972) 364.
12. R. Blumenthal, J.P. Changeux and R. Lefever, J. Membrane 2 (1970) 371.
13. A. Goldbeter and G. Nicolis, Biophysik 9 (1972) 212.
14. A. Babloyantz, Biopolymers 11 (1972) 2349.
15. I. Prigogine, G. Nicolis and A. Babloyantz, Physics Today 25 (1972) no. 12.
16. A. Babloyantz and J. Hiernaux (submitted to Nature).
17. T. Erneux, Mémoire de Licence (U.L.B., 1974).
18. L. Wolpert, J. Theor. Biol. 25 (1969) 1.
19. F. Crick, Nature 225 (1970) 420.
20. A. Babloyantz and J. Hiernaux, Proc. Nat. Acad. Sci. U.S. 71 (1974) 1530.
21. A. Babloyantz and J. Hiernaux (submitted to J. Theor. Biol.).
22 A. Gierer and H. Meinhardt, Kybernetik 12 (1972) 30.
23. H. Meinhardt and A. Gierer, J. Cell. Sci. 15 (1974) 321.

SUBJECT INDEX